S0-DKM-902

Abnormal Loading on Structures

Abnormal Loading on Structures

Experimental and numerical modelling

Edited by K. S. Virdi, R. S. Matthews,
J. L. Clarke and F. K. Garas

London and New York

Due to unforeseen circumstances the *Abnormal Loading on Structures* conference did not take place. However, it is hoped that this collection of papers will be a useful and authoratitive guide to state-of-the-art research in this key area of structural engineering.

First published 2000 by E & FN Spon
11 New Fetter Lane, London EC4P 4EE

Simultaneously published in the USA and Canada
by E & FN Spon
29 West 35th Street, New York, NY 10001

E & FN Spon is an imprint of the Taylor & Francis Group

Printed and bound in Great Britain by
T J International Ltd, Padstow, Cornwall

Publisher's Note

This book has been prepared from camera-ready copy provided by the authors.

British Library Cataloguing in Publication Data
A catalogue record for this book is available from the British Library

Library of Congress Cataloging in Publication Data

A catalogue record for this book has been requested

ISBN 0-419-25960-0

CONTENTS

INSTITUTION OF STRUCTURAL ENGINEERS' INFORMAL STUDY GROUP 'MODEL ANALYSIS AS A DESIGN TOOL'

The Group, which was formed in February 1977, operates under the auspices of the Institution of Structural Engineers and presently membership stands at over 500 covering some 40 different countries. Members come from a wide range of backgrounds including: research, design, engineering and contracting organisations, universities, government departments, local authorities and utility companies.

The primary objective of the Group is to create opportunities for members of the Institution and the profession to exchange information on the use of testing and model analysis to solve design problems. The scope of the Group encompasses the whole spectrum of structural engineering applications including: conventional structures, bridges, foundations, pressure vessels, offshore, harbour and coastal structures etc. It is intended to cover structures made of a wide range of materials and subjected to different loading conditions.

The Group's activities comprise the publication of a quarterly newsletter, organising international conferences, visits to test centres in the UK and Europe, sponsoring specialist lectures and holding an annual prize award competition for student dissertations on the application of physical modelling and testing in design.

Further information about the Group may be obtained from the Convenor, Prof F K Garas, FIStructE, 6 Amersham Gardens, High Wycombe, HP13 6QP, UK.

CONFERENCE ORGANISATION

Scientific Committee

Dr Ing K Brandes, BAM, Berlin, Germany
Dr Ing A Castoldi, ISMES SpA, Italy
Dr W G Corley, Construction Technology Laboratories, IncUSA
Dr J W Dougill, Institution of Structural Engineers, UK
Dr K J Eaton, Institution of Structural Engineers, UK
Prof J Eibl, Technische Universität Karlsruhe, Germany
Prof F K Garas, City University, UK
Prof H Gulvanessian, Building Research Establishment, UK
Prof Y Hasegawa, Waseda University, Tokyo, Japan
Prof T Krauthammer, Pennsylvania State University, USA
Dr J Kruppa, CTICM, France
Prof F M Mazzolani, Universita di Napoli Federico II, Italy
Prof P Marti, ETH Hönggerberg, Switzerland
Eng I Osman, The Arab Contractors, Egypt
Ir C De Pauw , Belgian Building Research Institute, Belgium
Dr A Pinto, Joint Research Centre, Ispra, Italy
Dr H G Russell, Russell Consultancy, USA
Prof G Somerville, Consultant, formerly British Cement Association, UK
Prof T P Tassios, National Technical University of Athens, Greece
Prof K S Virdi, City University, UK

Organising Committee

Prof F K Garas, City University
Prof J L Clarke, Concrete Society
Dr R Matthews, Taywood Engineering
Dr R Moss, Building Research Establishment
Prof K S Virdi, City University
Mr G S T Armer, Consultant
Dr S Doran, Institution of Structural Engineers
Mrs H Stevenson, Conference Secretary

PREFACE

Designing for hazardous and abnormal loads has become an important, even essential, requirement in the design process of most major civil engineering structures. The range of these structures includes tall buildings, bridges, conventional and nuclear power plants, chemical and other processing plants, oil and gas platforms, and harbour and coastal installations.

Hazard identification and risk assessment are of crucial importance when defining the extreme loading which a structure should be able to withstand. They are also vital elements in the establishment of appropriate protection for the user and the population at large in the event of a failure. The complementary requirement to maintain the cost of the structure at an acceptable level is, likewise, served by a clear understanding of its prospective service conditions.

Physical testing of full scale elements and structures combined with analytical modelling has played a significant role, over the years, in the study of their behaviour when subjected to hazardous loads. The experimental and numerical techniques have enabled designers to solve difficult engineering problems and thus improve the standards of design, safety, construction and in-service performance. In dealing with hazardous and abnormal loads, many problems are faced in the design, in undertaking and interpreting relevant experiment resultss, for example, establishing similitude requirements for material properties and behaviour, modelling the effect of time-varying loads and determination of natural frequency and proper boundary conditions.

Thirty four papers are reproduced in this book, whish were presented at the International Conference entitled 'Abnormal Loading on Structures - Experimental and Numerical Modeling', held at City University, London on 17 - 19 April 2000. The aim of the conference was to provide a forum for discussion and exchange of information on the relevant experience in the design and construction of structures subjected to abnormal loading, using physical tests and numerical modelling.

The following subjects are covered in the papers included here: loading on structures, including accidental loading, earthquakes and fire, effects of the methods of construction on structural response to abnormal loading, development of new materials and construction techniques, behaviour of engineering materials, inspection, monitoring, repair, and rehabilitation of structures, non-destructive testing for monitoring and assessment, structural safety and risk analysis, numerical simulation and modelling, and case studies of on-site testing of structures

The papers represent a state-of-the-art examination of structural design for hazardous and abnormal loads.

K.S. Virdi
R. Matthews
J.L. Clarke
F.K. Garas

ACKNOWLEDGEMENTS

The editors gratefully acknowledge the work of all the authors, and of the members of the Scientific and Organising Committees.

AVALANCHE SCENARIOS FOR ALPINE HIGHWAY BRIDGES

Avalanches on Alpine bridges

T. VOGEL
Institute of Structural Engineering, Swiss Institute of Technology (ETH), Zurich, Switzerland

Abstract
After the explanation of some basics of avalanches the calculation of velocities, flow depths and debris zones is shown. An example with a bridge of the St. Gotthard highway demonstrates how hazard scenarios are established and refined following actual events and what actions result from these premises.
Keywords: Avalanche, debris, dense flow, dry snow, friction, hazard scenario, highway bridge, pier, roughness, superstructure.

1 Dealing with avalanches

Avalanches have always threatened inhabitants and travellers in the Alps. Research on avalanches started in Switzerland in 1931 with the establishment of the Federal Committee on Snow and Avalanches in Bern. A further milestone was the foundation of the Swiss Federal Institute for Snow and Avalanche Research on the Weissfluhjoch at Davos in 1942.

Mountaineers fear most avalanches they trigger themselves. They have to judge the local situation, to act responsibly and sometimes to forego an intended tour. Research institutions like the one mentioned above provide a daily avalanche bulletin that judges the regional situation as accurately as possible.

For inhabitants of Alpine regions the situation is different. They are most threatened by avalanches running down to the valley floors. Avalanches may be triggered artificially in order to determine their time of run down and to prevent further accumulation of snow. More effective, however, are preventive measures like growing forests on the steep valley flanks and avoiding housing in endangered zones. The attempt to influence the flow of an avalanche by walls and dams is limited to some

Abnormal Loading on Structures edited by K. S. Virdi, R. S. Matthews, J. L. Clarke and F. K. Garas.
Published in 2000 by E & FN Spon, 11 New Fetter Lane, London EC4P 4EE, UK. ISBN 0 419 25960 0

favourable cases. Traffic routes often cannot avoid endangered zones and have to cross them through galleries or on bridges. These structures are designed to withstand the actions from avalanches.

1.1 Types of avalanches

Avalanches can be divided up into two types that behave in a different way and have different properties and characteristics (Table 1). Mixed type avalanches can also be observed.

Dense flow avalanches that consist of dry or wet snow have a granular structure and move similar to sand or gravel downhill. Their velocity depends above all on the inclination and the roughness of the terrain. They follow canyons and valleys and can be modelled as granular fluids.

New snow that has not yet consolidated can have entrained air once getting in motion. It forms a *dry snow avalanche*, like a gravity stream in water or a dispersion stream of a heavy gas. The entrainment of air and the erosion of snow form mechanisms that increase its mass with time. Such processes accelerate the avalanche and even more snow is dragged down. Dry snow avalanches move in the direction of the steepest slope and the friction to the ground matters in the run-out only. They are experimentally modelled as brine or suspensions in water or treated as two-phase flows.

Table 1. Characteristics of dense flow and dry snow avalanches (from [1])

	Unit	Dense flow avalanche	Dry snow avalanche
Typical velocity	[m/s]	30-60	50-100
Flow height	[m]	< 2-5	50-100
Density	[kg/m^3]	100-300	5

1.2 Partition of the avalanche path

Along the track of an avalanche three different parts can be assigned.

The *starting zone* includes all parts with an inclination between about 28 and 50 degrees, where snow packs can become unstable.

The *track* is the path formed by natural or artificial flanks that lead the avalanche downhill. Along the track the underlying snow can be captured thus increasing the volume in motion. Depending on the other relevant conditions the avalanche front speed may increase as well as decrease.

Where the inclination falls under a critical value given by the friction coefficient, the snow mass slows down and the *run-out* or *debris zone* begins.

1.3 Predictions and calculations

It is not the intention to give a complete manual for the calculations of avalanches. It shall be demonstrated, that with some few assumptions and estimations reasonable results can be achieved.

1.3.1 Dense flow avalanches

The model used for dense flow avalanches is based on the works of Voellmy [2] and Salm *et al* [3]. They assumed that the friction of the underlying snow can be described by a dry friction, linearly increasing with weight of the flowing snow layer and a dynamic drag proportional to the square of the avalanche velocity applying Eqn. 1.

$$v_0 = \sqrt{d_0 \xi (\sin\psi - \mu\cos\psi)} \tag{1}$$

The values needed are the roughness parameter ξ describing turbulent friction, a dry friction parameter μ denominating substratum friction and the angle of inclination ψ. Details, examples and all other formulas are given in [3]. In the meantime, more sophisticated models ([1], [4]) are available that have not greatly influenced practical work until now.

The size of the formation zone can easily be determined. More experience is needed to establish the depth d_0 of the snow pack that loses stability and influences the snow mass being involved. It depends on a basic value d_0^* denominating the possible increment of snow depth within three days and the inclination ψ. Values of d_0^* are available for various regions and return periods of 30, 100 and 300 years. Snowdrifts can increase d_0^* by 0.5 m.

The maximum flow Q can be estimated by multiplying d_0 with the maximum width of the formation zone B_0 and a representative velocity v_0 for a rectangular size (Eqn. 2), or by dividing the total snow volume K by the flow time Δt for any shape of the formation zone (Eqn. 3). In both cases a velocity is needed, that can be calculated, using Eqn. 1.

$$Q = B_0 d_0 v_0 \tag{2}$$

$$Q = K/\Delta t \qquad \Delta t = l/v_m \tag{3}$$

Knowing the flow Q, the velocity v can be calculated in all sections along the track, either by estimating the width B of the avalanche for flat slopes or by taking into account a hydraulic radius R for channelled avalanches. Again ξ, μ and ψ are needed. Changes of the inclination ψ cause acceleration or retardation respectively, determining the distance until steady conditions apply again.

The starting point of the run-out can easily be found when the inclination decreases suddenly beneath the critical value $\psi_k = \arctan\mu$ and has to be iterated in other cases. The calculation of the length of the run-out takes into account energy considerations only. Although the proposed values for ξ and μ vary over a range of factor two and depend on numerous parameters, the calculated run-out lengths agree well with observations.

1.3.2 Dry flow avalanches

Dry flow avalanches are more complicated to calculate and the observation of historic events helps to determine the endangered zones. The air pressures produced may be predominant more for facades and roofs of buildings than for engineering structures.

2 Structures and avalanches

In the formation zone appropriate structures are used to prevent snow packs from becoming unstable. They are not treated here.

Run-out zones are calculated with the given rules and by taking into account the local experience of the whole observation period. They are the basics for zones of forbidden or restricted housing, depending on the return period.

Structures forming part of the infrastructure, like roads and railway lines, normally cross the track of the avalanche or the run-out zone. Galleries are built to let avalanches run over them and are supposed to function even when the avalanche is occurring. Bridges should be built high enough to allow dense flow avalanches to pass underneath and withstand dry flow avalanches undamaged. Regarding their height, it is not possible to protect users on the bridge and a well-timed closure is required. Should a bridge always be in operation, it has to be constructed as a tunnel bridge.

3 Actions due to avalanches

The following three types of structures can be identified:

- Large obstacles like averting dams and walls that are loaded by dynamic pressures and friction forces. These are not treated further.
- Supporting structures like galleries that are subject to hydrostatic pressure, dynamic friction and possibly deviating forces during the rundown and carry the remaining snowpack and possible creep forces after the event.
- Small obstacles compared to the cross section of the avalanche like piers and girders of bridges that are subject to dynamic pressure and friction during the run down and may remain covered with debris of snow afterwards.

All actions are normally taken as static loads, neglecting dynamic amplification factors. Large blocks or logs may cause impact loads that exceed static loads by a factor of two. Avalanches with a return period of 30 years are regarded as an ordinary hazard scenario, i. e. the normal load factors apply. Avalanches with a return period of 100 or even 300 years are regarded as accidental events, load factors are reduced to 1.0 and accompanying actions are taken into account with a load factor ψ_{acc} which is equal to zero in most cases [5].

3.1 Actions on avalanche galleries

A guideline, issued by a working group of the Swiss Federal Highway Authority and the Swiss Federal Railways [6], establishes the procedure for determining the actions on avalanche galleries. An avalanche specialist has to specify flow depth, flow width and velocity of the design avalanche in a cross section up to 100 m above the railway or road to be protected. This enables the designer to choose the most favourable position and length of the gallery and to calculate the actions to be taken into account.

3.2 Actions on bridges

For bridges the avalanche specialist normally specifies directly the actions that have to be taken into account by the designer of a new bridge, or by the assessor of an existing bridge. An example is given in Section 5.

4 The St. Gotthard highway

The Northern access ramp of the St. Gotthard highway tunnel crosses the Swiss Canton of Uri over a length of 47 km and rises from 450 m to 1060 m above sea level. Due to the narrow valley the highway is carried by bridges or passes through tunnels over large stretches. The adjacent mountains rise to 3200 m above sea level.

4.1 The rehabilitation project

The highway was commissioned between 1971 and 1980. The first bridges, however, were begun in 1963 to serve as access to the remote construction sites of the various tunnels. In the meantime a growing use of deicing salt, increased traffic and axial loads as well as alpine accidents such as avalanches, rockfalls, mud flows and floods have considerably damaged the highway and especially the bridges. In 1990 a preservation programme was started to remove deficiencies and repair the damage of the first 25 years and to retrofit the structures for at least another 50 years.

The rehabilitation project prepared in 1998 covers the highest part of the highway called Group 4 from the Wassen exit to the entrance of the St. Gotthard tunnel. The knowledge gained until then on frequency, extent and characteristics of avalanches was taken into account, leading to design actions on piers and superstructures of bridges. Check calculations showed that all bridges could withstand the expected actions with a return period of 100 years.

5 Case study Reuss bridges Wattingen

The most interesting situation is now described in more detail based on the technical reports of the avalanche specialist in charge [7], [8].

5.1 The Rorbach avalanche

The Rorbach valley descending West-East produces an avalanche that is well documented since commissioning of the St. Gotthard railway line in 1882. 14 major events are recorded over 115 years, most of them causing damage to the infrastructure. The most severe event was a dry snow avalanche in 1981, that damaged the railway bridge, covered the roads and highways up to 4 m and produced a debris cone 8 m to 30 m deep, 400 m in lateral extent following the river Reuss for another 280 m (Fig. 1).

The starting zone can be divided into two parts. The upper one lies higher than 2000 m a.s.l. and covers an area of about 1 million m^2. Snow masses that begin to move up there must overcome a flat path 700 m in length to reach the valley floor. The lower part of the formation zone is steeper; each heavy snowfall causes snow flows that fill the narrow canyon and reduce the roughness, facilitating subsequent larger avalanches. Table 2 shows estimated and calculated values of two major events.

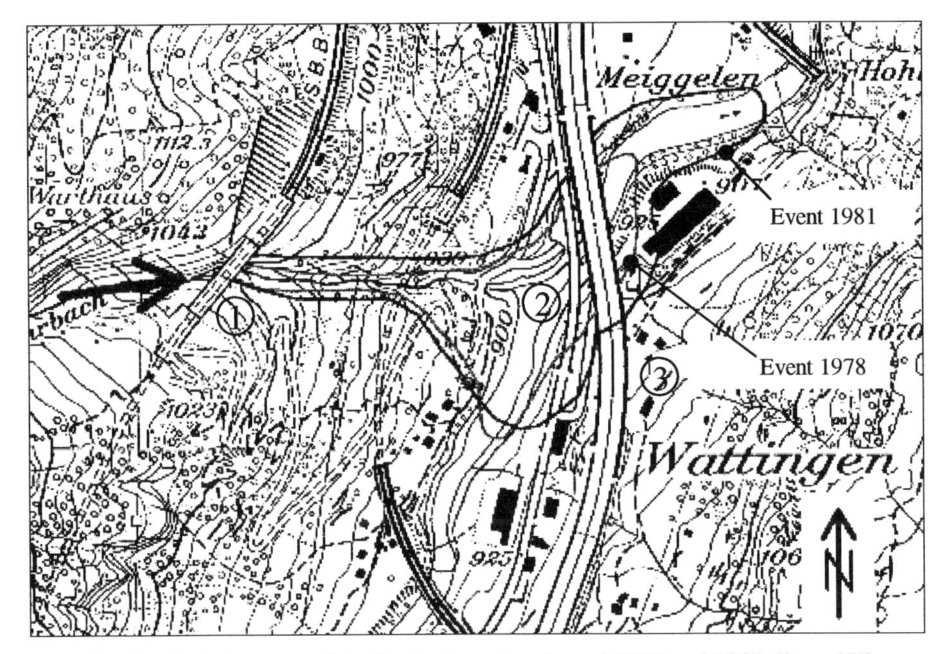

Fig. 1. Debris zones of the Rorbach avalanches of 1978 and 1981 (from [7])

Table 2. Estimated and calculated values of two run downs of the Rorbach avalanche

	Unit	February 2, 1978	January 6, 1981
Estimated debris volume	[m^3]	120,000	220,000
Estimated snow volume in the starting zone	[m^3]	300,000	550,000
Length of avalanche body	[m]	500	700
Duration of event	[s]	20	35

Such avalanches reach the valley floor with high speed and impact the opposite valley flank, losing most of their kinetic energy. The snow then flows down along the river Reuss with reduced speed.

5.2 Structures in the endangered area

5.2.1 The Rorbach bridge of the railway line

The railway line crosses the Rorbach valley on a bridge (marked 1 in Fig. 1) that has been subject to severe damage at every major event. In such critical situations the line had to be closed, because operational safety was not guaranteed, causing delays and distractions. In 1984 the original bridge was replaced by a tunnel bridge designed for a lateral distributed force of 10 kN/m^2, covering the flow pressure of a dry snow avalanche. If the canyon is not completely filled with snow, dense flow avalanches should pass underneath.

5.2.2 The Reuss bridges Wattingen

The St. Gotthard highway A2 as well as the local road cross the river Reuss at the same location, just where the Rorbach avalanche reaches the river (marked 2 and 3 in Fig. 1). Dense flow avalanches act on the piers, dry snow avalanches may blow snow-air mixture high up, producing a snow layer of some meters depth on all bridges.

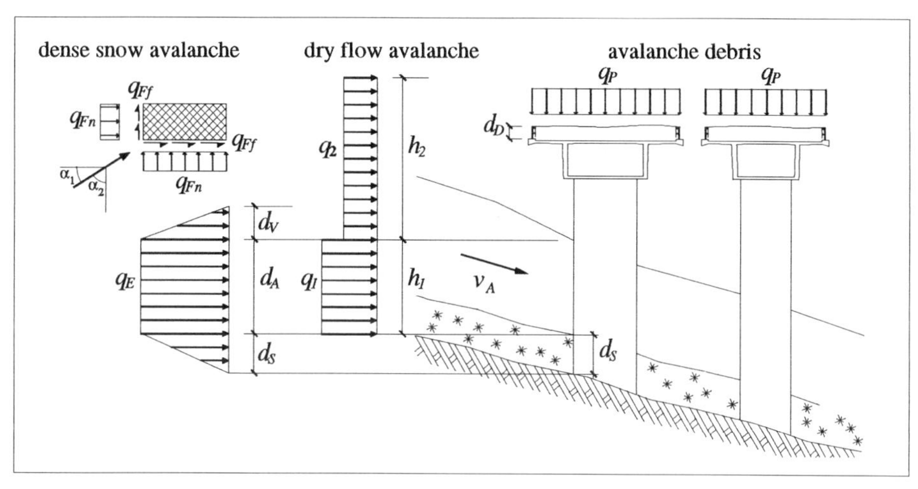

Fig. 2. Actions on piers and girders of the Reuss bridges A2 Wattingen, normal and accidental event

Table 3 Hazard scenarios for the Reuss bridges A2 Wattingen

Hazard scenario	Unit	Normal event	Accidental event
Return period	[years]	30	100
Avalanche debris on bridges			
Maximum depth d_D	[m]	2.5	4.5
Vertical load q_P	[kN/m^2]	7.5	15.8
Dry snow avalanche			
Air pressure q_1, q_2	[kN/m^2]	1.8, 1.3	2.5, 1.7
Altitude above h_1	[m a.s.l.]	920	930
Height h_2	[m]	20	20
Dense flow avalanche			
Avalanche velocity v_A at relevant pier LO2	[m/s]	11	14
Normal component of flow pressure q_{Fn} with most unfavourable angle α	[kN/m^2]	17.3	27.9
Parallel (friction) component of flow pressure q_{Ff} with most unfavourable angle α	[kN/m^2]	5.2	8.4

Dense flow avalanches produce a flow pressure q_F that is taken as constant over the depth of the avalanche d_A and linearly reduced to zero in the underlying snow depth d_S and in the dynamic depth d_V. Since dense snow avalanches can occur combined with a dry snow zone, the associated air pressure has to be taken into account. It is applied to two zones with the heights h_1 and h_2 and the pressures q_1 and q_2, respectively.

For piers the angle α between shaft surface and avalanche motion determines deviating and friction forces. The limited size and a possible favourable shape of the piers are taken into account by reduction factors.

The maximum actions taken into account for the check calculation of the rehabilitation project are given in Fig. 2 and Table 3.

5.3 The situation in February 1999

In the night from February 8 to February 10, 1999 a first dry snow avalanche associated with a snow powder flow ran down. The Rorbach bridge as well as the Reuss bridge Wattingen were covered with snow up to 0.2 m depth.

Since snow fall continued another large avalanche was expected, running on top of the already deposited one and thus being able to hit either the Rorbach bridge or the superstructures of the Reuss bridges Wattingen. Artificial triggering was ruled out due to unforeseeable consequences.

On February 20 the snow pack was finally triggered by natural causes and formed a dense snow avalanche that reached the Reuss bridge of the local road.

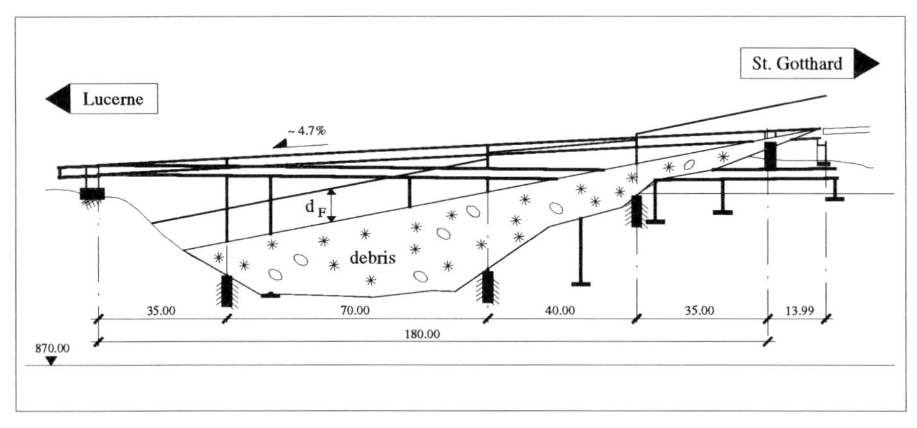

Fig. 3. Longitudinal section through the Reuss bridge A2 Wattingen (uphill track ROMEO)

5.4 Extreme hazard scenario

These experiences initiated the establishment of an extreme hazard scenario with a return period beyond 300 years, namely the filling up of the Rorbach canyon and the formation of a debris cone by a first avalanche and the subsequent release of a second avalanche. The calculation took into account the new topography changed by the debris of the first avalanche, resulting in steeper slopes and smaller depths and changing the direction of the track.

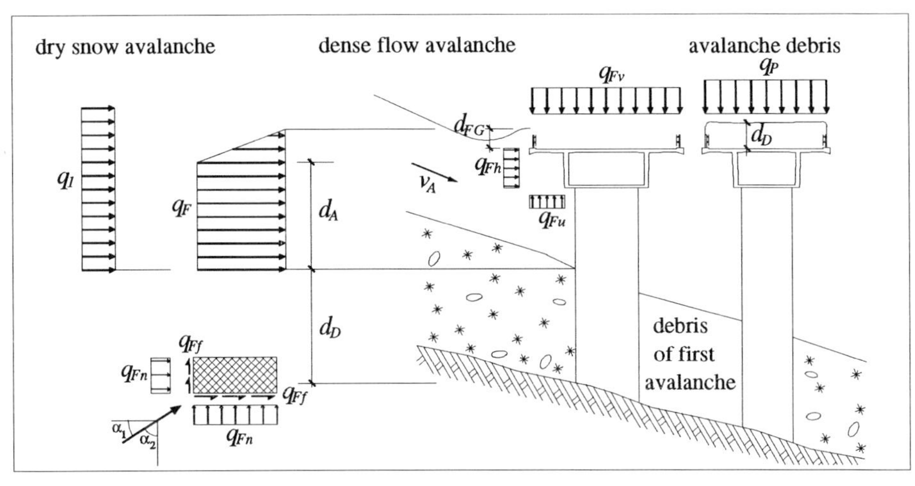

Fig. 4. Actions on piers and girders of the Reuss bridges A2 Wattingen, extreme hazard scenario

Table 4 Extreme hazard scenario for the Reuss bridge A2 Wattingen (uphill track ROMEO)

Extreme hazard scenario	Unit	Range of girder, number of pier			
Avalanche debris on bridges		c	b	a-2	a-1
Maximum depth d_D	[m]	2	3	6	8
Vertical load q_P	[kN/m^2]	5	7.5	21	28
Dry snow avalanche					
Air pressure q_l on total height	[kN/m^2]	2.0	3.0	2.0	
Dense flow avalanche on girder					
Flow depth d_{FG} on girder	[m]	1.5	2	3	4
Vertical load q_{Fv} on girder	[kN/m^2]	2.2	3	9	12
Horizontal pressure q_{Fh} on girder	[kN/m^2]	-	-	85 triangular	100 uniform
Uplift pressure q_{Fu} on cantilevering deck slab	[kN/m^2]	-	-	15	25
Dense flow avalanche on columns		RO4	RO5		RO6
Avalanche velocity v_A	[m/s]	10	16		21
Flow depth d_A on debris	[m/s]	4	8		6
Normal component of flow pressure with most unfavourable angle α	[kN/m^2]	12.5	36.5		74
Parallel (friction) component of flow pressure with most unfavourable angle α	[kN/m^2]	3.8	11		22

Nevertheless, the superstructure of the Reuss bridge A2 Wattingen is subject to either vertical loads caused by the debris of a dry snow avalanche or horizontal flow pressures due to a dense flow avalanche. The cantilever part of the deck slab would be loaded additionally by uplift forces.

All these forces are greatest at the southern end and decrease along the bridge due to the more favourable topography in the northern part (Fig. 3). The bridge carrying the uphill lanes called ROMEO was subject to a new check calculations applying the actions of Figure 4 and Table 4. The calculations showed, that the girder could withstand the vertical loads but would be heavily overstressed in the southern part by carrying the horizontal loads spanning from one abutment to the other.

Which measures will be taken to meet the additional requirements has not been decided yet. The most promising option is the enlargement of pier RO6 to form a stiff intermediate support for horizontal actions.

6 Conclusions

The presented methods of calculating actions due to avalanches are not very sophisticated. At least they have a physical background and fulfil the requirements for accuracy. A direct verification of actions by measurements is difficult, but the extent of the debris zone allows a back calculation of the event that caused them.

The case study shows that avalanche actions can heavily depend on previous events, because the debris accumulates over a whole winter period and can change the topography considerably.

7 Acknowledgements

The author wishes to thank the representative of the owner Heribert Huber, Cantonal Highway Department Uri, for his generous support and the consultants A. Burkard, Brig, R. Vögeli c/o Winkler & Partners Effretikon and D. Meister c/o A. Rotzetter&Partners, Zug for their agreeable co-operation.

8 References

1. Hutter, K. (1992) Lawinen-Dynamik (Dynamics of avalanches), *Schweizer Ingenieur und Architekt*, No. 13/1992, pp. 259-269.
2. Voellmy, A. (1955) Über die Zerstörungskraft von Lawinen (On the destruction force of avalanches), Schweizerische Bauzeitung, Vol. 73, No 12 pp. 159-162, No. 15 pp. 212-217, No. 17 pp. 246-249, No. 19 pp. 280-285.
3. Salm, B., Burkard, A. and Gubler, H. U. (1990) Berechnung von Fliesslawinen – Eine Anleitung für Praktiker mit Beispielen (Calculation of dense flow avalanches – a guideline for practitioners), *Reports of the Swiss Federal Institute for Snow and Avalanche Research*, No. 47, July 1999, 37 pp.

4. Norem, H., Irgens, F., Schieldrop, B. (1987) A continuum model for calculating snow avalanche velocities in *Avalanche Formation, Movement and Effects* (ed. Salm, B. and Gubler, H.), Proceedings of the Davos Symposium, September 1986, IAHS Publication No. 162, pp. 363-378.
5. Swiss Society of Engineers and Architects (1989), *Actions on structures*, SIA-Code 160(1989), Zurich, SIA, 1989, 104 pp.
6. Swiss Federal Department of Traffic and Energy (1994), Richtlinie Einwirkungen auf Lawinenschutzgalerien (Guideline for actions on avalanche galleries), EDMZ, Bern 1994, 15 pp.
7. Burkard A. (1998), Lawineneinwirkungen auf die Wattingerbrücke und den Lehnenviadukt Laubzug (Avalanche actions on the Wattingen bridges and the Laubzug viaduct), Technical Report, Brig, February 19, 1998, 29 pp.
8. Burkard A. (1999), Lawineneinwirkungen auf die Wattingerbrücken A2 – Extrem-Szenario (Avalanche actions on the Wattingen bridges – extreme scenario), Revised Technical Report with appendix, Brig, August 25, 1999, 11 pp/17 pp.

CONSIDERATION OF EXCEPTIONAL SNOW LOADS AS ACCIDENTAL ACTIONS

Exceptional snow loads

G. KÖNIG and D. SOUKHOV
Institute of Concrete Structures, University of Leipzig, Leipzig, Germany

Abstract

The investigation on German snow data show that for some German climatic stations (located in coastal regions of North Germany) the exceptional heavy snow falls can be identified. These snow loads can cause the essential damages of structures. Normally these exceptional values have a very large return period (1,000 or even 10,000 years) but can occur during relative short design working life of structure (50 or 100 years). According to ENV 1991-1 "Basis of Design" these exceptional snow falls can be considered as accidental actions. Based on German data statistical analysis is undertaken and procedure for codified design is considered.

Keywords: Accidental actions, characteristic value, exceptional value, probability distribution, snow loads, statistical analysis.

1 Introduction

Very heavy snow falls were observed in different parts of Europe, particularly in coastal regions. They cause snow loads which are significantly larger than the snow loads which normally occur in these regions.

As an example the city Schleswig (altitude 43 m above sea level) can be considered. It is located in North Germany, not far from the Danish border. It belongs to Snow Load Zone III according to ENV 1991-2-3 "Snow Loads" [1]. According to this standard, the characteristic value (with return period of 50 years) of snow load for this location and altitude, is 1.13 kN/m^2. But the maximum observed value, which occurred on 19th February 1979, was 2.37 kN/m^2. The ratio between these two values is equal to 2.1 and larger as the partial safety factor for snow load which is equal to 1.5. Including those high values together with the more regular snow events into the

Abnormal Loading on Structures edited by K. S. Virdi, R. S. Matthews, J. L. Clarke and F. K. Garas.
Published in 2000 by E & FN Spon, 11 New Fetter Lane, London EC4P 4EE, UK. ISBN 0 419 25960 0

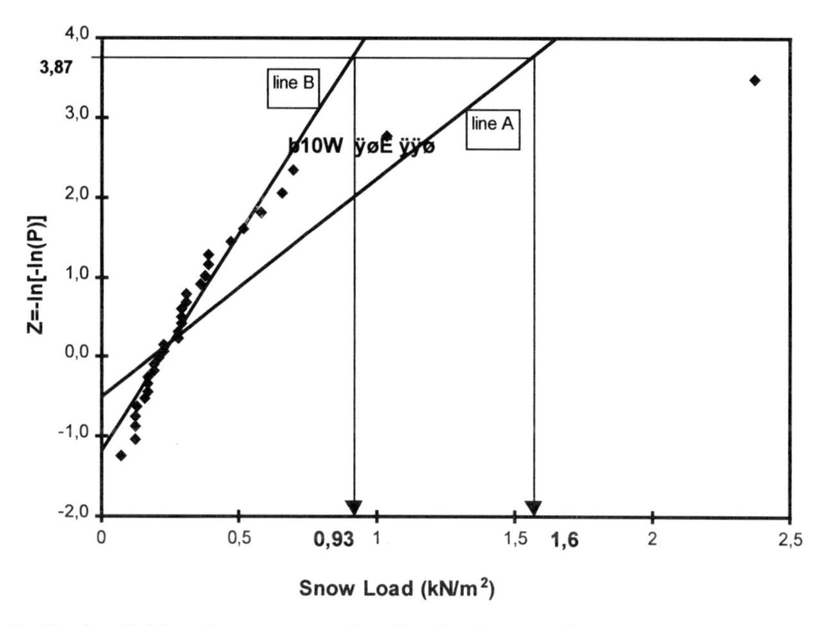

Fig. 1. Station Schleswig, extreme value distribution type I

sample of the annual maxima of snow load disturbs the statistical processing of the snow data. If the extreme value distribution type I for maxima is used for the fitting of data then using of probability paper leads to the plot shown in Fig. 1 (on this plot Z is the axis of reduced variate and P is probability that the corresponding value of snow load is not being exceeded). The parameters and therefore position of fitting line are defined by means of least squares method.

From this plot can be seen that including the exceptional value of 2.37 kN/m^2 in statistical processing (line A) leads to characteristic value of 1.6 kN/m^2 but the line does not fit the data. Excluding the exceptional value (line B) leads to characteristic value of 0.93 kN/m^2 and the line fits the data better (but also not quite well). This problem was also discussed in the background document to ENV 1991-2-3 [2] and in the final report of the "European Snow Loads Research Project" [3]. The main question of discussion is following: if the exceptional value is excluded from the sample considered how should this value be taken into account in codified design ?

2 Investigation on German snow data

During the work on "European Snow Loads Research Project" about 330 German climatic stations were investigated based on observed data of water equivalent (and/or snow depth). The record period of most stations was about 30 years. First of all the annual maxima of snow load were obtained and then the characteristic values (with return period of 50 years) were calculated based on extreme value distribution type I for maxima (but the log-normal and Weibull distributions were also discussed).

12 climatic stations were declared as the stations with the exceptional snow events. List of these stations can be seen from the Table 1.

Table 1. The list of German climatic stations with exceptional snow events

N	Station	Number Of record Years	Number Of years with snow	Exceptional value of snow load Q_{exc} (kN/m^2)	Date of occurrence of exceptional value	Next maximum snow load value after exceptional one (kN/m^2)
1	Norderney	18	15	1.56	22 Feb. 1979	0.70
2	Schleswig	33	32	2.37	19 Feb. 1979	1.04
3	Hamburg	33	32	1.82	15 Feb. 1979	0.67
4	Bederkesa	18	17	1.52	16 Feb. 1979	0.62
5	Bremen	32	30	1.53	18 Feb. 1979	0.58
6	Cuxhaven	31	24	1.48	01 Mar. 1979	0.61
7	Soltau	25	25	1.26	19 Feb. 1979	0.62
8	Hannover	33	31	1.23	24 Feb. 1979	0.56
9	Kiel	33	31	1.78	15 Feb. 1979	0.68
10	Norden	12	12	1.07	19 Feb. 1979	0.31
11	Tostedt	27	26	1.47	16 Feb. 1979	0.74
12	Visselhoevede	27	24	1.54	17 Feb. 1979	0.59

The location of all these stations is the north-west of Germany in the vicinity of North and/or Baltic sea. The date of occurrence of exceptional event for these stations is the second half of February 1979 when a very heavy snow falls were observed in north-west Germany which caused, for example, a damage of a lot of roofs in region of Hamburg.

With the help of extreme value distribution type I for maxima the characteristic (with return period 50 years) values were calculated for the case when the exceptional value is included in the statistical processing and for the case when the exceptional value is excluded from the consideration. The results can be seen in the Table 2. In this table the plot correlation coefficients are also given. These coefficients show how well the line (or other words the chosen distribution) fits the data. In ideal case the plot correlation coefficient will be equal to 1.0. This would mean that all data points lie on the same line.

The values of plot correlation coefficient from Table 2 show that extreme value distribution type I fits the data not well. Even for the case when exceptional value excluded the fitting is not satisfactory, as can be seen for line B on Fig. 1 for Schleswig. The same is observed for the most of other stations. Therefore it can be concluded that extreme value distribution seems to be not the right function for fitting the original data points. Thus other distributions should be considered.

Table 2. Characteristic values and plot correlation coefficients based on the extreme value distribution type I

N	Station	A: Exceptional value is included		B: Exceptional value is excluded		Ratio
		50 years return value S_A (kN/m^2)	Plot correlation coefficient	50 years return value S_B (kN/m^2)	Plot correlation coefficient	$k = Q_{exc} / S_B$
1	Norderney	1.51	0.8601	0.84	0.9097	1.86
2	Schleswig	1.60	0.8224	0.93	0.9673	2.55
3	Hamburg	1.23	0.8284	0.71	0.9759	2.56
4	Bederkesa	1.42	0.8860	0.81	0.9870	1.88
5	Bremen	1.05	0.8237	0.59	0.9811	2.59
6	Cuxhaven	1.11	0.8514	0.66	0.9905	2.24
7	Soltau	1.07	0.9168	0.75	0.9820	1.68
8	Hannover	0.90	0.8928	0.61	0.9946	2.02
9	Kiel	1.22	0.8220	0.71	0.9814	2.51
10	Norden	1.13	0.7704	0.39	0.9538	2.74
11	Tostedt	1.16	0.8769	0.75	0.9706	1.95
12	Visselhoevede	1.17	0.8373	0.66	0.9748	2.35

During the work "European Snow Loads Research Project" [3] it was pointed out that for Germany 3 probability distribution functions are to be considered as possible candidates: extreme value distribution type I, log-normal distribution and Weibull distribution. Which from these functions fits the data better depends on the local climatic conditions. It was found that for lowland of North Germany the log-normal distribution fits the data best. Let us look at the example of station Schleswig presented on the log-normal probability paper at Fig. 2. From this plot it is clear that line B (after excluding the exceptional value) fits the original data very well, essentially better as the line B on the probability paper for extreme distribution type I (compare with Fig. 1). The preference of the log-normal distribution (in comparison with the extreme value distribution type I) for the stations considered can be confirmed with the help of the plot correlation coefficients presented in Table 3.

The comparison of the results from the Table 1, Table 2 and probability plots of the stations (not given here, only station Schleswig is given as example) shows that for all 12 stations the log-normal distribution is the best fitting one if the exceptional value is included in the statistical consideration. But more interesting the case when the exceptional value is excluded. Then for 7 stations the log-normal is again the best fitting distribution. For 3 stations (Bederkesa, Cuxhaven and Hannover) both the distributions, extreme value type I and log-normal, give almost the same value of plot correlation coefficient and both fit the data well. For station Kiel extreme value distribution type I gives higher value of plot correlation coefficient as log-normal one but the both distributions fit the data not satisfactory. In this specific case the best candidate is Weibull distribution which gives the value of plot correlation coefficient equals 0.9849 and 50 year return value of snow load equals 0.64 kN/m^2. Almost the

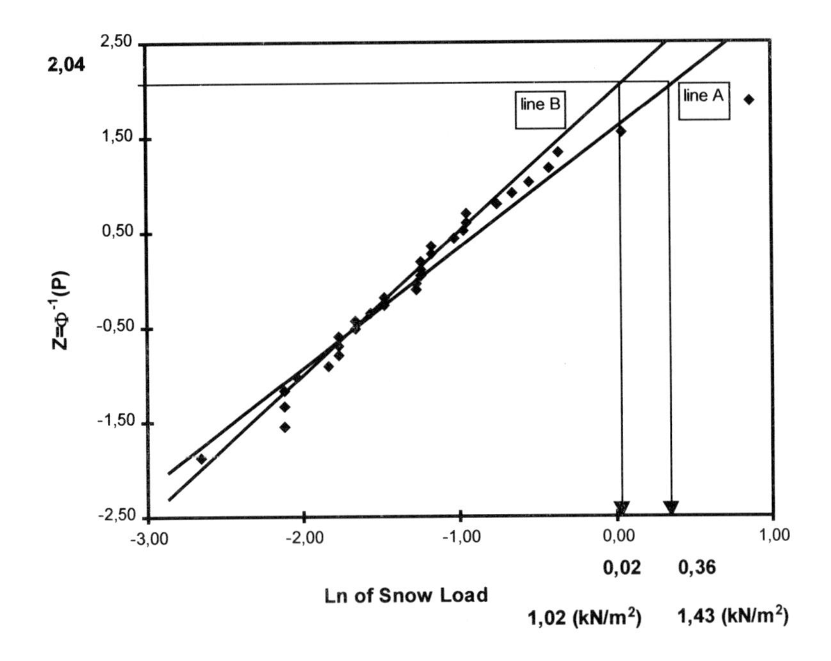

Fig. 2. Station Schleswig, log-normal distribution

Table 3. Characteristic values and plot correlation coefficients based on the log-normal distribution

N	Station	A: Exceptional value is included		B: Exceptional value is excluded		Ratio
		50 years return value S_A (kN/m^2)	Plot correlation coefficient	50 years return value S_B (kN/m^2)	Plot correlation coefficient	$k = Q_{exc} / S_B$
1	Norderney	1.72	0.9473	0.97	0.9567	1.61
2	Schleswig	1.43	0.9724	0.102	0.9924	2.32
3	Hamburg	1.13	0.9719	0.82	0.9883	2.22
4	Bederkesa	1.73	0.9866	1.16	0.9868	1.31
5	Bremen	0.94	0.9677	0.66	0.9878	2.32
6	Cuxhaven	1.06	0.9717	0.77	0.9856	1.92
7	Soltau	1.12	0.9802	0.87	0.9857	1.45
8	Hannover	0.92	0.9865	0.73	0.9925	1.68
9	Kiel	1.18	0.9653	0.89	0.9741	2.00
10	Norden	1.13	0.8814	0.45	0.9401	2.39
11	Tostedt	1.14	0.9744	0.84	0.9875	1.76
12	Visselhoevede	1.13	0.9724	0.78	0.9867	1.97

same situation with station Norden. Again Weibull is the best fitting distribution and gives for snow load the 50 year return value of 0.35 kN/m^2. But in this case the extreme value distribution type I fits the data also well and a little better as log-normal one (but this station has a shortest record period - only 12 years). Snow load value with 50 year return period is equal to 0.39 kN/m^2 and is a little larger as by Weibull distribution.

3 Presentation of exceptional snow load as accidental action

The only country among the CEN members where the problem of exceptional snow load is reflected in the building standard is France. According to the French Code of Practice N84 the snow value should be considered as exceptional one (based on snow depth d) if the following criteria is fulfilled:

$$d_{max} > 1.5\, d_{50} \tag{1}$$

where: d_{50} the 50 year return period value of snow depth if the maximum value of snow depth is excluded

d_{max} the maximum value of snow depth

The value of 1.5 based on the value of safety factor from French codes. The same value is set for partial safety factor for variable actions in ENV 1991-1 "Basis of Design" [4].

Because in the structural design the effect of snow is considered as loading the criteria similar to Eqn. (1) should be applied to snow load S and can be presented as:

$$S_{max} \geq k\, S_{50} \qquad \text{or} \qquad S_{max} \geq k\, S_k \tag{2}$$

Here the symbols have the same meaning as in Eqn. (1) but related to the snow load and S_k is called characteristic value. The constant factor k has to be defined. Two approaches can be considered for this purpose.

3.1 Statistical aspect

Firstly the statistical aspect is to be taken into consideration.

Because the extreme value distribution type I for maxima is used for the fitting of snow data almost in all CEN countries (except Denmark) let us consider the P-fractile based on this distribution:

$$X_p = u - \ln(-\ln P) / c \tag{3}$$

Parameters u and c can be determined using the method of moments and then fractile can be presented as in Eqn. (4) dependent on the mean value m and the coefficient of variation V of the sample:

$$X_p = m \{1 - 0.78\, V\, [\, 0.577 + \ln(-\ln P)\,]\, \} \tag{4}$$

The European building standards do not define directly a return period for accidental actions. Only the Background Document for ENV1991-1 [5] notes that this return period can be up to 10,000 years. The German Reactor Safety Rules for Nuclear Stations set the same value of 10,000 years. This corresponds to the fractile with the probability of not being exceeded during one year of 0.9999. The characteristic value for variable actions is defined as a value with a return period of 50 years (i.e. $P = 0.98$). Therefore the factor k can be defined as the ratio of these two values and it becomes dependent only on coefficient of variation V:

$$k = X_{0.9999} / X_{0.98} = (1 + 6.73\,V) / (1 + 2.59\,V) \tag{5}$$

As investigations on German snow data show [6], values of V vary with altitude. As mean value of V the value of 0.6 can be set. But if only stations with altitude not higher as 200m above sea level are taken into account (all 12 above considered stations in North Germany belong to this group), then value of 0.6 will be the lower bound for V, because for these stations V varies mainly between 0.6 and 1.4. For comparison a return period (T_{ret}) of 1,000 years is also considered, in this case the probability of not exceeding during one year is equal to 0.999. Using these values of T_{ret} and $V = 0.6$ (as a lower bound) will lead to following values of k:

$k \approx 2.0$ (T_{ret} = 10,000 years); $k = 1.55$ (T_{ret} = 1,000 years)

From the above shown calculations it can be concluded that the load values with $k \geq 2$ have a return period at least of 10,000 years and the load values with $k \geq 1.5$ have a return period at least of 1,000 years.

3.1 Normative aspect

As second possibility a normative procedure from the Eurocodes can be used to define the coefficient k.

The document ENV 1991 - 1 "Basis of Design" [4] defines in Section 4 "Actions and environmental influences" § 4.1 "Principal classifications" (2) snow loads as variable actions. But allowance is made in Clause (4): "Some actions, for example seismic actions and snow loads, can be considered as either accidental and/or variable actions, depending on the site location (see other Parts of ENV 1991)". This permits to consider the exceptional snow events as accidental loads.

Thus, if the snow event is identified as exceptional one, i.e. connected with the characteristic value by Eqn. (2), then the snow load should be taken into account in two design situations: persistent/transient (P/T) situations and accidental (A) situations. The coefficient k needs to be determined by taking into account not only consideration of the actions, but also the influence of the resistance.

According to ENV 1991 - 1 "Basis of Design" [4], Section 9 it shall be verified that:

$$E_d \leq R_d \tag{6}$$

where: E_d design value of the effect of action
R_d corresponding design value of resistance

For simplicity let us consider the case when snow load is the only variable action and there is only one permanent action (e.g. self-weight). Then according to Clause 9.4.2 "Combinations of actions" of ENV 1991-1 [4] there are two cases to be considered:

- persistent and transient design situations for ultimate limit states verification other than those relating to fatigue:

$$E_d = \gamma_G G_k + \gamma_Q Q_k \tag{7}$$

- accidental design situations:

$$E_d = \gamma_{GA} G_k + A_d \tag{8}$$

where:

$\gamma_G = 1.35$	the partial safety factor for permanent action
G_k	the characteristic value of permanent action
$\gamma_Q = 1.5$	the partial safety factor for variable action (snow load)
Q_k	the characteristic value of variable action (snow load)
$\gamma_{GA} = 1.0$	the partial safety factor for permanent action for accidental design situation
$A_d = \gamma_A \cdot A_k$	the design value of accidental action
$\gamma_A = 1.0$	the partial safety factor for accidental action(snow load)
A_k	the characteristic value of accidental action (snow load)

According to ENV 1992 "Design of Concrete Structures", Part 1-1 "General Rules and Rules for Buildings" [7]:

P/T: $R_d = R_{k,c} / \gamma_C$ for concrete (9)
$R_d = R_{k,s} / \gamma_S$ for steel reinforcement or prestressing tendons

A: $R_d = R_{k,c} / \gamma_{CA}$ for concrete (10)
$R_d = R_{k,s} / \gamma_{SA}$ for steel reinforcement or prestressing tendons

where:

$\gamma_C = 1.5$	the partial safety factor for concrete (P/T situations)
$R_{k,c}$	the characteristic value of concrete
$\gamma_S = 1.15$	the partial safety factor for steel reinforcement (P/T situations)
$R_{k,s}$	the characteristic value of steel reinforcement
$\gamma_{CA} = 1.3$	the partial safety factor for concrete (Accidental situations)
$\gamma_{SA} = 1.0$	the partial safety factor for steel reinforcement (Accidental situations)

Using Eqn. (6), (7) and (9) it is possible to write for P/T situations for concrete:

$$1.35\, G_k + 1.5\, Q_k \leq R_{k,c} / 1.5$$

Considering as an unfavourable case the ratio $G_k = 0.5\, Q_k$ and noting that $Q_k = S_k$ one can obtain:

$$3.26\, S_k \leq R_{k,c} \tag{11}$$

Similarly for accidental situations Eqn. (6), (8) and (10) and $A_k = S_{max} = k\, S_k$ from Eqn. (2) will lead to equation:

$$1.0\, G_k + k\, S_k \leq R_{k,c} / 1.3$$

Taking again as unfavourable case the ratio $G_k = 0.5\, Q_k$ it will be obtained:

$$(\, 0.65 + 1.3\, k\,)\, S_k \leq R_{k,c} \tag{12}$$

Because the right hand parts in Eqn. (11) and (12) are the same (characteristic value of concrete strength) the left hand parts (design value of action effect) shall be also the same, independently of the design situation. Then k can be calculated as:

$$k = (\, 3.26 - 0.65\,) / 1.3 = 2.0 \tag{13}$$

Other ratios of G_k / Q_k can also be taken into consideration. Assuming that $G_k = Q_k$ it will result in $k = 2.29$, and for the case when $G_k = 1.5\, Q_k$, k will be equal to 2.57.

Considering the case $G_k = 0.5\, Q_k$ as unfavourable, exceptional snow events with $k \geq 2.0$ should be fixed as accidental ones.

4 Analysis

The normative approach leads to values of k which are greater or equal to 2.0. The statistical approach leads to values of k which are greater or equal to 2.0 for return period of 10,000 years and values of k which are greater or equal to 1.5 for return period of 1,000 years. Thus snow events with k greater or equal to 1.5 can be considered as accidental ones.

Which from these two design situations (P/T or A) is more unfavourable (and therefore the decisive one) depends on the ratio of the characteristic value of permanent action to the characteristic value of the snow load.

Table 2 shows that if the extreme value distribution type I is used then all 12 considered climatic stations have the factor k greater than 1.5 and 8 station from these 12 ones have the values of k greater than 2.0. If log-normal distribution is used (see Table 3) then 10 stations have the factor k greater than 1.5 and 5 from these ones have the values of k greater than 2.0. Only two stations have the k factor less than 1.5. But for Soltau this factor is equal to 1.45 which is very near to 1.5. Therefore all stations (only Bederkesa is doubtful if the log-normal distribution is used) should be considered as ones with exceptional events and treated according to the procedure for accidental actions.

5 Conclusions

1. Analysis of 12 German climatic stations with exceptional values of snow load shows that the log-normal distribution fit the original snow data better as the extreme value distribution type I for maxima.

2. The exceptional values of snow loads can have the return period of 1,000 or even 10,000 years.

3. The largest snow load value from the sample should be identified as exceptional one if the ratio of this value to the characteristic value (with return period of 50 years) of snow load determined after excluding this largest value from statistical processing is greater or equal to 1.5.

4. The exceptional values of snow loads should be treated as an accidental actions and snow load, in this case, should be taken into account for two design situations, persistent / transient and accidental ones.

6 Acknowledgements

Authors would like to express their appreciation for financial support made by Commission of the European Communities DGIII-D3 to carry out the reported work, which forms the part of the prenormative research for Eurocode 1, Part 2-3 "Snow Loads".

7 References

1. European Committee for Standardisation. (1994) *Basis of Design and Actions on Structures. Part 2-3: Snow Loads.* CEN, Brussels. ENV 1991-2-3.
2. University of Pisa. (1995) *New European Code for Snow Loads. Background Document.* Proceedings of Department Structural Engineering University of Pisa.
3. Commission of the European Communities DGIII-D3. (1998) *Scientific Support Activity in the Field of Structural Stability of Civil Engineering Works. Snow Loads. Final Report.* University of Pisa. Department of Structural Engineering.
4. European Committee for Standardisation. (1994) *Basis of Design and Actions on Structures. Part 1: Basis of Design.* CEN, Brussels. ENV 1991-1.
5. European Convention for Constructional Steelwork. Joint Committee on Structural Safety (JCSS). (1996) *Background Documentation. Eurocode 1 (ENV 1991). Part 1: Basis of Design.* JCSS, Brussels. First Edition.
6. Soukhov, D. (1998) The Probability Distribution Function for Snow Load in Germany. *Leipzig Annual Civil Engineering Report (LACER)*, Vol. 3, pp. 263-274.

EVALUATION OF THE PEAK FORCES ON ROOF TILES UNDER STORMY CONDITIONS

Peak forces on roof tiles

B. PARMENTIER, S. SCHAERLAEKENS and J. VYNCKE
Belgian Building Research Institute, Limelette, Belgium

Abstract

The problem of determining the net forces on tiles and slates in domestic housing has never been resolved in a completely rigorous way, despite considerable safety risks and economic losses per annum due to tile and slate loss during stormy conditions. Instead, empirical rules have been introduced in normative documents [20] and guides of 'good practice' [4,5,6] aimed towards craftsmen. The most particular aspect with respect to the load on the elements resides in the fact that they are not only subject to the external pressure, but also to the internal pressure that builds up just underneath. The internal air pressure depends, among other aspects, on the geometry of the joints between the distinct elements, the global geometry of the roof, the boundary conditions at ridges and roof top, etc. An extra complication is introduced in that the external pressure on the roof is locally altered by the external shape of the elements. The Belgian Building Research Institute (BBRI), together with the Von Kàrmàn Institute (VKI) in Brussels, has initiated a research programme on this subject and this paper describes the scope of the research together with the set-up of the experimental and numerical work and some preliminary results.

Keywords: Full-scale experiments, numerical modelling, peak forces, roof tiles, wind loading, wind-tunnel tests.

1 Introduction: Tiling and slating practice in Belgium

It is common in Belgium to provide an under-roof under the tiles during the construction of a pitched roof. Generally, this under-roof consists of thin wooden boards, but thin plastic sheets are also used. Both are laid with sufficient overlaps to make sure that the under-roof is quite impermeable, at least in comparison with the permeability of the joints between the covering elements.

Abnormal Loading on Structures edited by K. S. Virdi, R. S. Matthews, J. L. Clarke and F. K. Garas.
Published in 2000 by E & FN Spon, 11 New Fetter Lane, London EC4P 4EE, UK. ISBN 0 419 25960 0

The frontal permeability is caused by the gaps in the joints between the tiles. Also, in the thin air layer underneath the tile cladding, air movement in two orthogonal directions is possible, which is characterised by the lateral permeability, that may easily be many times greater than the frontal permeability.

2 Scope and structure of the research programme

It is well known that the net forces induced by the wind on tiles and slates are the result of the (integrated) pressure differences between the external pressure (p_e) and the internal pressure (p_{ia}) that build up just under the elements in the thin air layer. Both external and internal pressures depend on several parameters that we will discuss further in this presentation. Two of the major parameters are the position of the tile on the roof and the wind direction. There are indeed some places locally where the pressure coefficient (Eq. 1) can be much higher than the average value on the roof. This can be explained by the development of 'delta wing vortices' along the roof edges (Fig. 1).

$$C_p = C_{pe} - C_{pia} = \frac{p_e - p_{ia}}{1/2.\rho.V^2} \tag{1}$$

where ρ is density of air (1.225 kg/m³) and V is the wind velocity.

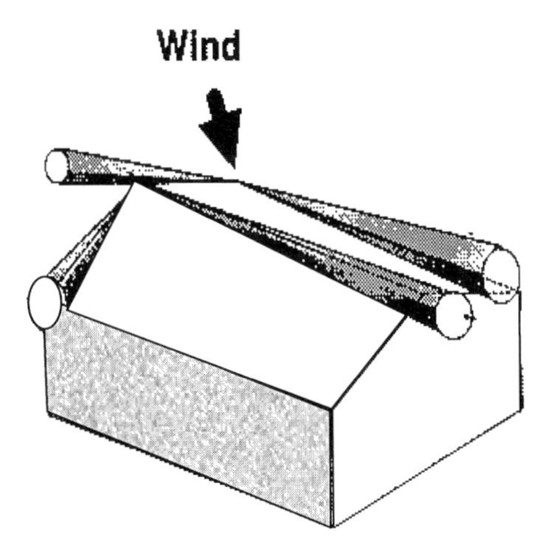

Fig. 1. Delta-wing vortex pairs on duopitch roof [11].

The centres of these vortices are regions of high negative pressure. This is the principal cause of high mean peripheral uplift on roofs, and causes much of the damage reviewed in literature [10].

To this purpose, the Belgian recommendations propose empirical rules for fixing tiles (with nails or other types of fixations) to keep the tiles from taking off. Three classes of fixation are considered: none, one out of every two, one out of every four tiles must be fixed [6]. The extensive damage due to windstorms in Belgium in 1990 has made it evident that the assumptions made for C_{pia} in the existing Belgian Wind Standard NBN B03-002-1 [7], for example, that 2/3 of C_{pe}, in function of the percentage of frontal openings, are not always a correct representation of the reality. It turns out that this approach is a very rough simplification, but until now, there has been no alternative. It was felt that the contribution of the fluctuating internal pressure field might be much more important than thought before. Therefore, the start of an extensive research programme leading to a better knowledge of the internal pressures was estimated necessary.

To perform this research, all the parameters that play a role in determining the net forces on tiles have to be taken into account: 1. structure of the oncoming wind flow: mean values and atmospheric turbulence characteristics, 2. global geometry of the building: plan, shape of the roof, overhangs, parapets, 3. frontal and lateral permeability values of the cladding system: details of roof construction, type of tiles, etc, and 4. local geometry of the cladding: external shape of the tiles, overlap length, etc.

The research programme first concentrates on tiles. Later on, slates, façade elements and external ballast blocks for flat roofs may be treated following the same methodology. The aim is to evaluate the internal pressures underneath the tiles, resulting from a fluctuating, in time and space, external flow field, and given the rest of the above-mentioned parameters. For this purpose a computer program based on finite volumes has been developed. Permeability values are measured in the lab; the influence of the external shape of the elements results from stationary Computational Fluid Dynamics (CFD) calculations. The validation of the numerical method and necessary input data required the set-up of a full-scale experiment, which will be described below. Other external data fields are provided by parametric studies from wind-tunnel tests.

3 Numerical model for the internal pressure

The numerical model used in this study is a fairly simple one, based on the work of CSTB (Centre Scientifique et Technique du Bâtiment, France), expressing the fact that in a small volume under isentropic quasi-stationary conditions, the input in air flow leads to an internal pressure rise and vice versa [1,8,22]. The governing algebraic expressions are (with $\gamma = 1.4$ and Δp the pressure difference between two neighbouring volumes):

$$\frac{dp}{dt} = \frac{\gamma . P_0}{V} . q(t) \qquad (2)$$

$$q(t) = q_x(t) + q_y(t) + q_z(t) \qquad (3)$$

$$q_x = \Phi_x . S_x . \sqrt{\left(\frac{2.\Delta P}{\rho}\right)} \tag{4}$$

where Φ_x is the permeability coefficient through the surface S_x in the direction x.

In adopting a finite volume representation, these equations can be translated to the following expression:

$$\begin{aligned} \frac{dp_{i,j}}{dt} &= K_{i,j}.\sqrt{P_{ei,j}(t) - p_{i,j}(t)} \\ &+ XK.(\sqrt{p_{i+1,j}(t) - p_{i,j}(t)} + \sqrt{p_{i-1,j}(t) - p_{i,j}(t)}) \\ &+ YK.(\sqrt{p_{i,j+1}(t) - p_{i,j}(t)} + \sqrt{p_{i,j-1}(t) - p_{i,j}(t)}) \end{aligned} \tag{5}$$

in which XK,YK and ZK are factors that incorporate respectively the frontal permeability, the lateral permeability, in the parallel and perpendicular direction to the ridge of the roof.

This relation characterises the variation of pressure for a time-interval dt , in an element, numbered [i,j], of the grid, based of the values of the pressure of the neighbouring volumes. Of course, boundary and initial conditions have to be fed into the program.

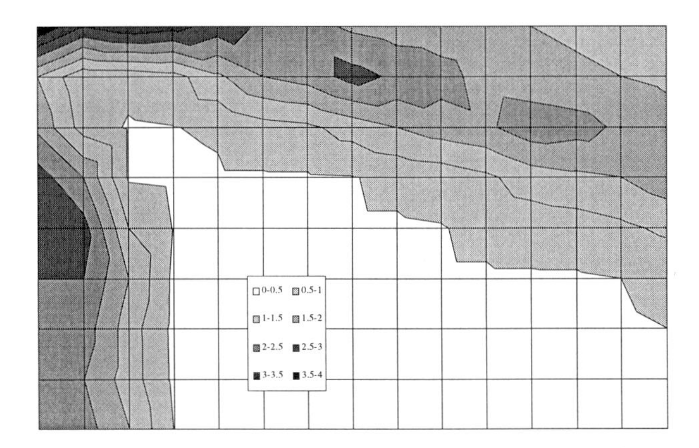

Fig. 2. Results obtained by BBRI using a numerical model to determine the pressure distribution at the upwind corner of the rooftop of a building.

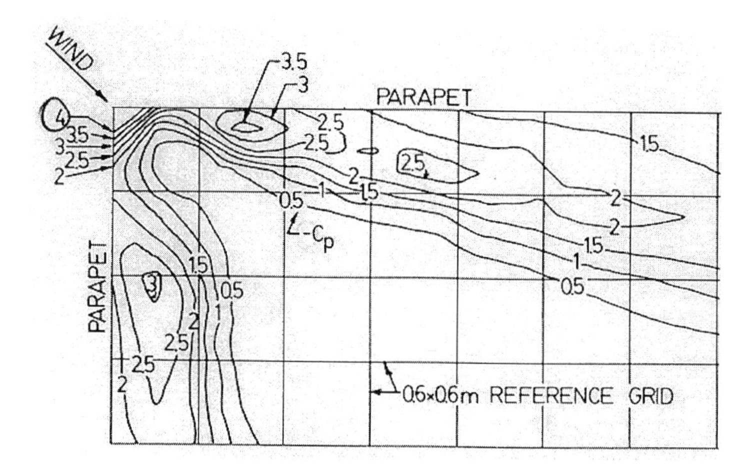

Fig. 3. Results obtained by Kind (1994) at the upwind corner of the rooftop of a building.

In CSTB, the non-linear set of coupled equations (Eqs 2-4) was solved with a 4th order Runge-Kutta algorithm. At BBRI, a much simpler direct iteration scheme is used (Eq. 5). In steady state, the program has been able to reproduce tests from literature e.g. results obtained by Kind [18] using the Laplace's equations (Figs 2-3).

In these tests, only time-average pressures were used. Kind considered that peak pressures could be dealt with in the same way because the fluctuating pressures appear to behave in a quasi-steady manner for typical energy-containing gust frequencies. The present research programme will give more precision as to the validity of this assumption, since, the numerical behaviour under fluctuating input data could be validated against field measurements.

In parallel, the VKI is working on a representation/transformation of permeable cladding as an analogous electrical circuit i.e. the internal space underneath the tiles as a capacitance (C), the porous outer cladding as a resistance (R) combined with an inductance (L, delay). The volume is connected with neighbouring volumes (capacitances) through resistance and inductance elements.

4 Experimental programme

4.1 Frontal and lateral permeabilities

The experimental programme is divided into two parts.

In the first part, the frontal and lateral permeabilities have been determined. The test programme has revealed a clear difference between the generally more air-open ceramic tiles and the more air-tight concrete tiles (which was to be expected due to the smaller fabrication tolerances). This can be seen in Fig. 4. The results indicate also very distinct frontal and lateral permeabilities (due to the geometry of the under-roof) as can be seen in Figs. 4-5.

In the second part, tests are being conducted to investigate the repeatability/reproducibility of the tests.

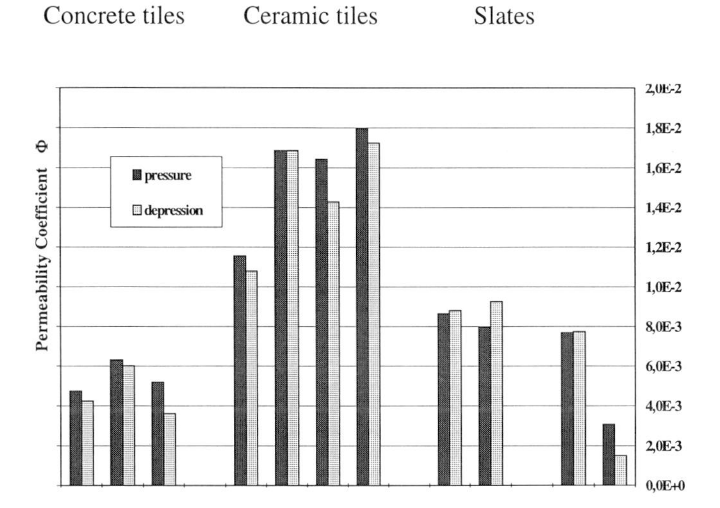

Fig. 4. Frontal permeability of several types of tiles.

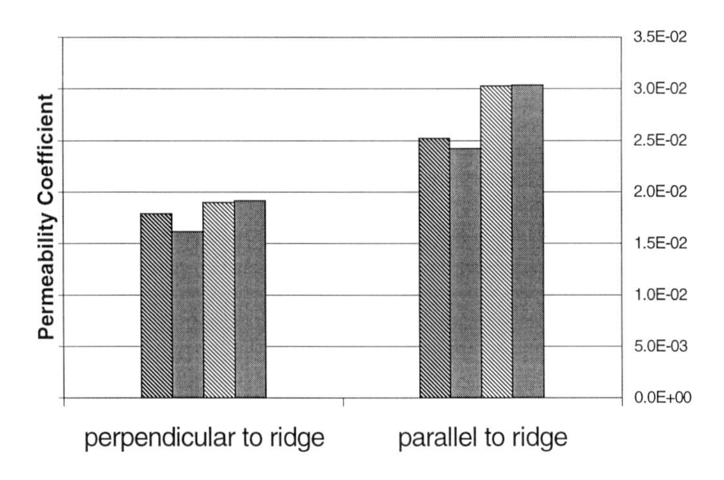

Fig. 5. Lateral permeability of several types of tiles.

Finally, some non-stationary experiments will be performed to investigate whether there is a delay factor in the air movement. In that case, a more explicit identification of this term must be investigated since numerical results will be dependant on the delay between the external pressure (input) and the internal pressure (output) when the wind moves through the tiles. At present, the most important delay appears to be due to the interaction between internal elements themselves in the volume modelling and not due to the connections (joints) of the volume elements with the external pressure.

4.2 Full scale experiment

An important part of the research programme consists of pressure measurements on a full-scale single-family dwelling equipped with a pitched roof and covered with ceramic tiles. The construction of the roof is of the type discussed above. The external and internal pressures are measured on a regular grid.

Additionally, in the zones where delta wing vortices arise, a more dense measuring system has been adjusted. This low-rise house has been conceived to rest on a rotary base, to allow any angle of the incident flow against the roof to be investigated. This house has been constructed on the experimental site of BBRI in Limelette (20 km from Brussels) during April 1999.

After preliminary tests, the set-up of the full-scale measuring system was realised. Data acquisition in the experimental house is automatically triggered by analysis (mean value and direction) of the last 5 minutes wind speed data recorded at 10 Hz. If the mean wind velocity over this period is beyond 7.5 m/s, some 100 pressure transducers and meteorological instruments are sampled at 20 Hz during 30 minutes (Fig. 6).

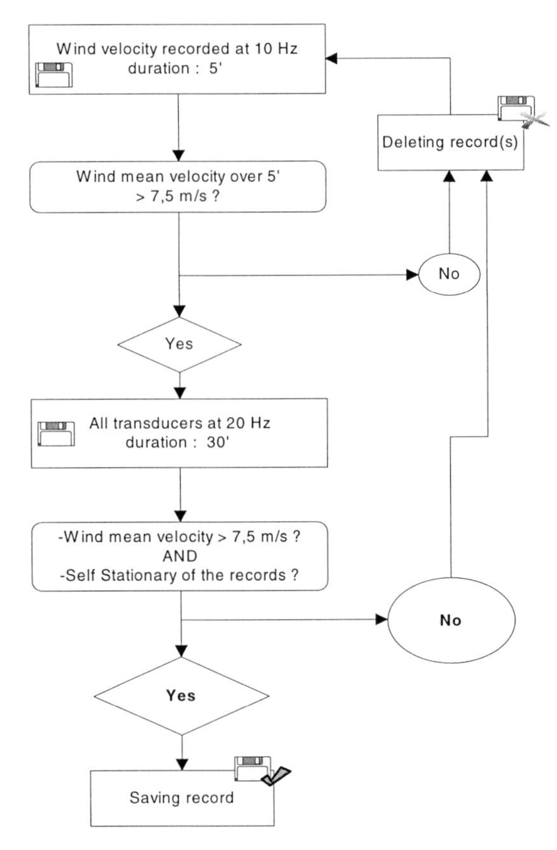

Fig. 6. Scheme of data analysis programme

After the end of each record, tests about wind velocity as well as consistency are automatically performed by the so called 'reverse arrangement' and 'run' tests [2] by Matlab® routines. If one or both tests show inconsistency for the meteorological data, all the data relative to all transducers are erased.

The architecture of the pressure measurement system is shown in Fig. 7. Each tapping point is connected to the pressure transducer through a 3–way valve. This enables a calibration of the transducers before each session of recording. The entire acquisition system, including calibration and analysis, is managed by a Visual Basic® program.

For meteorological data, a directional vane and two 3-cup anemometers (5 m and 10 m above ground) have been installed on a 10 m meteorological mast. Near the mast, a ground reservoir is put in the ground. This reservoir, located outside the zone where the static pressure field is influenced by the house [19], allows reference static pressure for all the pressure transducers to be obtained. This reference pressure is transmitted to the test building via a PVC pipe. This pipe has a large enough diameter to prevent the possibility of blockage due to water condensation, infiltration of small animals, etc.

At this moment (August 1999), some tests are performed to calculate the time-constant of the system. Hence, it will be possible to determine its sensitivity for dynamic amplification. The first measurements have been started (checks of the system) and several useful records are expected in the future (provided we can count on some heavy weather).

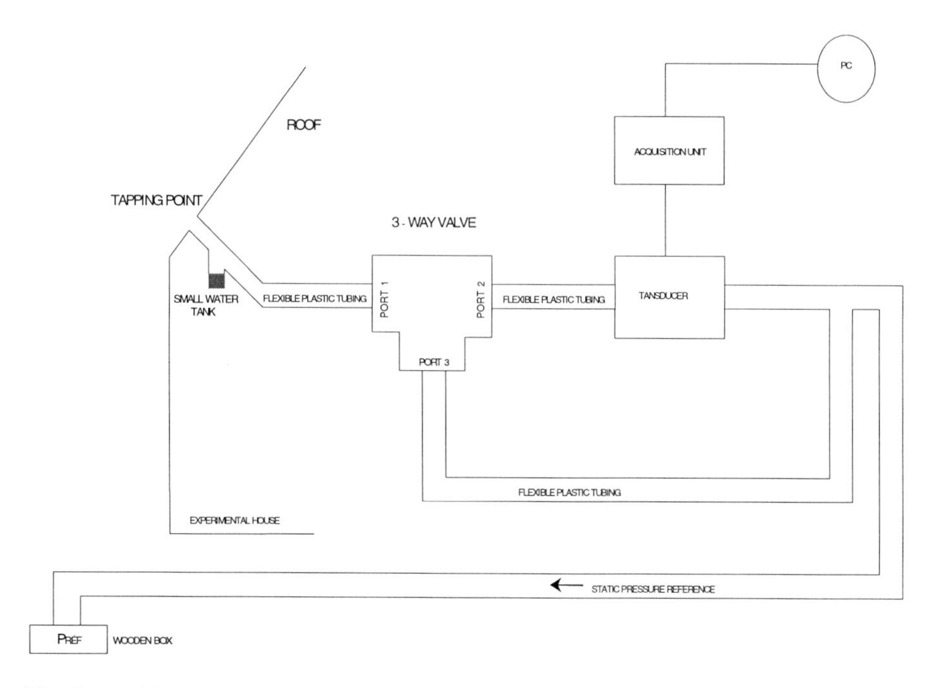

Fig. 7. Architecture of the pressure measurement system.

5 Wind-tunnel tests

The site where the experimental house is being constructed has, like every possible site, very specific topographic and roughness features. It is planned, therefore, to perform a parametric study in the atmospheric boundary layer (ABL) Wind Tunnel of VKI to obtain additional input data for the numerical model (on other geometric shapes, other roughness coefficients, etc.). Also, in a further stage of the research, some tests may be performed on a model of full-scale dwelling covered with thin (PE), porous sheets to simulate the permeable tile covering. This could result in a procedure for wind-tunnel testing of permeable building façades.

6 Conclusion

This research is divided into two parts: a theoretical and an experimental approach.

The theoretical approach is based on the results of CFD (Computational Fluid Dynamics) calculations, which have illustrated the influence of the external shape of the elements and provided a numerical characterisation of frontal permeability.

A numerical 'finite volume' model has been developed. This model has been validated with respect to literature; fluctuating pressures still have to be compared with the measurement results from the full-scale test building. Finally, an electric analogy model (for closed and open box) was developed in terms of resistance (R) , inductance (L) and capacitance (C).

The experimental approach is based on laboratory tests for measuring the delay (time filter) characteristics of the cladding in addition to frontal and lateral permeabilities. Small wind-tunnel tests have also been performed for studying the influence of the external shapes of the elements. ABL wind-tunnel tests are performed in the laboratories of the VKI by modelling the porous outer surface of the permeable covering with thin metal sheets with small openings. A full scale test building is erected to measure pressures on roof for all interesting wind directions.

Full scale tests are necessary in order to validate the numerical models and the ABL wind-tunnel approach. Parametric studies with electrical analogy, numerical model and ABL wind-tunnel tests shall later be realised to propose some adaptations to NBN B03-002-1 (Belgian Standard on Wind Loads).

7 Acknowledgement

The support received from the Belgian Federal Government (IWONL/IRSIA and Ministry of Economic Affairs) under grants CI 1/4-8827/208 and CC CIF-407 is gratefully acknowledged.

8 References

1. Amano, T. et al. (1988) Wind loads on permeable roof-blocks in roof insulation systems. *J. of Wind Eng. And Ind. Aerodyn*, No. 29.
2. Bendat, J.S. & A.G. Piersol (1996) *Random data. Analysis and measurements procedures*, J. Wiley & sons, New-York.
3. Bienkiewicz, B. & Y.Sun (1997) Wind loading and resistance of loose-laid roof paver systems. *J. of Wind Eng. And Ind. Aerodyn*, No. 72.
4. Belgian Building Research Institute. (1989) *Daken met pannen in gebakken aarde. Opbouw - Uitvoering*. BBRI, Brussel, TV 174.
5. Belgian Building Research Institute. (1995) *Daken met natuurleien. Deel 1. Opbouw en uitvoering*. BBRI, Brussel, BBRI, TV195.
6. Belgian Building Research Institute. (1996) *Daken met betonpannen. Opbouw en uitvoering*. BBRI, Brussel TV202.
7. Belgian Standards Institution (BIN-IBN). (1988) *Windbelasting op bouwwerken*. BIN-IBN, Brussel, NBN B03-002.
8. Chaibi, A. (1991) *Analyse des charges instantanées induites par le vent sur une paroi semi-perméable*, PhD thesis, CSTB, Nantes.
9. Cheung, J.C.K. & W.H.Melbourne (1988) Wind loading on a porous roof. *J. of Wind Eng. And Ind. Aerodyn.*, No. 29.
10. Cook, N.J. (1985) *The designer's guide to wind loading of buildings structures, part 1*, BRE, Garston.
11. Cook, N.J. (1990) *The designer's guide to wind loading of buildings structures, part 2: Static structures*, BRE, Garston.
12. Gerhardt, H.J. & C. Kramer (1990) Wind loads on permeable roofing systems. *J. of Wind Eng. And Ind. Aerodyn.*, No. 36.
13. Gerhardt, H.J. & C.Kramer (1983) Wind loads on wind-permeable building façades. *J. of Wind Eng. And Ind. Aerodyn.* No. 11.
14. Gerhardt, H.J. et al. (1979) On the wind-loading mechanism of roofing elements. *J. of Wind Eng. And Ind. Aerodyn.*, No. 4.
15. Gerhardt, H.J.et al. (1990) Wind loading on loosely-laid pavers and insulation boards for flat roofs. *J. of Wind Eng. And Ind. Aerodyn.*, No. 36.
16. Hazelwood, R.A. (1980) Principles of wind loading on tiled roofs and their application in the British Standard BS 5534. *J. of Wind Eng. And Ind. Aerodyn.*, No. 8.
17. Kind, R.J. & R.L..Wardlaw (1982) Failure mechanisms of loose-laid roof-insulation systems. *J. of Wind Eng. And Ind. Aerodyn.*, No. 9.
18. Kind, R.J. (1994) Predicting pressure distribution underneath loose-laid roof cladding systems. *J. of Wind Eng. And Ind. Aerodyn.*, No. 51.
19. Levitan, M.L. (1993) *Analysis of reference pressure systems used in field measurements of wind loads*. Ph.D thesis, Texas Tech University, Lubbock.
20. Ministerie van Verkeer en Infrastructuur. (1971) *Dakbedekkingen. Eerste Deel. Pannen- en leiendaken.*. Eengemaakte Technische Specificaties, Ministerie van Verkeer en Infrastructuur, Brussel, STS 34.
21. Simiu, E. & R.H. Scanlan (1996) *Wind effects on structures*, J. Wiley & sons, New-York.
22. Solliec, C. et al. (1993) Procédure de dimensionnement au vent des bardages et vêtures – approches expérimentales et numériques. *J.of Wind Eng.And Ind.Aerodyn.*, No. 48.

WIND LOADING ON PITCHED ROOFS WITH COMPLEX GEOMETRIC SHAPES

Wind loading on pitched roofs

N. TARANU, E. AXINTE, D. ISOPESCU and I. ENTUC
Department of Civil Engineering
Technical University of Iasi, ROMANIA

Abstract

The assessment of wind loading on building roofs requires good knowledge of the interaction between meteorological, aerodynamic and structural aspects of the problem. The paper presents some useful experimental results obtained from wind tunnel tests performed in the Wind Engineering Laboratory (WEL) at the Technical University of Iasi concerning the particular values of the pressure coefficients on pitched roofs with complex geometric shapes utilised in the third climatic zone. The experimental programme has been carried out in the WEL to evaluate how various factors influence the wind action on pitched roofs. Experimental measurements have been performed to determine the effects of the following factors:

- general slope and local change in slope,
- wind direction with respect to the building main axes,
- coupling of two or more sections in different configurations,
- local pressure distribution at ridges eaves and overhangs,
- geometrical discontinuities due to special architectural features,
- Height regime of the surrounding buildings, etc.

The experimental values of the pressure coefficients determined in the WEL for the special conditions mentioned above clearly indicate the need for special fixing conditions of the roof coverings and for additional checking of the secondary roof load bearing elements.

Keywords: Pitched roofs, pressure coefficients, wind loading and wind tunnel tests.

Abnormal Loading on Structures edited by K. S. Virdi, R. S. Matthews, J. L. Clarke and F. K. Garas.
Published in 2000 by E & FN Spon, 11 New Fetter Lane, London EC4P 4EE, UK. ISBN 0 419 25960 0

1 Introduction

Wind is a dynamic natural phenomenon and for this reason it is also a very complex problem from a structural point of view. In structural design of common buildings the basic approach to wind load analysis is to treat this dynamic phenomenon as a static load problem using the Bernoulli equation to translate wind speed into wind pressure.

Wind forces are influenced by the geometry of the structure (height, width, depth, plan and elevation shape) as well as the surrounding landscape.

New types of roofs have been used after 1980 in residential areas replacing the flat terraces by pitched roofs with complex shapes. These new types of roofs require accurate knowledge of the pressure (values and distribution) which provides a good basis for wind forces evaluation and structural design of roof elements.

2 Wind action on pitched roofs

The basic wind pressure (g_w) against a flat surface perpendicular to the direction of wind is determined by Bernoulli's equation:

$$g_w = \frac{1}{2} \rho v^2 \tag{1}$$

in which

ρ = the density of the air;
v = the speed of the wind.

The influence of the geometry of the building, the topography of the surrounding landscape, and the turbulent character of the wind are further described through the empirical wind pressure equation [2]:

$$p_n^n = \beta \, c_{h(z)} \, c_{n_i} \, g_w \tag{2}$$

where:

p_n^n = the nominal intensity of the wind component normal to the exposed surface;
β = the wind gust factor, whose magnitude is a reflection of the wind turbulence as well as the interaction between the wind and the structure and its elements;
$c_{h(z)}$ = coefficient of variation of the basic dynamic pressure which depends on the height of the building;
g_w = the basic dynamic wind pressure, calculated at a height of 10 m above open level terrain;
c_{n_i} = the pressure coefficient or shape factor that reflects the influence of wind on the various parts of a structure.

The pressure coefficient is the most difficult to understand and evaluate. It may have positive (i.e. pressure) or negative (i.e. suction) values and depends on the shape of the structure (plan and elevation) relative to the wind direction as well as whether the building is open or closed (airtight). The pressure coefficient relates to the flow of the wind around the structure and numerous tests have been run to determine the value of $\mathbf{c}_{ni}$ for different geometric characteristics. For complex building shapes it is always advisable to run wind tunnel tests to determine the wind loads and their distribution.

Individual components and local areas may have local pressure concentrations due to local disturbance of the airflow. Roof eaves and other projections are areas of concentrated wind pressure. A complex distribution of design pressures with higher values of the pressure coefficients is recommended by almost all national standards around ridges, eaves and overhangs. These localised distributions affect the strength of local supporting elements and the strength of fixtures, but they do not affect the overall resistance requirements of the structure.

3 The experimental programme

The experimental work has been carried out in the laboratory for physical modelling of wind-construction environment interaction provided with a boundary layer wind tunnel of the open circuit type.

The natural wind has been modelled [1] to account for the variation of mean wind speed with height above ground, Fig. 1a, and for the intensity and scale of the turbulence, Fig. 1b, appropriate to the terrain site. (In Fig. 1. Z_g is the gradient height, V_g is the wind speed at the gradient height, and V is the wind speed at the height Z).

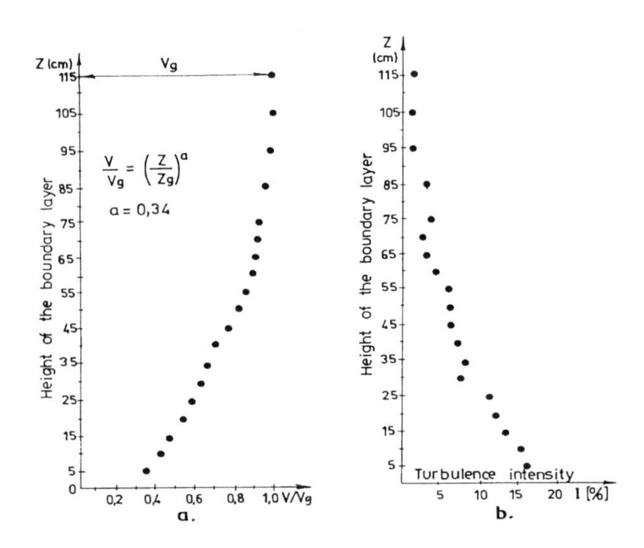

Fig. 1. Mean wind speed and turbulence intensity inside experimental chamber.

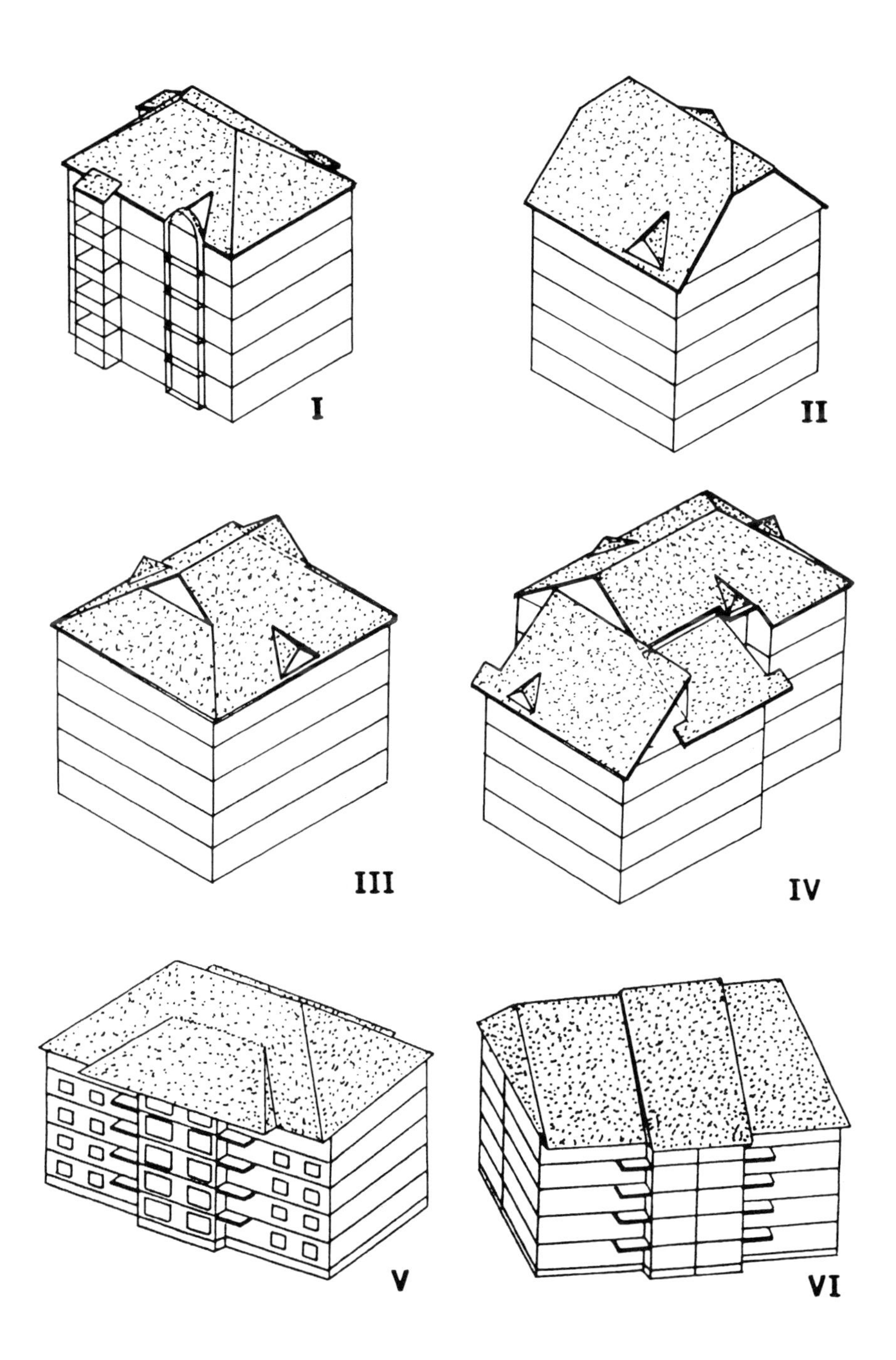

Fig. 2. Types of pitched roofs tested in the wind tunnel.

The experimental chamber of the wind tunnel has a cross section of 19600 cm^2 suitable for experiments concerning usual buildings up to five stories and urban areas of limited extent. Classical measuring devices and a TSI system have been utilised to determine pressures and mean velocities, standard deviation and the turbulence intensity.

The model scale (1/200) has been selected in such a way that the specimen does not obstruct more than 10% of the tunnel active cross-section. Six different types of pitched roofs, Fig. 2., have been tested. Each roof has been designed with three different slopes corresponding to those required for ceramic tiles and zinc coated steel plates.

In order to obtain a detailed characterisation of the airflow, a mesh of measuring points have been positioned on the roof surfaces, Fig. 3.

The test models have been immersed in a turbulent boundary layer with a height equal to 115 cm and a Reynolds number Re = 120000. A round mobile table has been utilised to modify the wind velocity direction with respect to the building axes.

Measurements have been performed for each section and layout in 12 positions corresponding to an angular sector of 30^0. The pressure coefficients have been determined in all measuring points for each angular position.

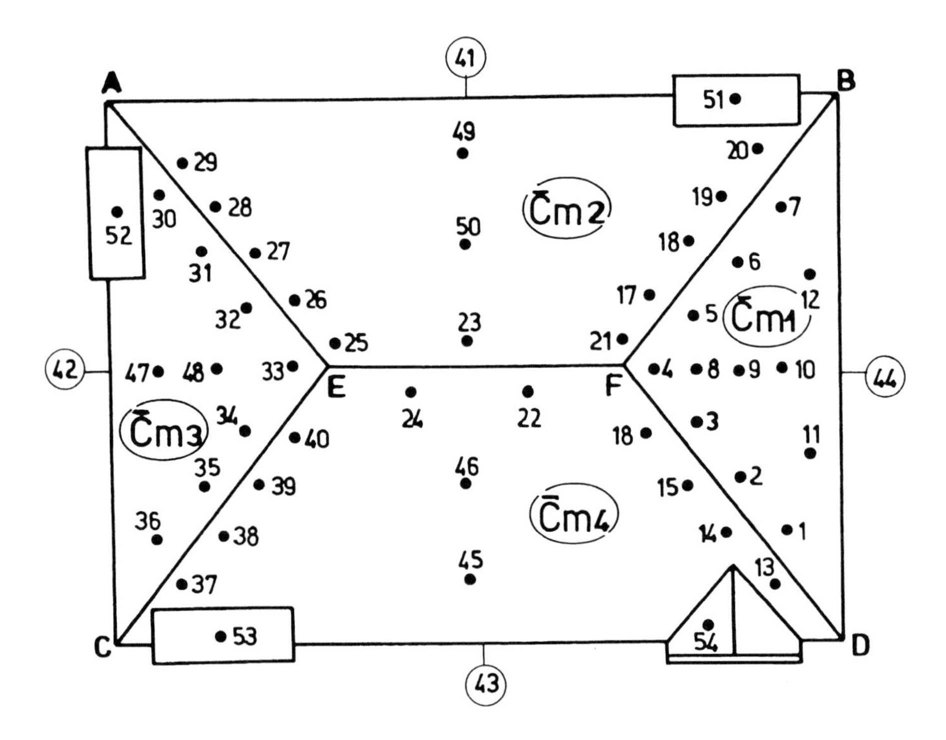

Fig. 3. Measuring points on and characteristic areas for roof type I

Each model has been tested individually and also immersed in some typical urban planning schemes. A five story building scale model with a pitched roof type I (Fig. 2.) located as shown in Fig. 4. has been tested to reveal the influence of the built environment on the local pressure coefficients [2]. Each urban planning scheme has a different pressure field depending on the relative position of the building model against neighbouring construction and their height, and also on the street network in the urban area [3]. In addition, the local pressure coefficients on roofs have been evaluated with respect to:

- variation of the general slope,
- local change in slope,
- air flow with respect to the principal axes of the building model,
- pressure distribution along ridges and eaves,
- geometrical discontinuities due to some particular features, and
- coupling of two sections of the same roof types.

The maximum values of the individual pressure coefficients have been identified for each experimental set up, layout and wind direction.

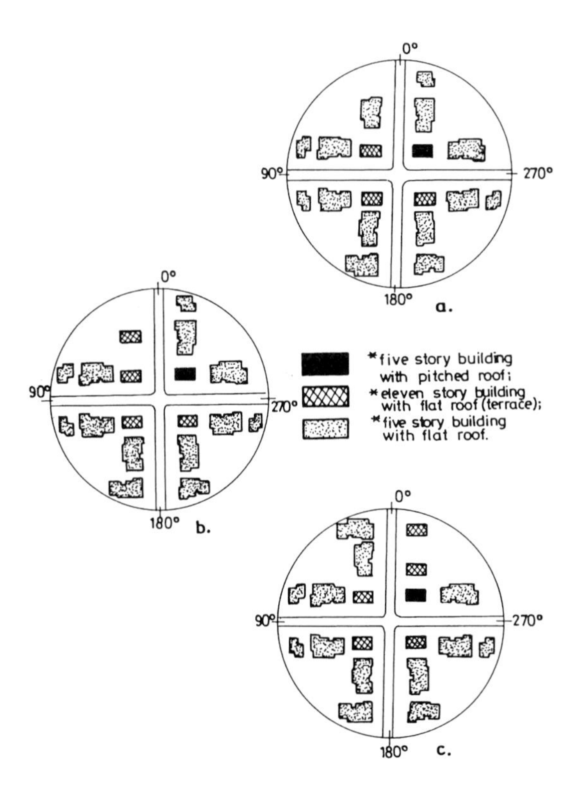

Fig. 4. Lay-out schemes analysed in the experimental programme

4 Experimental results

Based on the experimental data measured in the wind tunnel tests, the pressure coefficients have been determined with the following formula:

$$c_n = \frac{p_t - p_0}{p_d} \tag{3}$$

where

p_o=	the upstream static pressure measured in an area which is not disturbed by the building under study;
p_t =	the local pressure in a point on the roof surface;
p_d =	the basic dynamic pressure at the height of the measuring point.

A special computer programme "PRESS" has been written to process the individual pressure coefficients and to evaluate the average pressure coefficients $\overline{C}_{m1}, \overline{C}_{m2}, \overline{C}_{m3}$ and $\overline{C}_{m4}$ on the characteristic areas (DFB, ABFE, AEC and CEFD respectively) illustrated in Fig. 3.

The extreme values of the individual pressure coefficients (positive and negative) have been identified for each physical model and experimental condition.

The wind pressure distributions on pitched roofs with complex geometric shapes cannot usually be determined using the presentation existing in national codes and standards. Therefore wind tunnel tests based on physical modelling are needed to obtain an accurate evaluation of the wind loads. Some of the most important results of the experimental programme are:

- The local pressure coefficients are significantly influenced by the complexity of the roof geometry and by the characteristics of the surrounding buildings;
- The initial low pitches increase upward drafts much beyond standard prescriptions;
- Planning details can increase locally the wind forces on the roof surfaces;
- The most unfavourable directions do not coincide with the principal axes of buildings;
- All values of the average pressure coefficients determined for the lay-out schemes shown in Fig. 4 are negative (Fig. 5) although the standard prescriptions would suggest positive values for slopes higher than 30°. This result emphasises the need for proper anchoring of the roof structure to the main building structure;
- High positive values (1.4 …1.7) have been identified on vertical positions of the roof type I, II, and IV, requiring special strength cheeks of the roof elements;
- For sharp edges, ridges and eaves high negative pressure coefficients (-3.5 … -5.15) have been identified on larger strips than those indicated in national standards.

The experimental results indicate the need for special fixing conditions of the roof coverings and for special checking of supporting roof elements.

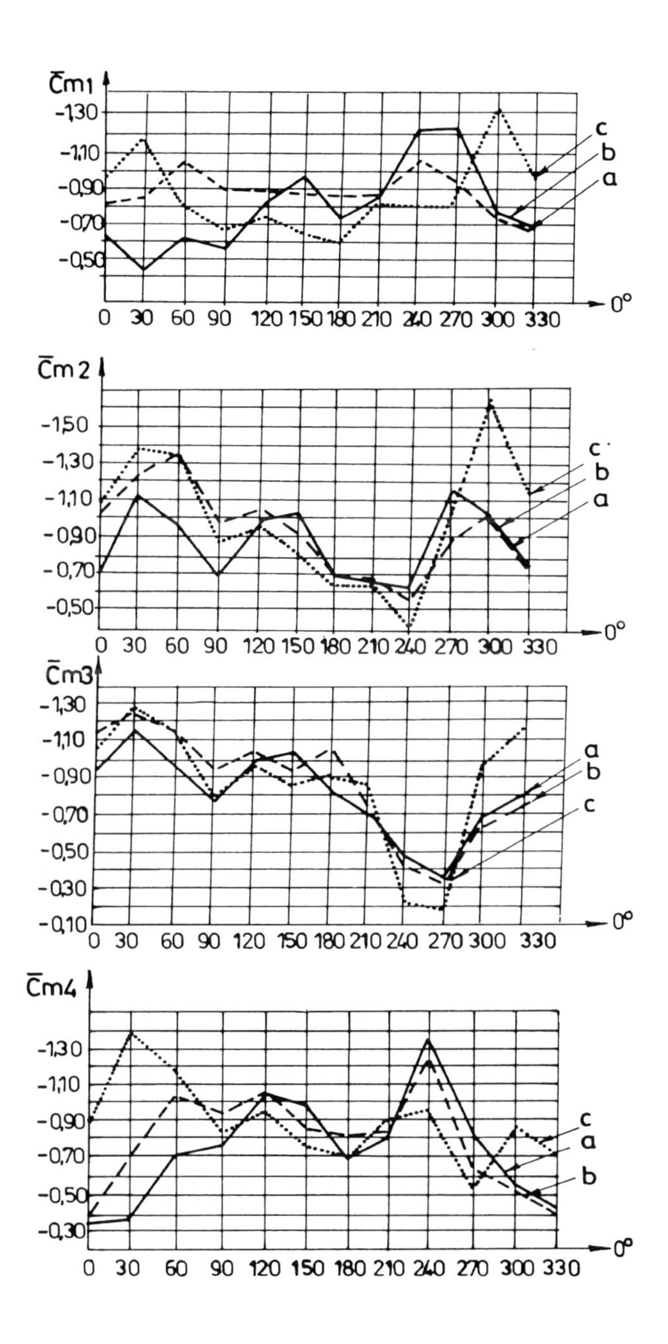

Fig. 5. Angular dependence of the average pressure coefficients on the characteristic areas

5 References

1. Cook, N.J. (1985) *The designer's guide to wind loading on building structures,* Butterworths, London.
2. Taranu, N., Axinte, E. and Isopescu, D. (1993) Optimizarea constructiilor si sistematizarea localitatilor pe baza interactiunii intre actiunea vantului si mediul construit, *Contract, Techn. Univ. "Gh. Asachi", Iasi,* MLPAT, 1309.
3. Axinte, E., Taranu, N. and Isopescu, D. (1997) Influence of the height regime of surrounding buildings on pressure field induced by wind on pitched roofs, *Buletinul Institutului Politehnic Iasi*, Tom XLIII, Fasc. 1 - 2, pp. 61 – 67.

TOWARDS A STATISTICAL HUMAN BODY MODEL FOR DETERMINING JUMP DYNAMIC FORCES

Human body stiffness

T.D.G. CANISIUS
Structural Reliability and Risk Analysis Unit, Building Research Establishment Ltd, Watford, UK

Abstract

This paper presents preliminary results from research on developing a statistical human body model for use in calculating jump effects on scaffold boards. This work, which takes forward the work of a previous research programme, is aimed at using available measurements to predict jump effects under conditions not considered previously.

Keywords: Human body model, jump effects, scaffold boards, structural dynamics

1 Introduction

Scaffold boards, which form part of temporary access platforms, are commonly subject to forces that result from human jumps. In order to obtain information on these impact forces, through the sponsorship of the Health and Safety Executive, BRE have conducted an investigation aimed at quantifying such forces. The results of earlier work were published in two previous papers by Canisius *et.al.* [1] [2]. The work reported in this paper is aimed at using the available measurements to predict jump dynamic effects under conditions not considered previously.

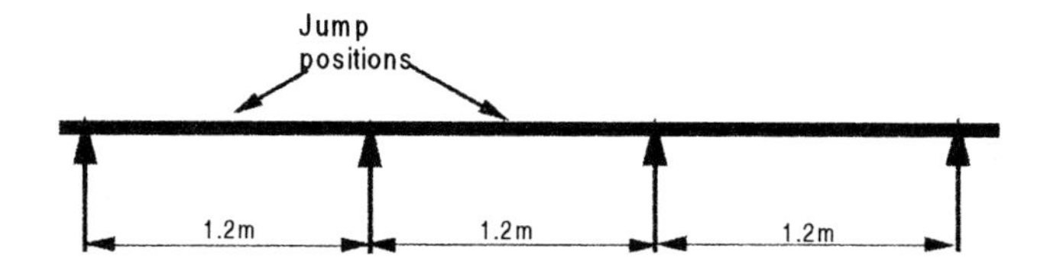

Fig. 1. Scaffold board test set-up.

Abnormal Loading on Structures edited by K. S. Virdi, R. S. Matthews, J. L. Clarke and F. K. Garas.
Published in 2000 by E & FN Spon, 11 New Fetter Lane, London EC4P 4EE, UK. ISBN 0 419 25960 0

1.1 Previous work on scaffold boards

Fig. 1 shows the arrangement of scaffold board supports and jump positions used during the study of Canisius *et.al.* [1] [2]. In this case the scaffold boards were simply placed on the supports, allowing lift-off where relevant.

Fig. 2 shows typical time history curves for the deflection of the scaffold board at the jump positions. There, the results for two types of jumps, *viz.* soft and hard jumps, are shown. A soft jump is identified where the jumper flexes the knees to reduce the impact while a hard jump is where flexing of the knees is not done voluntarily. During this work it was established that soft and hard jumps, respectively, occur on the balls and heels of the feet.

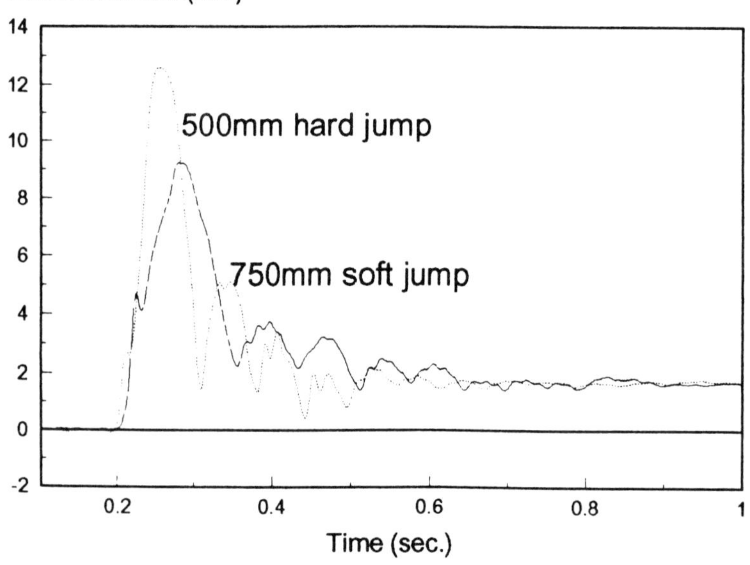

Fig. 2. Typical time histories of board deflection at jump position.

The work reported by Canisius *et.al.* [1] [2] used impact load factors (ILFs) to express human jump load effects on timber scaffold boards. The ILFs were defined such that their product with the weight of the jumper provided the equivalent static forces on the board (as described by the corresponding peak board deflections at the jump position). During the tests it was found that when jumps are repeated on the same board position by the same person, the ILFs do not remain the same. The reason for this is that the purposeful flexing of knees of a soft jump could vary from jump to jump, while a 'hard' jump may contain varying amounts of involuntary knee flexing. An example of this variation can be seen in Fig. 3 where the effects of repeated jumps of a single person, on to the same board, are shown.

In the above work the ILFs were classified according to the jump height and then, due to their natural variations, were described in terms of probability distributions. Also for ease, and justifiably for the purpose, the results from jumps at the mid-span and outer spans were combined together, and this provided an average value for the ILF, which was nearly independent of the board stiffness. However, some slight variation with board stiffness, generally with opposite sense was identifiable when considered separately. In addition, the variation of ILF fractiles with jump height was expressed using an approximate analytical formula that was calibrated against experimental results. Fig. 4 shows the ILF statistics for the case of soft jumps expressed in terms of percentiles of the probability distribution and the jump height.

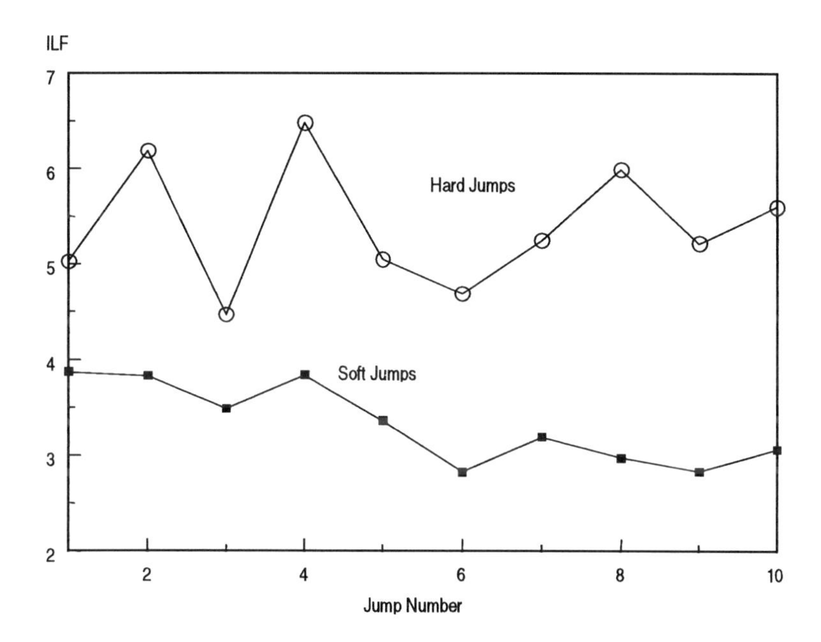

Figure 3: Variation of jump force (as ILF) from repeated jumps.

1.2 Current work and the need for new developments

The previous work mentioned above had several limitations. As the impact effects were described only as ILFs, the results were valid only within the scaffold board stiffness range used in the tests. Also, they were applicable only for boards 'simply placed' on the supports. Thus, the impact forces on platforms (and even floors) with stiffness outside this range, and on scaffold boards where the boards are tied to the supports, cannot be predicted from these ILF formulae developed for other circumstances.

This paper presents preliminary results from a method being developed for predicting human impact forces on platforms (of differing stiffnesses) not considered in the above test programme. This method is also capable of predicting impact forces and platform responses under jumps from heights not considered in the tests. However, as before, the relations to be developed will be constrained to be applicable only to jumps by people having a similar physical build to that of the 'representative' person who carried out the bulk of the test jumps.

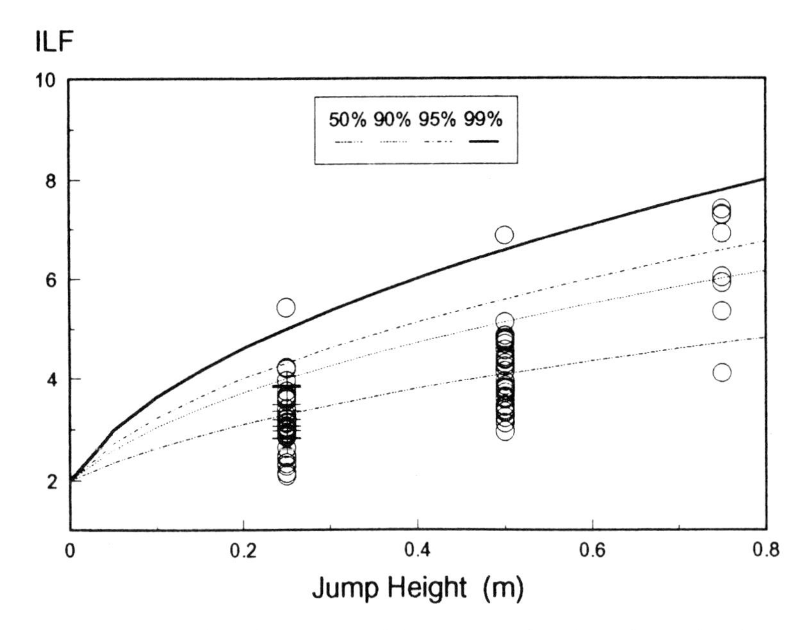

Fig. 4. Fractiles of ILFs for soft jumps, as a function of jump height.

The method proposed would ultimately express human body stiffness in a probabilistic manner so that, either voluntary or involuntary, variations in jumps can be considered in evaluating jump forces. This would aid in the expression of jump forces probabilistically as required by modern reliability-based design methods.

2 Basic Considerations

The work presented here assumes that the human body can be represented as a single degree of freedom (SDOF) dynamic system as shown in Fig. 5. Of course, it is possible to model a body with increasing numbers of degrees of freedom, but this is not considered to be necessary here. Also, damping is considered to be of no significance to this first-peak impact problem.

The mass of the human body SDOF system can be considered as the mass of the jumper. Then, the non-constant nature of ILFs for repeated jumps indicates that the spring stiffness to be used in the human body model, which is the only variable system parameter, cannot be taken as constant. The effect of this control, which may be voluntary or involuntary and may give rise to soft and hard jumps, should be bounded both from below and above. For example, no human control of the jump effect may normally be possible during a high-speed impact from a very high position. Another reason for variations in ILFs can be fluctuations in the way jumps are initiated by a jumper; the effect of this random variation is expected to be indirectly captured by the final statistical evaluation of results proposed here.

The spring constants for different jumps are obtained here by considering the motion of the SDOF system's impact with the scaffold board. The scaffold board is assumed to be massless and is represented by its generalised stiffness and an independent degree of freedom. The human body spring stiffness necessary to provide the experienced peak board deflection is determined here as the primary item of interest. The calculated time to peak displacement is checked, but only to make sure that it is not wildly off that measured. On having carried out these calculations for each individual jump, it is intended to evaluate the human body spring stiffnesses statistically to provide a probability distribution that may be used in calculating impact forces and times-to-peak deflections in other situations.

3 Theory and implementation

Consider the impact of a human of mass M, who jumps on to a scaffold board from a height h. Assuming that the scaffold board is without mass, the dynamical system can be idealised as shown in Fig. 5. The generalised stiffness of the mass-less scaffold board can be obtained by considering its deflection under a static load placed at the jump position. It is expected to improve the model by incorporating scaffold board mass effects.

Using the notation denoted in Fig. 5, for a jump from a height h, the time to peak scaffold board deflection T can be calculated from standard structural dynamics theory as:

$$T = \tan(-\varphi)\sqrt{\frac{M(K_0+K_b)}{K_0K_b}} \tag{1}$$

where

$$\varphi = \sqrt{\frac{2hK_0^2}{gM}\frac{(K_0+K_b)}{K_0K_b}} \tag{2}$$

Similarly, the maximum board deflection δ_b can be calculated as

$$\delta_b = \frac{K_0}{K_b}\delta_0 \tag{3}$$

where

$$\delta_0 = \sqrt{\frac{2Mgh(K_0+K_b)}{K_0K_b}}\sin\theta + \frac{Mg}{K_0}(1-\cos\theta) \tag{4}$$

with

$$\theta = T\sqrt{\frac{K_0K_b}{M(K_b+K_0)}} \tag{5}$$

In the above g is the acceleration due to gravity. The equivalent human body spring stiffness that provides the experimental maximum board deflection is iteratively obtained for each jump from the above equations.

4 Results

The measurements from the repeated jump case reported in [2], and described above, are used in the preliminary analysis reported here. These included 10 soft jumps and 10 hard jumps each by three persons, individually, from a height of 0.25m above the scaffold board. For the presentation of results these have been denoted as follows:

- Jumps 1 to 10: Soft jumps by Jumper A.
- Jumps 11 to 20: Hard jumps by Jumper A.
- Jumps 21 to 30: Soft jumps by Jumper B.
- Jumps 31 to 40: Hard jumps by Jumper B.
- Jumps 41 to 50: Soft jumps by Jumper C.
- Jumps 51 to 60: Hard jumps by Jumper C.

Of the three jumpers, jumper A had a tall and lean build. The others had stocky builds, with jumper B being the stockier.

Fig. 6 presents equivalent human body stiffness calculated for each jump. The x-axis legend of the figure provides the jump number, the type of jump (soft or hard) and the jumper (A, B and C). Due to the different amount of voluntary knee flexing, there are variations in the equivalent body stiffness determined for soft jumps. As some involuntary slight flexing of knees occurs during hard jumps, the equivalent stiffness for them too is not unique for a given jumper, although they may be expected to be the same for nominally identical jumps.

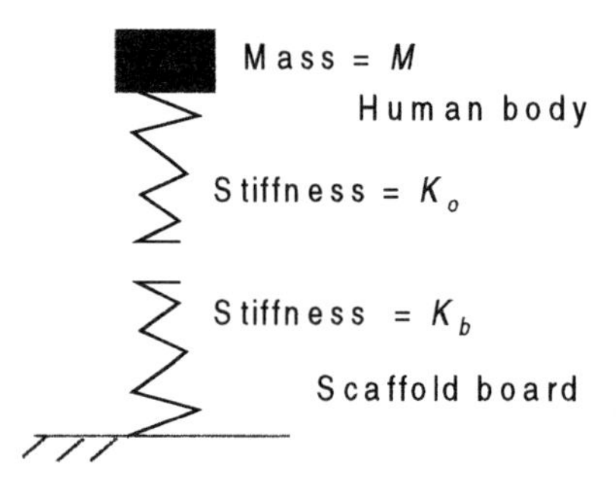

Fig. 5. Idealisation of the human body and the scaffold board.

The results clearly indicate that the equivalent body stiffness is lower, except on one occasion, for the soft jumps than for hard jumps. In fact, this is to be expected. The soft jumps also show a jumper's tendency to 'learn', *i.e.* the person attempts to soften the impact further as the series of jumps progresses. This indicates that only the first soft jump from among repeated jumps should be considered as relevant for design calculation purposes.

Fig. 7 presents the measured and calculated times from contact of the body to the peak deflection of the board. As can be seen, the calculated times are greater than the measured times. The reason for this could be the fact that the human body is not a single degree of freedom system, and its equivalent stiffness varies during impact. This is supported by the fact that it is the soft jumps, where more effort is made by the jumpers to soften the impact, that provide larger differences between the measured and calculated times. Fig. 8 presents the differences in measured and calculated times as percentages of the measured values. Although the maximum error is nearly 550%, they are generally lower than 300%. Even at the maximum error, the change is not sufficiently great to significantly affect strain-rate related calculations relevant for design purposes *e.g.* the determination of material strength. Thus, the method presented here, on being expressed statistically, can be seen as a way forward for quantifying human body dynamic characteristics for jump effect evaluations.

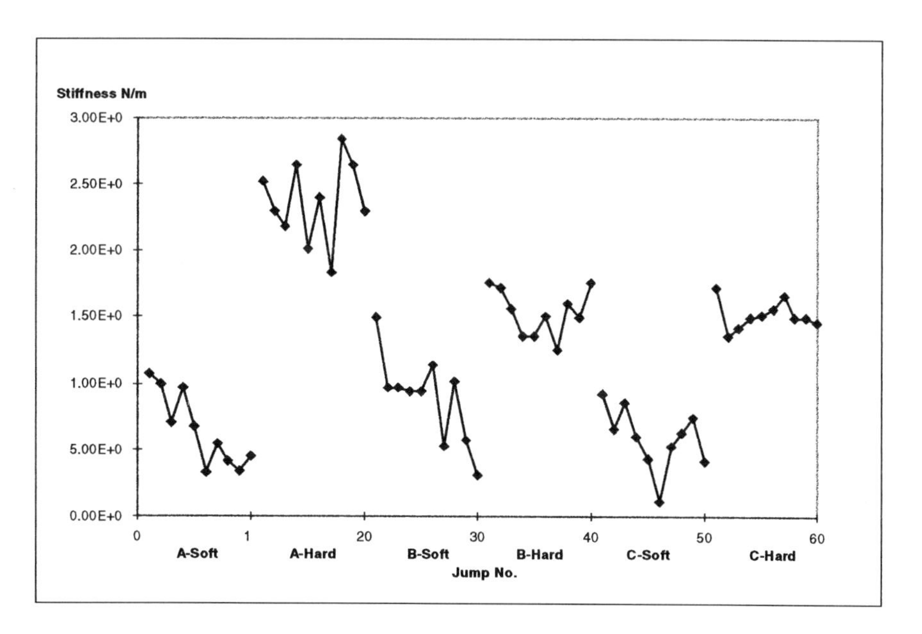

Figure 6: The Equivalent Body Stiffness for the six sets of repeated jumps by three persons.

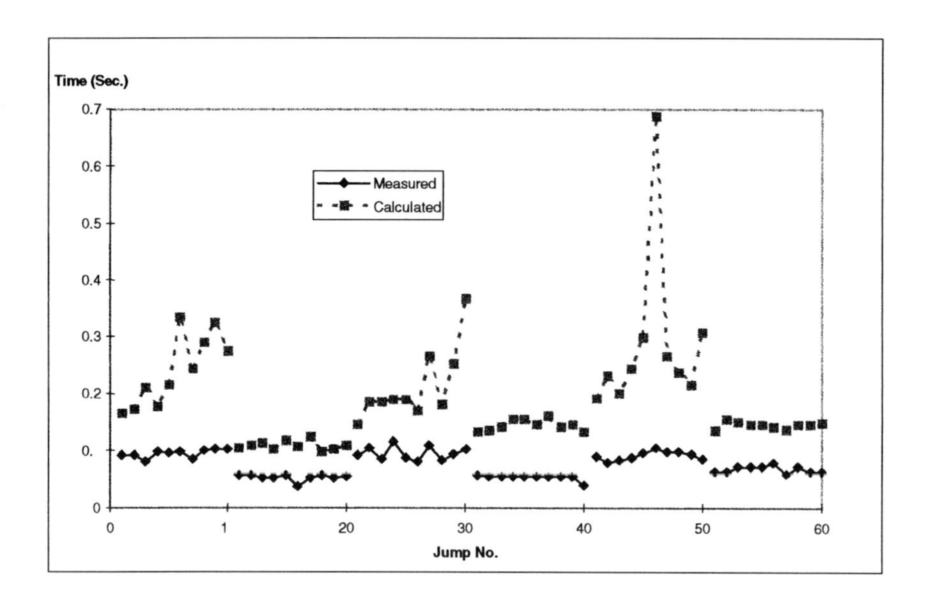

Figure 7: The calculated and measured times to peak displacement.

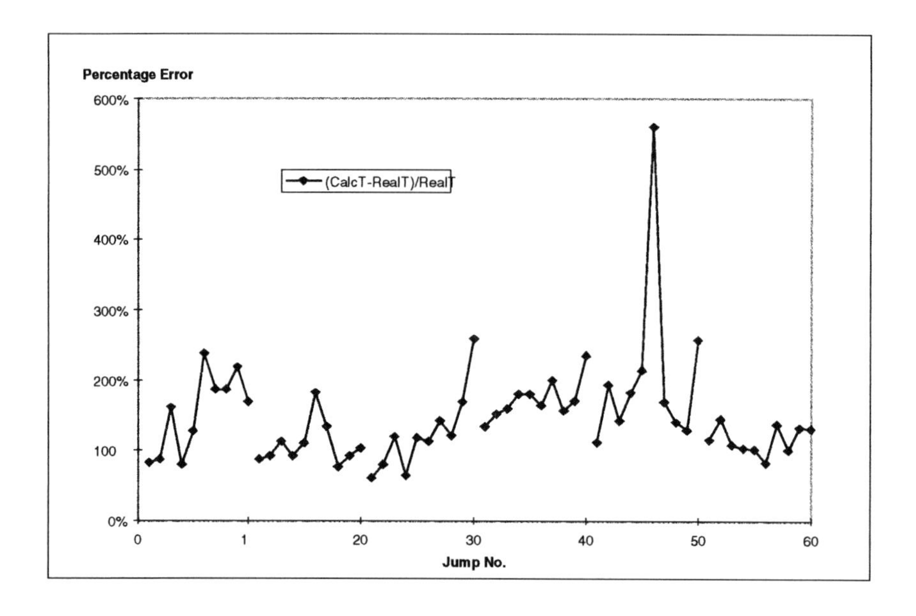

Figure 8: The difference between calculated and measured time to peak displacement for each jump.

5 Conclusion

This paper presents a method for obtaining an estimate of the equivalent human body stiffness to be used in jump effect calculations. This could help calculate impact forces on surfaces outside the scope of the current work.

The body stiffness values, determined so as to provide the measured peak deflections and the effective forces, produce a sufficiently accurate prediction of the measured times to the peak deflections. This confirms the validity of the proposed human body model for practical calculations.

6 Acknowledgement

The author wishes to acknowledge

- the invaluable contribution of Mr. A. Bougard of BRE who prepared the data used in the analyses; and
- the support of the Science and Technology Division of the Health and Safety Executive of the UK which sponsored the original investigations.

7 References

1. Canisius, T.D.G., Bougard, A.J. and Ellis, B.R. (1997) *Human impact loads on timber scaffold boards,* Proc. of Seminar 'Working at Height', Maitra, A. (*Editor*), Thomas Telford, London (in press).

2 Canisius T.D.G, Bougard A.J. and Ellis B.R. (1998); *Performance Timber Scaffold Boards Under Human Jump Load Effects*, Proc. ICE, Buildings and Structures, pp. 332-341, November.

THE DEVELOPMENT OF A SEISMIC TEST FACILITY FOR FULL-SCALE STRUCTURES

Facility for full-scale seismic test

P.A.MOTE
President, Nevada Testing Institute Inc. Las Vegas, USA
P.R. GEFKEN
SRI International, Menlo Park, USA
G.S.T. ARMER
GA Consultants, Boston, UK

Abstract
The Nevada Testing Institute (NeTI) is developing a full-scale seismic testing centre at the Nevada Test Site (NTS) to allow full-scale testing that in turn will reduce loss of life and property damage that result from medium and large earthquakes near metropolitan areas.
Keywords: Earthquake, full-scale testing.

1 Introduction

The disasters at Northridge, California, and Kobe, Japan, are testimony to the need for testing earthquake resistant designs in full scale. A testing centre, capable of hosting full-scale structures and testing them by shaking the ground they rest on, does not presently exist. However, the Nevada Testing Institute (NeTI) is developing a full-scale test centre at the Nevada Test Site (NTS), 105 km north west of Las Vegas, Nevada. The goal of the Centre is to allow full-scale testing to assess the safety of structures, that in turn will reduce loss of life and property damage that result from medium and large earthquakes near metropolitan areas.

The need for full-scale seismic testing was identified over 25 years ago at a National Research Council, National Academy of Engineering, and National Science Foundation (NSF) Workshop on Simulation of Earthquake Effects on Structures in San Francisco. However, building a full-scale testing facility has not been practical up to now because of the inherent limitations posed by traditional shaking table technology. The weight and dimensions of a full-scale, multistorey structure would require such a massive test facility that costs alone have prohibited its development.

However, a new technology has been demonstrated that promises affordable testing at full scale. NeTI was founded to apply this technology at the NTS, to facilitate the

Abnormal Loading on Structures edited by K. S. Virdi, R. S. Matthews, J. L. Clarke and F. K. Garas.
Published in 2000 by E & FN Spon, 11 New Fetter Lane, London EC4P 4EE, UK. ISBN 0 419 25960 0

use of the site by the international science and engineering research communities, and to enhance their efforts to achieve the goals of the national earthquake hazard mitigation agenda. NeTI's primary objective is to develop advanced and unique testing capabilities for earthquake hazard mitigation designs and technologies related to civil structures and their foundations.

2 Testing development

NeTI and its partner, SRI International (SRI), are developing the ground motion generation centre to deliver a wide range of tailored strong ground motions to test beds on which large and full-scale multi-storey civil engineering test structures can be erected. The Centre makes use of the fundamentals of the Repeatable Earth Shaking by Controlled Underground Expansion (RESCUE) technique, conceived by SRI over a decade ago and funded by the National Science Foundation (NSF). Development activities have centred around the production and evaluation of ground motion on a 1/7-scale basis. The plan is to begin generating full-scale motion by 2000.

The RESCUE technique produces ground motions by simultaneous expansion of a planar array of buried vertical sources. The current source design, shown in Fig. 1, comprises a rubber bladder around a rectangular mandrel. Propellant is burned in steel canisters to produce high-pressure gas, which is subsequently vented in a controlled manner into the source module to produce low-pressure gas (up to 1MPa). This causes the rubber bladder to expand and move the soil, thus providing input motion to the structure being tested.

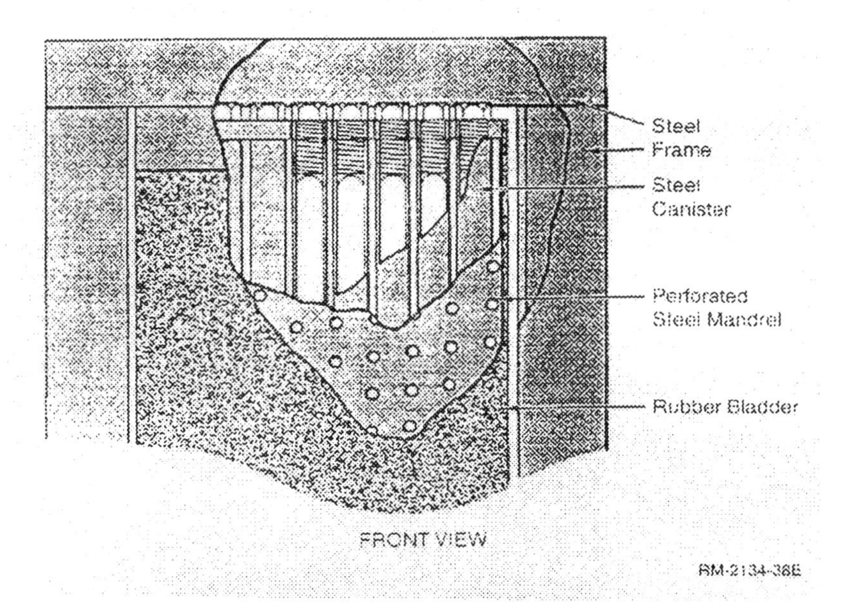

Fig. 1. SRI's Repeatable Earth Shaking by Controlled Underground Expansion (RESCUE) source design.

Because the sources generate low pressure, they do minimal damage to the soil test bed, allowing for sequential pulses to be applied to the soil for multiple tests. Each source design calls for up to 20 canisters, to produce 20 applied-pressure pulses over a time span of 10 to 60 seconds. Each pressure pulse typically produces two cycles of ground motion because of the elastic rebounding response of the soil medium. Large ground motions, particularly with respect to ground displacements, are created by surrounding the test bed with trenches.

Since 1996, proof-of-concept (POC) analysis and testing on a scaled ground motion source module which produces 1/7-scale strong ground motion, have been conducted by the NeTI test team, which includes SRI, NeTI, Bechtel Corp., Bechtel Nevada, Defense Special Weapons Agency (now part of the Defence Threat Reduction Agency), Department of Energy/Nevada Operations, and the Lawrence Livermore National Laboratory (LLNL). POC testing was completed during the summer of 1998 at the NTS. The results of these tests show that the RESCUE technique can produce strong, tailored ground motion generating pressure pulses to simulate specific response spectra, ground displacements, and horizontal accelerations and velocities required by the structural engineering research community. For the POC test series conducted at the NTS, NeTI formed working partnerships with LLNL and the Universities of California at Irvine (UCI), Nevada at Las Vegas (UNIV), and Nevada at Reno (UNR), to assist in the evaluation of SRI's RESCUE technique as well as to provide test support for university and other research programmes. In particular, the ground motion generated at NTS was used to enhance the partners' research programmes in the areas of dynamic soil behaviour and the dynamic response of electrical isolation equipment.

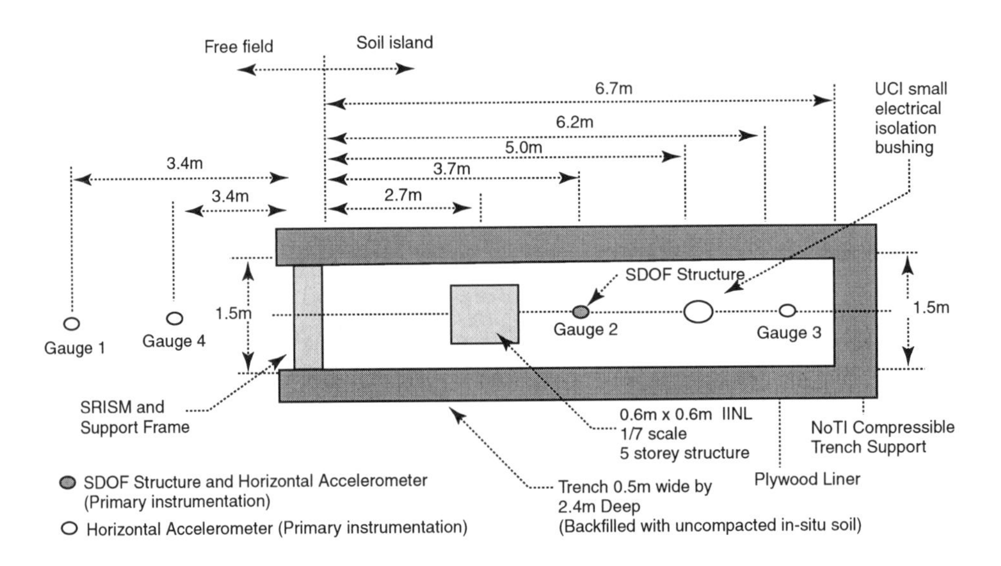

Fig.2. 1/7-scale soil island test bed at NTS with RESCUE source module

The evaluation tests illustrated the following capabilities of the RESCUE technique: tailoring of source peak pressure and pulse shape, source load and ground motion reproducibility, large ground motions, and hardware durability.

Fig. 2 shows the overall testing arrangement at NTS, which consisted of a single 1/7-scale RESCUE SRI source module loading one side of a soil island. This scaled source module contained two propellant canisters, giving two applied-pressure pulses to the soil island.

Fig.3 illustrates the ability to tailor the applied-pressure pulse shape by adjusting the timing and firing sequence. The longer pulse is ideal for generating long-period type motion, whereas the short pulse is ideal for generating vibration motion when pulses are produced sequentially and propagated in forward and reverse directions. The magnitude of the peak pressure can be varied by simply altering the propellant quantity.

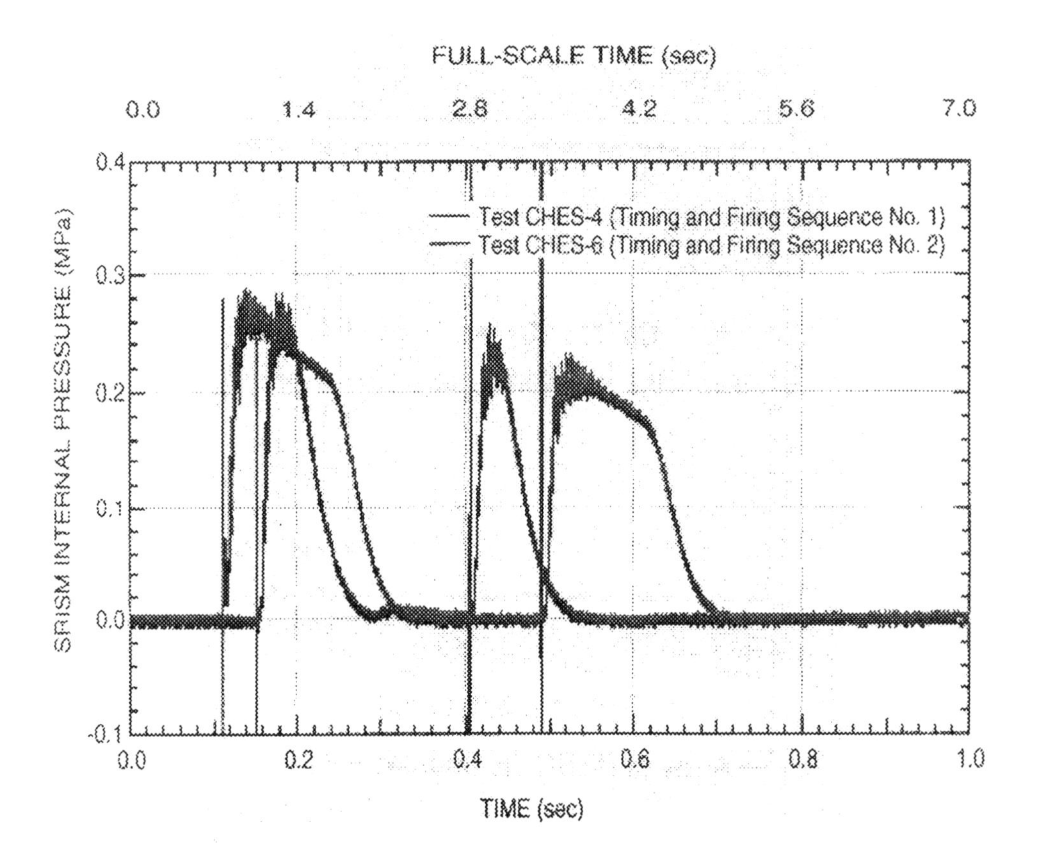

Fig.3 RESCUE source applied pressure pulse tailoring capability

Twelve tests were performed with the source module. Seven of the tests were carried out with the source module in a stiff clay type soil at SRI's remote test site in Tracy, California. After these tests, the source module was shipped to NTS in Nevada, where five additional tests were performed in a very weak, silty sand-type soil. The source module sustained no damage during these tests and remains completely functional.

The NeTI minimum response spectrum encompasses the mean of the horizontal ground motion response spectra from the Loma Prieta, Cape Mendocino, Landers, Northridge, and Kobe earthquakes. The ground motion generated at NTS using the RESCUE technique meets or exceeds the NeTI minimum response spectrum for a wide range of periods.

The results from tests conducted in California and at NTS indicate that an applied RESCUE source pressure of 1 MPa will produce full-scale accelerations of 1 to 2 g's, velocities of 100 cm/s to 140 cm/s, and displacements of 60 cm to 100 cm.

Fig 4 shows a NIKE2D finite element model of the full-scale soil island developed by LLNL. This model is used to investigate the response of the soil island to different applied-pressure pulse sequences.

Fig.5 shows the predicted horizontal response spectrum for the full-scale soil island and a series of four 1 MPa pressure pulses. This response spectrum is near the Japanese NUREG 3XS2 response spectrum for the design of nuclear power plants.

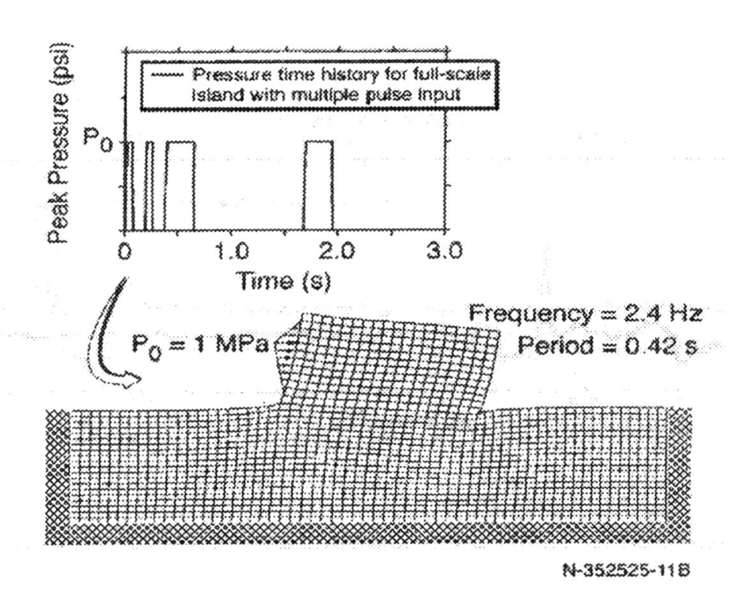

Fig. 4. NIKE2D model of NeTI full-scale soil island test bed

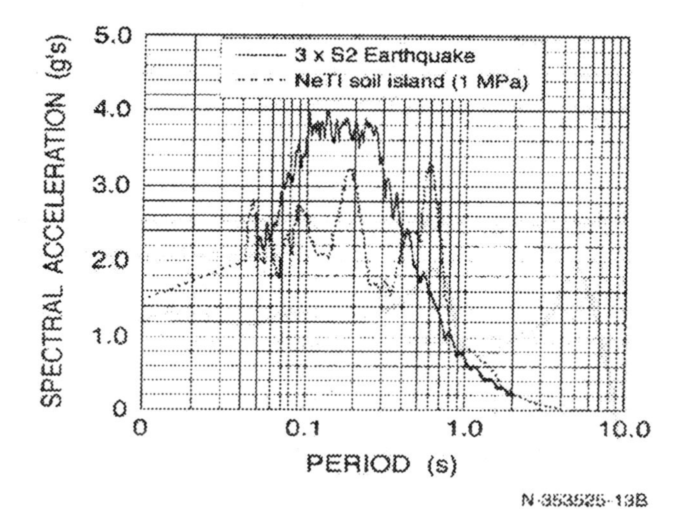

Fig. 5 Calculated NeTI full-scale test bed response spectrum

3 Recent (1999) test series

Another test series was planned for the summer of 1999 at the NTS. The 1/7-scale tests already performed were designed to be pulsive so that the impact of each propellant detonation could be clearly defined and analyzed. The motions were not intended to represent a specific type of strong ground motion. The 1999 work, the results of which may be presented at the time of the conference, will demonstrate the capability of producing two types of strong ground motions:

a. Motions representative of near-fault motion, which may be the result of an accumulation of waves emanating from the fault's rupture zone at an earthquake's epicentre that produces a strong long-period pulse called the "fling effect."
b. Vibrations with both positive and negative displacement excursions.

The two RESCUE pressure pulses generated in the tests to illustrate these types of motions will be 1/7-scale versions of the pressure pulses used in the finite element computations described previously.

The long-period motion requires changing the timing and firing sequence to cause an overlapping of the applied pressure pulses. To illustrate this, the timing of the arrival of the second acceleration pulse was changed, so that the second starts soon after the first. Up to three such long period motion tests will be performed.

To produce vibrations, a second source module will be placed at the opposite end of the test bed from the existing source module shown in Fig 2. The testing and firing sequence will control each source module so that they fire out of phase of each other, producing vibratory motion.

4 Test operations and partnerships

The national earthquake engineering research agenda falls primarily under the nation's private and public universities with major support from the Federal Emergency Management Agency (FEMA) National Institute of' Standards and Technology (NIST), NSF, and U.S. Geological Survey (USGS), the primary National Earthquake Hazard Reduction Program implementation agencies.

For the Centre to support the national research agenda and activities, earthquake engineering research universities must take a strong role in test operations. Accordingly, NeTI has been working with the California Institute of Technology, UCI, the University of California at Los Angeles, the University of California at San Diego, UNLV, UNR, the University of Southern California, the University of Texas at Austin, the University of Missouri-Rolla, and the University of Maryland at College Park, to form a collaborative operations partnership. These universities are part of the National Earthquake Engineering Research Centre and are well positioned to represent the national earthquake research institutions and national research agenda

The government's role in the Centre's operations includes working with NeTI to construct, operate, and maintain the Centre at the NTS. These activities provide the government contractor an excellent opportunity to exercise many aspects of its testing capabilities, with the potential of offsetting operations funding requirements.

NeTI and the Civil Engineering Research Foundation (CERF) have agreed to work together to involve industrial and international users in the Centre. NeTI and CERF will work through CERF Innovation Centres to bring its nationally recognised test and performance evaluation process to key stakeholders in each sector of the construction industry. Test programmes integrated into CERF Innovation Centres could be used to pre-approve technologies to expedite their introduction and acceptance into practice.

The research performed to date has demonstrated that the RESCUE technique produces large ground motions that can be used to excite full-scale test structures at a full-scale testing centre. The Centre, therefore, when more fully developed, will provide multiple tests with the same soil island test bed over an extended period, up to 40 applied-pressure pulses over 60 seconds, and load tailoring ability to meet specific ground motion requirements.

Longer term development activities include design and construction of the first of several full-scale soil islands over the next two years, with an anticipated completion of the full development of the NTS Strong Ground Motion Test Centre in 2004.

Each of the soil test beds (see Fig. 6) will have the dimensions and potentially different soil properties to meet a wide array of testing needs. This allows researchers to concurrently build structures at one location and conduct tests at another. Multiple testing beds overcome the inefficiencies of tying up a limited testing facility with one long-term research project involving fabrication and multiple testing of a structure.
The Centre concept offers additional efficiencies with a limited shared supply of modular ground motion sources, stored nearby and transported to a soil island when needed, and one test-control point, with one set of data collection and diagnostics capabilities for all soil island test beds.

The Centre will be able to accommodate large- to full-scale testing of typical multi-story buildings, nuclear reactor structures, highway overpass structures, fluid storage tanks, electrical switchyard equipment, and buried pipelines. The Centre will have the capability of fragility testing for all of these structures.

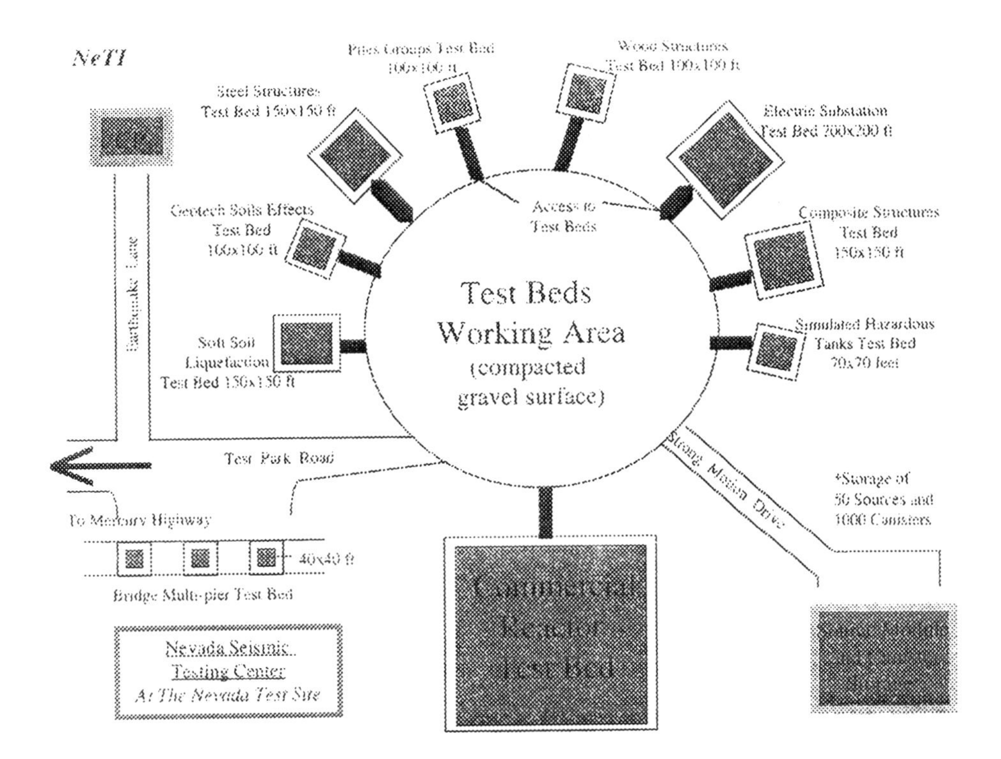

Fig. 6. The Nevada Seismic Testing Centre as it will be in 2004

Full development of the Centre is projected to be a five-year process that includes final-design performance testing of a full-scale ground motion source module, construction of the first full-scale test bed in 2000, and final development of the Centre through 2004.

5 Conclusion: Uses for the Centre

The Centre will provide several full-scale soil test beds for various structural types that cannot be accommodated currently. High-rise buildings, full-scale storage tanks containing simulated hazardous fluids, and pipelines crossing active faults are examples of appropriate test specimens.

Other important uses include: validation of numerical design codes based on full-scale testing, enhancing confidence in code performance, reducing the need for testing, enhanced infrastructure for research and education through the university partnership, broad dissemination of data through the university partnership, publish reports and papers in technical journals, public presentations at meetings and conferences, and site visits and tours conducted during the test series.

The potential benefits to society are: improved understanding of building behaviour during strong seismic shaking for better protection of occupants improved understanding of structural failure mechanisms, leading to engineering design that reduces the potential for damage and property losses. improved understanding of foundation performance during strong seismic shaking, leading to design improvement and occupant safety improved performance of storage vessels for environmental safety.

In summary, we expect that the Centre will provide an important new tool to the engineering community for testing earthquake-resistant designs.

THE BEHAVIOUR OF CONNECTIONS IN STEEL MR-FRAMES UNDER HIGH-INTENSITY EARTHQUAKE LOADING

Steel connections under high intensity earthquakes

G. DE MATTEIS and F. M. MAZZOLANI
Department of Structural Analysis and Design, University of Naples Federico II, Italy
R. LANDOLFO
DSSAR. University of Chieti G. D'Annunzio, Italy

Abstract

Recent earthquakes have clearly shown as the seismic behaviour of steel moment-resisting (MR) frames is affected by beam-to-column connections, which may promote the premature collapse of the structure, exhibiting brittle failure modes. Based on field experience arisen from such events, the primary failure causes are analysed and discussed in current paper. Then, a general overview on the suggested upgrading procedures for connection behaviour, emphasising the main aspects enhancing plastic rotation capacity, is provided. Finally, the effect of accurate structural modelling on the assessment of rotation demand is investigated.
Keywords: Beam-to-column connections, Earthquake, Seismic design, Steel frames, Strain rate, Structural modelling.

1 Introduction

Earthquake events recently occurred worldwide, namely Llolleo-Chile (1985), Michoacan-Mexico City (1985), Loma Prieta (1989), Landers (1992), Northridge (1994), Kobe (1995), Umbria-Italy (1997), Izmit-Turkey (1999), Athens-Greece (1999), Taiwan (1999), Mexico-Puerto Escondido (1999) have produced disastrous human and economical losses, destroying several residential, industrial and bridge structures. They clearly emphasise that the natural hazard related to earthquakes is very high and point to the need for revision in codification, design, construction process technique, repair and retrofit of existing buildings, and material fabrication. The present paper is addresses steel frame buildings, seismic risk, design rules and modelling of such structures.

For many years, steel moment-resisting (MR) frames have been unquestionably considered as a reliable and favourable seismic resisting system [1]. Credits to steel were mainly based on the inherent material ductility, which could result in high structural ductile response, allowing for a large capacity for energy dissipation [2]. The 1985 earthquake of magnitude 7.3 in Michoacan (Mexico City) evidenced the first collapse of an important high-rise steel building. Also, during the magnitude 6.7

Abnormal Loading on Structures edited by K. S. Virdi, R. S. Matthews, J. L. Clarke and F. K. Garas.
Published in 2000 by E & FN Spon, 11 New Fetter Lane, London EC4P 4EE, UK. ISBN 0 419 25960 0

Northridge (Los Angeles) earthquake several MR-frames suffered brittle failure of welded beam-to-column connections. Most of such structures were built within the preceding 10 years and therefore were designed according to recent design and constructional methods for seismic-prone countries. Similarly, in the magnitude 7.2 Hyogoken-Nanbu (Kobe) earthquake severe damage was detected in a number of modern steel buildings. For the first time in the Japanese history of large earthquakes, the image of steel structures was compromised. Again, damage occurred in a number of welded beam-to-column joints.

These events clearly shown that a high ductility material like steel could suffer brittle failure modes as well. As a matter of the fact, the international scientific community is aware of the need of understanding the reason of what happened. A review of current seismic provisions for steel structures is, therefore, in progress. At present, the main tasks under investigation are aimed at: (a) the prediction of the actual behaviour of all structural components of steel frames under severe dynamic loading conditions; and (b) a better hazard assessment of earthquake ground motions.

The former task deals with structural aspects, which are related to the capacity supplied by the system. Therefore, it interests especially structural engineers, who must identify the primary sources of misunderstanding in customary design procedures. In particular, efforts are addressed to the correct evaluation of strength demands and availability of required ductility ratio of structural members and connections. Another source of misunderstanding in standard design analyses is due to the improper accounting for non-structural elements, e.g. facade building panels.

Task (b) concerns with the surveying of the action shaking the structure, and therefore with the seismic demand. The related interest is mainly for seismologists, whose aim is to establish methodologies for predicting strong ground motions for future scenario earthquakes, in order to mitigate damage in urbanised area surrounded by active faults. However, structural engineers have to study the earthquake effects on structures and therefore either accelerograms or response spectra to be used for numerical simulations and design have to account for actual earthquake characteristics hypothetically occurring at the site of building erection. Aiming at a better mindfulness of damage potential of earthquake ground motions, it is a new trend to distinguish the effects of far-field and near-field earthquakes, the latter being characterised by impulsive motions and higher ground velocities. Anyway, design spectra suggested by codes do not account for all significant seismic hazard parameters (duration of earthquake, peak ground velocity, input energy content, etc.). As a consequence, the seismic risk is essentially based on the sole peak ground acceleration and the shape of response spectra is dependent on the soil characteristic only.

Some recent outcomes concerned with seismic behaviour of steel MR-frames are presented in this paper, emphasising the major aspects which condition their response to high-intensity earthquakes. In particular, the effects of beam-to-column joint response on the global performance of the whole structure are investigated, assessing the possibility to develop new and more suitable design procedures.

2 Field-observed behaviour

Majority of damages to steel members and connections were detected and identified following the Northridge and Kobe earthquakes. The former seismic event produced damage to several MR-frame structures. More than 100 were identified by inspection (concerning both low- and high-rise building types) and many more were suspected to have occurred in non-inspected structures [3]. The latter event caused damage to about 680 steel buildings, but the number concerned with failure of beam-to-column connections was about 30 [4]. Despite of some existing analogy between the above outcomes, exhibited damage was different due to peculiar connection detailing customarily used in U.S.A. and Japan.

Beam-to-column connection typology widely used in U.S.A. before 1994 is shown in Fig. 1a. It was aimed at fully developing the flexural strength of the connected beam. Deep H-shape beam flanges, which develop beam bending moment, were field-welded to the heavy H-shape column face by means of groove welds, after the beam web was field-bolted to a single plate shear tab, which was shop-welded to the column flange to resist the beam shear. In same cases, additional welds were provided to the shear tab, in order to increase the moment capacity of the connection. Usually, a backup bar was arranged at the lower side of the beam flange, left in place after joint completion. Web copes were required to allow for the bevel welds at the top side of the beam flanges. Also, in some cases continuity plates were used to face on relative flange column stiffness and doubler plates provided to increase column panel zone shear capacity.

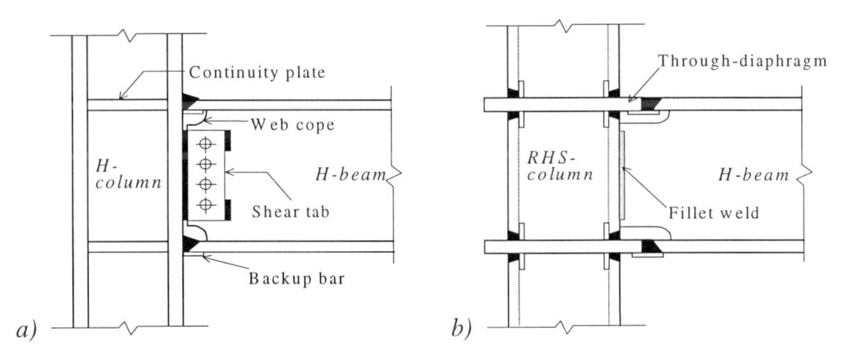

Fig. 1. Connection detailing.

Several damage typologies were observed. Frequently, occurred damage was due to cracks developing at beam bottom flange, in the vicinity of the groove weld. The fracture was either contained within the weld thickness or propagated throughout the column flange, by one of the following patterns: (1) developing horizontally behind the flange heat affected zone; (2) propagating longitudinally along the flange; and (3) progressing within the web of the column [3]. In very few cases, the complete rupture of the column section was observed as well. The above damage was sometimes coupled with the failure of beam shear tabs due to either vertical fracture of the net section or bolt shearing. Eventually, in some instances the yielding of the panel zone was noticed too. It is worth noting that in all cases connections did not develop significant plastic deformation, with flange local buckling and damage at beam top flange location being noticed rarely. Also, no steel building collapsed completely during the earthquake. Damage or failures were mainly localised, they being also difficult to inspect.

Beam-to-column connection typology widely used in Japanese practice before 1995 is shown in Fig. 1b. It consisted of using wide flange H-shape beams either field- or shop-welded to cold-formed rectangular hollow column sections. In case of shop-welding, the column tree configuration was adopted. It involved the shop-welding of stub-beam to the column and the field-bolting of the remaining beam part to the stub-beam. Continuity plates (diaphragms) were inserted into the box section, they being either through-, interior- or exterior-type. Also, beam web-copes and weld-tabs were used to facilitate the welding process.

In the case of shop-welding type, the main observed fracture mode involved the base metal of the bottom beam flange, initiated at the toe of the web cope. Besides, the brittle fracture of either the diaphragm, initiated at the edge of the butt weld, or the butt welded joint itself, initiated at the end tab weld, was noticed in some instances. Field welding type connections exhibited a similar but even worse behaviour, which involved also the fracture of the beam bottom flange heat affected zone and the fracture of the full penetration weld metal between the beam bottom flange and the through-diaphragm [4]. In some instances, fractures modes occurred in column members as well, due to cracks of column skin plates located above or below the top or the bottom diaphragm. This resulted in complete overturning and collapse of the structure.

Generally speaking, in the Northridge earthquake, residual storey drifts of damaged buildings were not significant. In fact, damage to interior and exterior finishes were limited and fractures occurred mostly at beam bottom flange only. On the other hand, in the Kobe earthquake, beam hinging due to remarkable yielding and local buckling, were often observed, and fracture occurred after a notable amount of plastic deformations were developed. Also, the columns remained almost elastic [5].

3 Related projects

In recent years several research projects dealing with steel structures under earthquake loading have been undertaken worldwide. Due to the dramatic consequences of the Northridge Earthquake, in U.S.A. the SAC Joint Venture was established, grouping the Structural Engineers Association of California (SEAOC), the Applied Technology Council (ATC) and the California Universities for Research in Earthquake Engineering (CUREe). The main aim was to focus on the seismic performance of welded steel moment connections and to develop interim recommendation for professional practice. The study was therefore addressed to the inspection of damage of earthquake affected buildings, to their repair, and to the upgrading of existing buildings to improve their future performance. The main findings of the research were presented in a number of detailed reports and used for the development of Interim Guidelines [6].

In a similar manner, after Hygoken-Nambu Earthquake, several Special Task Committees were organised to assess the impact of this event on steel building frames. Firstly, the Steel Committee of Kinki Branch of the Architectural Institute of Japan (AIJ) reported in detail the seismic damage that had occurred to steel buildings [7]. Also, the Japanese Society of Steel Construction (JSSC) investigated the fracture that had occurred to steel beam-to-column connections. Instances of damage were detected and the related causes investigated. Improved connection details and welding procedures were suggested [4].

In Europe, several research projects have been stated as well. Copernicus-RECOS project (1996-1999) was sponsored by EC and was co-ordinated by University of Naples. It brought together knowledge and experience of different specialists from several earthquake-prone European Countries, aiming at investigation of the influence of joints on the seismic behaviour of steel frames. Several areas of concern were analysed, comprising the evaluation of seismic hazard, the influence due to ground motion type, the experimental response of various types of beam-to-column connections subjected to both static and dynamic loading, the hysteretic modelling of connections, the influence of connections on the global response of steel MR-frames, and the effects of building asymmetry. Preliminary results are presented in [8], while an extensive Application Document is currently being processed.

The STEELQUAKE (1996-1998) research project sponsored by EC and co-ordinated by ISMES (Italy) in co-operation with several European partners concerned both experimental and numerical investigations on steel MR-frames. A large series of tests on sub-assemblages was carried out emphasising the effect of welding detailing and strain rate on the seismic response of beam-to-column connections [9].

The European Consortium of Shaking Tables (ECOEST) is working on another research project called Innovative CONcepts in Seismic engineering (ICONS) (1997-2000), involving co-operation between University of Liege (B), Imperial College London (UK), University of Darmstadt (D) and University of Madrid (S). Quasi-static, pseudo-dynamic and dynamic tests are being carried out on frame structures and beam-to-column connections to investigate the ductility of composite sections in bending subjected to dynamic loading. The research is still in progress.

4 Connection rotation capacity

4.1 General remarks

Seismic input essentially means energy input to structure, which becomes apparent through displacement demand, whose possible measure is top sway or interstorey drift. In case of high-intensity earthquakes, the main concept which recent code provisions are based upon is that the actual strength capacity of structures (F_y) is lower than the one that could be developed if the structure itself behaved linearly elastic (F_e). Structures resist earthquakes due to structural ductility allowing for some amount of energy dissipation. This permits a reduction factor $R_\mu = F_{el}/F_y$ to be used for defining design earthquake loading, which is a function of available ductility (μ).

The main essence of the reduction factor is that it allows a simplified elastic analysis to be performed instead of complicated inelastic ones. R_μ is provided by seismic codes, it being essentially based on empirical observations. In the case of single degree of freedom (SDoF) systems, R_μ may be evaluated for any natural period and for several values of available system ductility and damping ratios. Obtained spectra [10] show that it is period-dependent only in the short-period (acceleration-controlled) range, where inelastic displacements are higher than equivalent elastic ones. In particular, for very short period R_μ approaches the value 1, independent of the available ductility (μ). In contrast, for medium- and large-period range R_μ may be assumed to be equal to the available ductility, the inelastic displacements being very similar to the equivalent elastic ones.

In case of multi degree of freedom (MdoF) structures, the problem appears to be much more complicated. In fact, the response of the whole structure is affected by many factors: higher mode effects, torsional effects, structural irregularities, structural redundancy. The actual seismic response of MR-frames should be, therefore, obtained via inelastic time-history analyses, providing information on the distribution of inelastic deformations within the structure. In particular, components exhibiting plastic deformations should be able to dissipate energy without significant strength degradation, this requiring that particular care in detailing is taken. Besides, in order to favour the maximum energy adsorption capacity, the weak-beam strong-column (WBSC) criterion is recommended, this allowing storey mechanisms to be prevented and the maximum number of dissipative zones to be promoted [11]. Also, in order to avoid brittle failure modes to occur, standards for seismic design require that connection strength be greater than the beam strength. Anyway, collapse may not be defined in terms of strength, but it must be intended when the structure does not have the possibility to dissipate energy. It is apparent that the performance of the system may be therefore assessed in terms of local criteria, therefore on the basis of ductility demand to ductility capacity ratio of energy-dissipative zones.

The seismic analysis of MR-frames is usually based upon the separate evaluation of the rotation demand and rotation capacity of beam-to-column connections, where plastic hinges are likely to occur. The former is strongly influenced by the structural typology and seismic input, and relative results are highly sensitive to structural modelling. The latter depends essentially on the structural component itself, but is also related to the typology of the applied loading (impulsive- or cyclic-earthquake type).

4.2 Experimental evidence

The evaluation of available ductility is a difficult matter. The ductility of plastic hinges under monotonic loading is, in fact, strongly reduced under seismic loading, due to: (a) cyclic action, which, on one side, induces a deterioration of mechanical properties as a function of number of plastic excursions and their amplitude, and, on the other side, due to isotropic hardening, may produce the yielding of more than one component (beam section, column section and panel zone) at a give joint; and (b) dynamic effects, which due to high-strain rate may promote fracture modes.

Several tests were conducted especially in U.S.A. to determine experimentally the available rotation capacity of connection details used in practice. Excellent overviews on the experimental evidence are presented in [12, 13]. Before Northridge earthquake, investigations under static cyclic loading undertaken since the early 70's, evidenced contradictory results, some tests providing a satisfactory performance, other tests showing a very poor behaviour. Nonetheless, it was sustained that the current connection configuration, made of bolted web/welded flanges, was able to develop plastic rotation in the order of 0.02 rad, and this was the acceptable level of performance provided by prescriptive provisions. After 1994, a more precise examination of both exiting data and that obtained by using specimens perfectly equivalent to those adopted in buildings collapsed during the earthquake clearly show that the pre-Northridge connection typology behaves very poorly, exhibiting rotation in the order of 0.005 to 0.02 radians with premature failure after few cycles. Often, fracture occurred without developing any plastic deformation. Thus, the available evidence actually confirmed the majority of the fractures observed during Northridge earthquake, which had initiated at beam flange groove welds.

On the other hand, numerical investigations carried out on the same lay-outs of building struck by the Northridge earthquake, analysed with reference to the actual acceleration records evidenced that the majority of connections damaged by the strong motions were not subjected to large deformations indeed. Generally, rotation demands were limited, they being lower than 0.01 rad, and in several cases no inelastic deformations were noticed at the same joint location which had actually failed during the earthquake. This outcome shows that the adopted connections behaved worse than that could be predicted by careful and faithful examination of laboratory evidence.

So far as Kobe earthquake outcomes are concerned, it was observed that often the fracture of connection using cold-formed box columns was due to a ductile crack owing to strain concentration near welding. Also in this case, the fracture process was similar to that observed in laboratory tests [14]. Dynamic analyses of buildings that exhibited fracture at beam-to-column connections were also carried out [5]. First, they indicated that input motion concerned with the actual earthquake had been higher than the one predicted by current code provisions. Besides, high interstorey drift were recognised, they being in the order of 2%, which likewise should correspond to the limit value of connection rotation capacity. Also in this case there was an evident contradiction. In fact, the field observation show that, in many cases, connections revealed serious damage, but this was usually associated with minimal interstorey drift, which resulted in limited damages to claddings and interior partitions.

5 Causes for connection failures

5.1 U.S.A. occurrences

The large amount of experimental analyses carried out allows some of the main reasons producing brittle failure of connections to be detected. The general conclusion of research in U.S.A. is that there are several factors influencing the poor response of connection configurations used in practice [13, 15]. Surely, quality of welding strongly influences the performance of the connections: many failures were detected when welds had inadequate fusion or excessive porosity. Besides, tests on weld metal taken from damaged building showed very low values of toughness, which promoted brittle failure modes. Several investigations concluded that backup bars, when these were left in place, could negatively affect the response of the connection. In fact, fusion of the backup bar to the column flange results in a sharp notch at the root of the weld. Similar notches may occur due to continuity plates, when fillet welds are used. These notches act as initiators for cracking development. Other experiments showed that high stress concentration appeared at flange welding prior to developing full moment capacity of the beam. The main explanation of such a behaviour is the scarce aptitude of bolted web connection to contribute to joint moment resistance. Besides, a large proportion of the shear is carried by flange welds rather than by web tabs. Also, the absence of continuity plates could produce high concentration of flange stresses close to column web.

Sources for poor behaviour of connections are also due to the anisotropy of rolled steel shapes. In fact, owing to fabrication process, mechanical properties of steel have been found to be poorer in the through-thickness direction, i.e. perpendicular to the rolling direction. In particular, material strength and ductility is strongly reduced as the thickness of the elements composing the member section increases. This results in flanges with lower strength than webs and in a lower capacity of members with larger sections. Also, considering that at the beam-to-column connections, a triaxial stress state exists, reduction of through-thickness properties may penalise the performance of whole joint.

Connections are designed in such a way so as to be stronger than the connected elements. Test results show that the flexural strength of beam flange $F_{uf} = Z_f f_u$, where Z_f is the plastic modulus of beam flange alone and f_u is the material ultimate strength, predicts quite accurately the ultimate moment capacity of welded/flange bolted/web beam-to-column connections [12]. The flexural plastic strength of the connected beam may be evaluated as $F_{yb} = \alpha\, Z f_y$, where Z is the plastic modulus of beam cross-section, f_y is the material yielding strength and α is a factor accounting for strain hardening, usually assumed to be in the order of 1.2-1.3. It is clear that for a fixed factor α and a given beam section geometry (Z_f / Z), connection to beam strength ratio is ruled by the strength material ratio f_u / f_y. Ratios of ultimate to yield strengths for steel sections belonging to ordinary production have been examined in [12, 15]. Randomness of this ratio due to actual values compared to nominal ones may be quite large. Cases with low ultimate to yield strength ratio may result in the development of beam flange fractures prior to the attainment of cross-section plastic capacity.

Another important matter resulting in poor performance of joint is that, usually, the actual yield stress of steel materials is much larger that the nominal values. In U.S.A. it is customary to combine beams made of ASTM A-36 steel (nominal yield strength equal to 248 MPa) with columns made of ASTM A-572 grade 50 steel (nominal yield strength equal to 344 MPa). Statistical evaluation of mechanical properties evidence that difference between nominal and actual yield strength are higher for A-36 (about 35%) than for A-572 (about 15-20%), so that yield strength of the former tends to approach and sometimes exceed that of the latter. Obviously, in this case WBSC criterion is no longer assured, thus yielding affecting column section rather than the beam section, and the panel zone is subjected to higher stress than that could be predicted. This has been estimated to be relevant in 10% of cases [15].

Lastly, it is important to mention that available rotation capacity may be very limited in case of high beam depth, where plastic strains at flange are decidedly higher than the ones corresponding to the same absolute rotation in smaller sections. Similar conclusion may be reached if the contribution of the slab is accounted for in composite sections. The reduction of the number of moment resisting frames within a building (perimeter frame solution) and the reduction of moment connection in the frame (dual schemes) had been a common practice before 1994. The consequence was the reduction of redundancy in the structure, which means reduction of global over strength, and the use of larger beam and column sections for moment connections, which means reduction of available ductility [16].

5.2 Japanese occurrences

In the case of Kobe Earthquake, causes of connection failure have been mainly attributed to the ground motion which, in terms of input energy, was significantly stronger than that was prescribed by Japanese Standards. In fact, beam bottom flange connections fractured after developing considerable plastic deformations. No significant weld defects were identified to be the reason for initiating such a failure mode, but discontinuities in the weld zone beneath the web of the beam were reckoned to be partly responsible. The ultrasonic detection technique, consequently, is now being strongly recommended in Japan [7]. Nevertheless, several other factors have been considered and deeply investigated by scientists and researchers, focussing on the ductile fracture modes observed on the field.

The majority of steel multi-storey building frames in Japan are made of box-columns. They are widely used because: (1) they allow a simple structural organisation, avoiding beam weak axis connections; (2) WBSC criterion is easily realised; and (3) they assure a very good performance under biaxial bending. When fully welded connections are employed, however, the transmission of stresses from the web of the beam is not smoothly developed, it requiring the out-of-plane bending of the column flange. This has been also shown by numerical FEM simulations [4]. The seismic performance of moment resisting frames has been demonstrated to be strongly influenced by this transmission efficiency.

Stress concentration in beam flanges to column welded connections caused a considerable reduction of the inelastic deformation capacity of beams. The fact that the column diaphragm is often thicker than beam flange explains why in the majority of cases crack propagation interested beams rather than columns [17]. Because of the geometric configuration, strain concentration occurs around the toe of the weld access hole. Besides, such a concentration is increased when tack-weld of the backing bars overlaps the weld access hole itself. Suggestions for the improvement of welding procedures and connection details are provided in [4]. Likewise, strain concentrations appear owing to geometrical restraints exerted by through-diaphragm and weld access hole at beam-flange connections. Also in this case, floor slabs contribute with beam cross-section to carry out normal stresses. At beam-to-column joints, neutral axis approaches the top flange and as a consequence the bottom flange sustains higher strains than those corresponding to symmetrical cross-sections.

Problems arising in such connection typology are also due to the material used for box-sections. In fact, they are manufactured by cold-forming, which produces a reduction in ductility and toughness. It is apparent that in this condition the stress concentration at the weld toes of through-diaphragm to beam-flange connection more easily may promote cracks due to tensile fracture. Kuwamura [18] has shown by tests that this problem is relevant in the case of width-to-thickness (b/t) ratios less than 20, and it could be avoided by using hot-rolled box-columns.

5.3 The effect of strain rate and temperature

One aspect connected with high damage level in steel structures is certainly common to both Northridge and Kobe experience. In fact, both earthquakes were near-source type, i.e. the urban region shaken by the earthquake is within a few kilometers of the fault rupture zone. Such events are characterised by few long-duration velocity pulses in the fault-normal direction, which generally produce very high values of peak ground velocities (maximum recorded values for Northridge and Kobe earthquakes were 177 cm/s and 176 cm/s, respectively). In both cases, such pulses have quite large predominant period, they being larger than 0.7 s and 1.0 s for Northridge and Kobe, respectively.

Due to the great velocity of the seismic action, rate loading on structural members leading to fracture is much higher (it may be estimated in the order of 10^{-1}/s) than the one typically applied in laboratory quasi-static tests (about 10^{-5}/s). On the other side, it is well known that materials exhibit higher yielding and ultimate strength as far as strain rate increase. Moreover, since the yield strength is more strain rate-sensitive than the ultimate strength, high strain rate strongly reduces f_u/f_y ratio. A similar effect is due to low temperatures. The latter concern has been particularly important in Kobe, where temperature was very low at the time of the earthquake occurrence. Both aspects potentially alter the rotational ductility of member sections and connections. Since the material ductility factor is reduced, this results in a negative trend. Also, since the applied loading is repeated, number of cycles needed to failure could be reduced due loading velocity.

Some experimental tests on material are presented in [19]. Several other investigations have been performed after recent earthquakes, especially in Japan [4] and Europe [20], on both basic material and complete sub-assemblages. Analytical constitutive models are presented accounting for the strain rate effect. Obtained outcomes show that the lower the initial yield strength, the higher the strain rate-sensitivity of the material yielding. Also, the f_u/f_y ratio approaches the value 1 when strain rate increases. Analytical studies presented in [21] clearly show that these conditions may be deleterious for both ductility and low-cycle fatigue capacity of structural components.

It is worth noting that available experimental evidence is still questionable for what concerns the behaviour of connections under high-velocity cyclic loading. Fig. 2 illustrates the effect of strain rate (*f*) by some experimental tests performed on typical Japanese beam-to-box-column connections, presenting welding access holes at beam web (specimen TS) or not (specimen TN) and at different temperature (-50 °C: specimen 5; -20 °C: specimen 2 and +13 °C: specimen RT) [22]. It is apparent that strain rate considerably influences the monotonic dynamic response of the system in terms of both yielding and ultimate strength (Fig. 2a), but this effect is very similar to the one produced by low temperatures (Fig. 2b). On the contrary, stain rate seems not to be detrimental to low-cycle fatigue, the number of cycles to failure (N) being slightly and not always affected by loading velocity (Fig. 2c). On the other side, low temperatures strongly affect the cyclic dynamic response, brittle failure modes occurring in such cases (Fig. 2d). Similar conclusions were drawn by Canadian researchers, who did not observe any significant effect of strain rate on the seismic response of typical beam-reduced section (dog-bone) connections under cyclic dynamic loading [23]. Contradictory results have also been obtained in [20], where tests on typical European rigid full-strength extended end plate connections revealed a substantial decrease (34%) of number of cycles to failure, when high loading velocity is applied to low-strength steel (f_y<300 MPa), but the trend is opposite for high strength steel (f_y<300 MPa) [20].

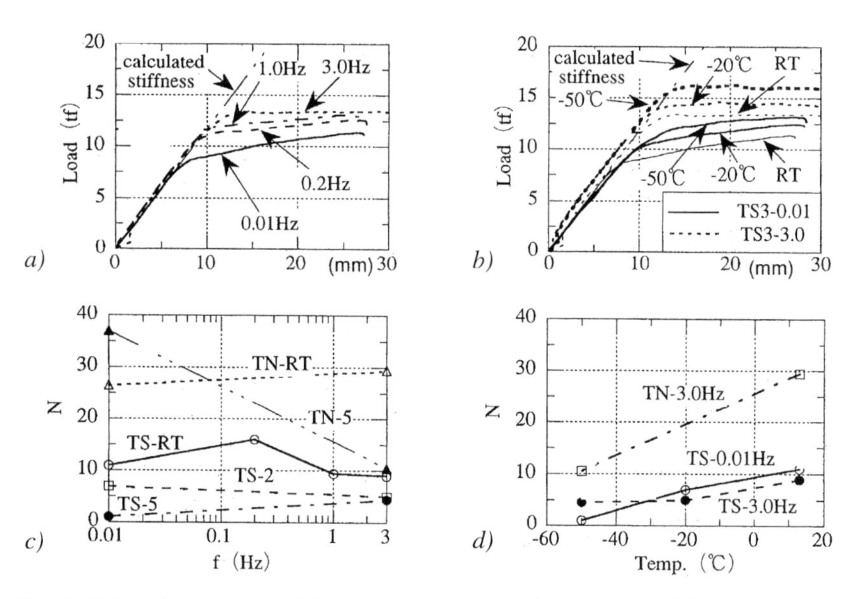

Fig. 2. Effect of strain rate and temperature on connection response [22].

6 Procedures for upgrading connection behaviour

After Northridge and Kobe earthquake many efforts to improve connection behaviour have been made. Several modifications in terms of connection detailing and welding procedures have been proposed and implemented in recent construction practice and code provisions. In particular, recent specifications give more importance to the toughness of the material, in order to reduce the potential of brittle failure modes due to strain concentration, welding effects, high strain rates, low temperature, and so forth. Test results indicate that improved connection details should be based on toughness values, determined according to Charpy V-notch impact test, equal to 27 J at a temperature of -29°C [13]. This value is not so distant to that prescribed by Eurocode 3 for the highest steel quality "D" (27 J at –20°C). Also, it is recommended that in common practice backup bars and weld tabs are removed after completion of the connection assemblage, while continuity plates should be used in all ductile moment connections. Furthermore, welded web connections are preferred to bolted ones, because they provide better stress patterns avoiding unexpected strain concentrations at beam-flange to column-face connections.

The current trend is to increase required design strength for connections by increasing the design factor α (see Section 5.1), which relates this strength to the one of the connected member. Factor α is determined in such a way so as to account for the randomness of material yielding and ultimate stresses, execution of connection details and of welding procedures [4]. The objective is to avoid high strain concentration in beam flanges as well as to favour the Weak Beam Strong Column criterion.

Generally, direct experimentation on full-scale specimens is required to either assess or verify the cyclic behaviour of new connection types to be adopted in high-seismicity zones. In particular, a new limit of available rotation capacity for connections is explicitly defined by AISC [24], which requires maximum rotation at least equal to 0.03 rad before failure, this being higher than previously suggested benchmark values. This change is significant because existing data show that pre-Northridge connection configurations are not able to comply with such plastic rotation limit. Therefore, in addition to the above detailing improvement, some substantial modifications in connection configuration have been proposed.

Tests on these new typologies are presented in [12, 13]. They show that the new rotation target is obtainable following two different strategies: by weakening the beam cross section at the joint or by strengthening the joint itself. Both are based on the concept to move the plastic hinge that is going to be formed away from the face of the column, to favour ductile plastic mechanisms. The first solution, which corresponds to the well know dog-bone configuration [25], is obtained by reducing the beam flange width starting from a certain distance (L_i) from the column face for a length equal to L_m that is able to accommodate the plastic hinge. Strengthening of the joint may be easily developed by using ribs, haunches, tapered cover plates, side plates and so forth, to be welded to bottom and top beam flanges. For the sake of an example, the comparison between the experimental cyclic response of a typical pre-Northridge and a dog-bone connection type, having very similar geometry and details, is shown in Fig. 3 [13]. The better performance of the latter is unmistakable. The influence of these newly proposed connection typologies on the seismic performance of moment resisting frames is presented in [26] for different frame configurations and ground motion typologies. The analyses indicate the superiority of dog-bone and reinforced connection types as compared with classical moment connection typologies, particularly because they strongly reduce the susceptibility of the structure to exhibit storey mechanisms rather than global collapse mechanism types.

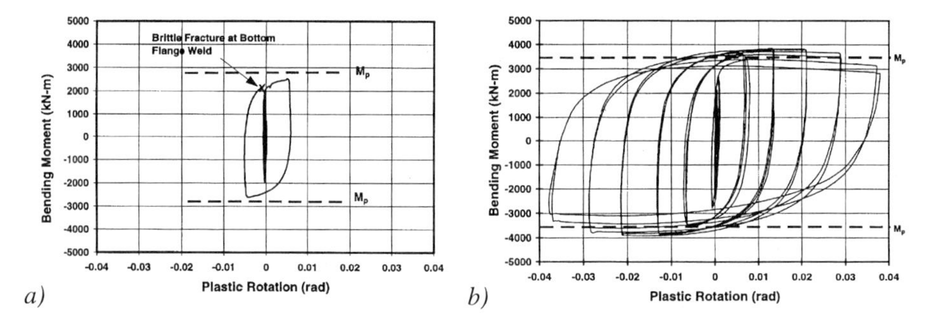

Fig. 3. Response of a typical pre-Northridge connection (a) and dog-bon connection (b) [13].

7 Connections modelling for prediction of rotation demand

According to current research developments, in order to assess plastic rotation demand for members and connections, inelastic time-history analyses of steel moment resisting frames under high-intensity earthquake conditions should be carried out considering explicitly the actual beam-to-column joint features [16]. Existing relevant experimental tests show a number of complex aspects that could have a major impact on seismic demands and that therefore should be accounted for in accurate structural system modelling. In particular, strong non-linearity, kinematic hardening of the monotonic restoring force characteristic, cyclic hardening, cyclic damage of mechanical properties and pinching of the hysteretic cycle (essentially for bolted joints) are the essential aspects arising from laboratory tests.

There are several methods available to predict the monotonic behaviour of joints: (1) empirical methods based upon test results; (2) mechanical methods mainly based upon the well known component method (for instance EC3-Annex J); and (3) mathematical methods based on the representation of the joint response through some basic parameters selected through either self-confidence or test results. A great deal of research effort is still under way for suitable prediction and modelling of the cyclic response of joints. On the other hand, it is necessary to recognise the usefulness of a preliminary assessment of the phenomenological aspects producing indeed a significant influence on the frame response and that therefore are worthy of accurate modelling and further investigations. To this purpose, the most effective approach is the semi-empirical one, based on the mathematical formulation of moment (M)-rotation (ϕ) relationship describing the joint behaviour. Such an approach has been proposed by the authors in [27, 28], where the influence of joint on both ductility and hysteretic energy demand is presented through a wide parametrical study relating to both SDoF and MDoF systems. In particular, with reference to MDoF, a five storey-two bay moment resisting frame, designed according to European Standards has been analysed. A number of conclusions were drawn. For example, in Fig. 4, which reports the maximum joint rotation demand, the effect of pinching modelling and strength degradation due to low-cycle fatigue is highlighted with reference to Kobe (impulsive-type) and El Centro (cyclic-types) earthquakes, scaled at three PGA values. The pinching effect is accounted for by the ratio (t_1/t_2) between two model parameters ruling the cycle-shape and the energy dissipated per cycle. Moment capacity degradation throughout the deformation history is schematised by a purposely defined damage model, whose parameter β represents the strength deterioration rate. The relative importance of connection modelling and earthquake typology is evident. In particular, strong pinching in hysteretic behaviours should not be disregarded (Fig. 4a), while strength deterioration is important particularly in the case of high-intensity and impulsive type ground motions (Fig. 4b).

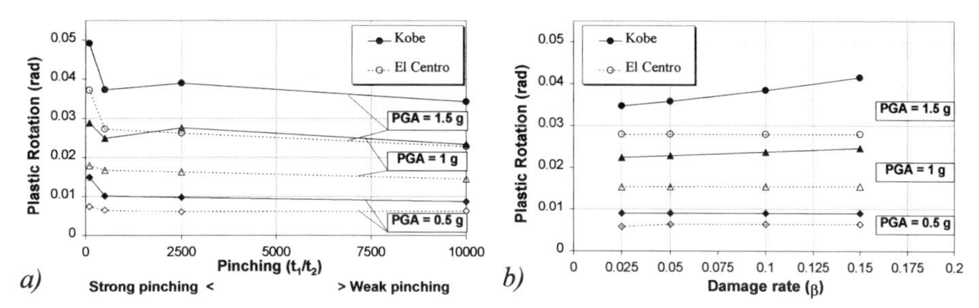

Fig. 4. Influence of connection modelling on ductility demand.

8 Final remarks

Current studies show that the aseismic design of steel moment resisting frame is not completely solved. Due to Northridge and Kobe earthquakes, several aspects have been discovered to be very influential with respect to customary design assumptions and constructional practice. In particular, inadequate beam-to-column connections may have a detrimental effect on the performance of the whole structure, exhibiting premature brittle failures. On the other side, several comprehensive research programs have been undertaken worldwide, aimed at a better understanding of the reasons why steel structures failed in these earthquakes. A number of aspects have been clarified, promoting detailing and design improvements to enhance the connection performance under cyclic dynamic loading. Nonetheless, more effort appears to be necessary, especially for what concerns the assessment of seismic risk for earthquake-prone regions and the effects that high-intensity impulsive-type ground motions may produce in terms of both seismic demand and structural capacity.

9 References

1. Mazzolani, F.M. and Piluso, V. (1996) *Theory and Design of Seismic Resistant Steel Frames*, E & FN Spon, London.
2. Mazzolani, F.M. (1999) Principles of Design of Seismic Resistant Steel Structures, in *4th National Conference on Steel Structures*, Ljubljana (Slovenia), May, 27-42.
3. Bruneau, M., Uang, C.M. and Whittaker, A. (1998) *Ductile Design of Steel Structures*, McGraw-Hill, New York.

4. Japanese Society of Steel Construction. (1997) *Kobe Earthquake Damage to Steel Moment Connections and Suggested Improvements*. JSSC, Technical Report No. 39-1996.
5. Nakashima, M., Yamao, K. and Minami, T. (1997) Post-earthquake Analysis of Steel Buildings Damaged during the 1995 Hyogoken-Nanbu Earthquake, in *Proc. of 5th Int. Coll. on Stability and Ductility of Steel Structures*, (ed T. Usami), Nagoya, Japan, 723-730.
6. Federal Emergency Management Agency. (1997) *Program to Reduce the Earthquake Hazards of Steel Moment Frames Structures*. Interim Guidelines, FEMA-267A.
7. Architectural Institute of Japan. (1995) *Reconnaissance Report on Damage to Steel Building Structures observed from the 1995 Hyogoken-Nanbu (Hanshin-Awaji) Earthquake.* AIJ, Steel Committee of Kinki Branch.
8. Mazzolani, F.M. (1999) Reliability of Moment Resistant Connections of Steel Building Frames in Seismic Areas: the First Year of Activity of the RECOS Project. In *2nd European Conference on Steel Structures* (Eurosteel 1999), Praha, Czech Republic, May 26-29.
9. Plumier, A., Agatino, R., Castellani, A., Castiglioni, C.A., Chesi, C. (1998) Resistance of Steel Connections to Low Cycle Fatigue, in *11th European Conf. on Earthquake Engineering*, Balkema.
10. Miranda, E and Bertero, V.V. (1994) Evaluation of Strength Reduction Factor for Earthquake Resistant Design. *Earthquake Spectra*, Vol. 10, 357-379.
11. Mazzolani, F.M. (1998) Design of Steel Structures in Seismic Regions: the Paramount Influence of Connections, in *Intern. Conf. Control of the Semi-rigid Behaviour of Steel Engineering Structural Connections,* (ed. R. Maquoi), Liége, 17-19 Sept., 371-384.
12. Tsai, K.C and Popov, E.P. (1997) Seismic Steel Beam-Column Connections, *in Report No. SAC-95-09, FEMA Beckground Report N. 288*, IV/1-IV/39.
13. Engelhardt, M.D and Sabol T.A. (1997) Seismic-Resistant Steel Moment Connections: Developments since the 1994 Northridge Earthquake. *Progress in Structural Engineering and Materials*, Vol. 1, No. 1, 68-77.
14. Kuwamura, H., and Akiyama H. (1994) Brittle Fracture under Repeated High Stresses. *Journal of Constructional Steel Research*, Vol. 29, 5-19.
15. Hamburger, R.O. and Frank, K. (1994) Performance of Welded Steel Moment Connections: Issues Related to Material and Mechanical Properties. In *Invitational Workshop on Steel Seismic Issues*, Los Angeles, September 8-9, SAC Rep. No. 94-09, 72-78.
16. Krawinkler, H. (1997) System Behaviour of Structural Steel Frames subjected to Earthquake Ground Motions, *in Report No. SAC-95-09, FEMA Beckground Report N. 288*, VI/1-VI/42.
17. Akyiama, H. and Yamada, S. (1997) Seismic Input and Damage of Steel Moment Frames, in *Behaviour of Steel Structures in Seismic Areas* (STESSA '97), (ed. F.M. Mazzolani, H. Akiyama), Kyoto, Japan, 789-800.
18. Kuwamura, H (1997) Ductility of Steel Members Susceptible to Brittle Failure, in *Proc. of 5th Int. Coll. on Stability and Ductility of Steel Structures*, (ed T. Usami), Nagoya, Japan, 925-932.

19. Soroushian, P. and Choi, K.B. (1985) Steel Mechanics Properties at Different Strain Rates. *Journal of Structural Engineering*, ASCE, Vol. 113 No. 4, 663-672.
20. Beg, D., and Plumier, A. (1999) Cyclic Behaviour of Beam-to-Column Connections: The Influence of Strain Rate, in *RECOS-Copernicus final report* (in press).
21. Gioncu, V. and Mateescu, G. (1999) Prediction of Available Ductility by means of Local Plastic Mechanism Method DUCTROT Computer Program, in *RECOS-Copernicus final report* (in press).
22. Terada, T, Yoshitaka, Y., Mase, S., Sakamoto, S. and Toshio, U. (1997) Structural Behaviour of Steel Beam-to-column Connections subjected to Dynamic Loads, in *Behaviour of Steel Structures in Seismic Areas (STESSA '97)*, (ed. F.M. Mazzolani, H. Akiyama), Kyoto, Japan, 656-663.
23. Tremblay, R., Tchebotarev, N. and Filiatrault, A. (1997) Seismic Performance of RBS Connections for Steel Moment Resisting Frames: Influence of Loading Rate and Floor Slab, in *Behaviour of Steel Structures in Seismic Areas (STESSA '97)*, (ed. F.M. Mazzolani, H. Akiyama), Kyoto, Japan, 664-671.
24. American Institute of Steel Construction. (1997) *Seismic Provisions for Structural Steel Buildings*. AISC, Chicago.
25. Plumier, A. (1990) A New Idea for Safe Structure in Seismic Zones, in *IABSE Symposium*, Brussels.
26. Anastasiadis, A., Mateescu, G., Gioncu, V. and Mazzolani, F.M. (1999) Reliability of Joint Systems for Improving the Ductility of MR- Frames, in *Stability and Ductility of Steel Structures* (SDSS'99), (Ed. D. Dubina and M. Ivanyi), Elsevier, Oxford, 259-268.
27. Della Corte, G, De Matteis, G. and Landolfo, R (1999) Modellazione di Nodi Trave-Colonna e Risposta Sismica di Telai di Acciaio, in *9° Convegno Nazionale ANIDIS*, Torino (Italy).
28. Della Corte, G, De Matteis, G. and Landolfo, R. (2000) Influence of Different Hysteretic Behaviours on Seismic Response of SDoF Systems, in *12th World Conference on Earthquake Engineering*, Auckland, New Zealand, 30 January- 4 February 2000 (in press).

NON-LINEAR ANALYSIS OF A REINFORCED CONCRETE STRUCTURE UNDER SEISMIC LOADING

Non-linear seismic analysis of a concrete structure

A. RUSHTON, J.E. SARGENT and N.E. KIRK
WS Atkins Consultants Ltd, Bristol, UK

Abstract

A non-linear analysis has been used to remove some of the conservatisms associated with assessing a safety-critical reinforced concrete structure under seismic loading. The structure consists of multiple concrete columns and beams supporting a heavy item of plant, which runs on rails supported by the beams. The explicit finite element code LS-DYNA has been used. The parameters of a concrete material model have been developed, by comparing several modes of structural failure with published experimental results. The ability of the concrete material model to cope with cyclic loading has been demonstrated. The reinforced concrete structure has been modelled using a simple mesh but with explicit representation of the reinforcement bars. Seismic loading has been applied to the model and the response of the structure and plant has been calculated. The integrity of the structure has been demonstrated, whereas previous design code based linear elastic assessments had indicated significant overstress. A safety case has been made which has removed the need for difficult and expensive modifications.

Keywords: Assessment, LS-DYNA, non-linear, reinforced concrete, safety case, seismic analysis.

1 Introduction

Increasingly stringent safety requirements for nuclear power plant make it necessary to demonstrate their integrity during a rare, but credible, seismic event. One such large piece of mechanical plant weighing 500 tonnes rests on reinforced concrete beams supported by concrete columns. The columns are significantly under-reinforced with respect to bending. Their reinforcement ratios are approximately 0.4%, compared with 1-2% for typical reinforced concrete columns.

Abnormal Loading on Structures edited by K. S. Virdi, R. S. Matthews, J. L. Clarke and F. K. Garas.
Published in 2000 by E & FN Spon, 11 New Fetter Lane, London EC4P 4EE, UK. ISBN 0 419 25960 0

Since this structure is in continual use, structural modifications would be very difficult, costly and intrusive. They could stop the plant working for a sufficient period to threaten the economics of continued operation.

Previous linear analysis, based on conventional design-code limits, showed that the columns would be significantly overloaded during an earthquake. The assumed linear elastic response gave bending moments much greater than allowed.

It was recognised that failure to obey the design code did not imply functional failure. However, significant departure from linear behaviour made it difficult to predict the forces in the concrete structure, and impractical to apply conventional static design-code criteria. A non-linear analysis of the structure, working beyond first yield of steel reinforcement and using fundamental failure criteria for concrete, was required to gain confidence that the structure would support the plant during a credible seismic event.

2 Approach

2.1 Modelling Procedure

The computer program LS-DYNA [1] was selected to model the plant and supporting structure in a relatively quick and inexpensive manner. The finite element based technique can handle complex loading, and can apply fundamental failure criteria for reinforced concrete material in place of the restrictive design code limits.

The reinforced concrete structures are modelled using a combination of eight-node brick elements for the concrete and fully integrated beam elements for the steel reinforcement bars. Each reinforcement bar in the columns was explicitly modelled.

The steel and concrete elements are connected only at nodal points. The bond between the steel and concrete between these nodes is not modelled.

2.2 Seismic Loading

The loading on the structure consists of earthquake displacement histories of ten seconds' duration. Synthetic time histories were derived from risk-based UK design spectra. The seismic loading was applied as a prescribed displacement via rigid elements attached to the bases of the columns.

2.3 Material Models

Current design codes for concrete use elastic criteria for uniaxial bending strength. They employ simple conservative empirical formulations for shear strength, in terms of global load combinations. These formulations are based upon tests of simple structural elements, for instance, simply supported beams. Whilst the codes give good estimates of ultimate strength for these simple structures, the linear code-based approach has the following practical limitations.

1. A reinforced concrete structural element reaches its elastic limit well before it loses the capacity to carry load. The codes do not predict strength beyond this limit.
2. The strength of concrete is properly related to its triaxial stress state. It is only crudely estimated in the design codes. Their use in complex load cases is necessarily conservative.

The concrete was modelled using a LS-DYNA material model in which the concrete stress is represented by the octahedral formulation. The formulation describes the stress state in an element by three scalars: the 'hydrostatic' stress (or pressure), the 'von Mises' (or deviatoric) stress, and the direction in which the deviatoric stress acts. It represents the rupture threshold as a surface possessing 3-fold rotational symmetry about the hydrostatic axis [2]. LS-DYNA employs an axisymmetric approximation of this surface. It is defined by a unique, conservative relationship between deviatoric stress and pressure.

The effect of this curve was explored on the model's predictions for the set of simple test cases. A conservative rupture curve, which gave the best overall fit to the empirical data, was adopted. The curve was scaled solely on the basis of concrete cylinder strength, which is related to cube strength.

Rupture of concrete does not imply structural failure. Concrete that has failed in tension will continue to bear compressive load and will sustain shear stress across flexural cracks provided sufficient compression is restored. However, cracked concrete under hydrostatic tension cannot transmit deviatoric stress in any direction.

Thus, the material model was able to represent the following characteristics of concrete that were important for this analysis:

- tensile cracking;
- compressive capacity after tensile cracking under cyclic loading; and
- shear capacity after tensile cracking under restored compressive stress.

There were two aspects of material behaviour that the model was not able to represent.

1. The bond between concrete and steel reinforcement is not modelled, so bond failure is not predicted. However, it was established that no reinforcement element in the structure was strained significantly beyond yield, either in tension or compression. The condition where bond failure might be expected, i.e. when reinforcement bars are stretched, is not approached.
2. Several pours of concrete will have been made during construction. The interfaces between successive pours represent potential areas of concrete weakness, particularly in tension. No allowance was made for such interfaces in the model. Construction interfaces in the structure are likely to be found at the bases of the columns and at the joints between the columns and the beams. However the model predicts that columns would, anyway, suffer extensive flexural cracking in these zones. The flexural cracks would mask any weakness at construction interfaces.

The steel reinforcement was modelled using a simple bilinear stress-strain relationship, characterised by an elastic modulus to first yield and a tangent modulus to fracture.

2.4 Model Testing and Development

Test cases involving simple structures, selected to represent principal failure modes for reinforced concrete, were run in order to qualify the reinforced concrete material model. The results from the finite element test analyses were compared with established experimental data and also with design code predictions. The effects of scaling the material properties, and of changing the number and size of elements, on the predicted global behaviour of each test specimen was explored. Then the plant support structure model was developed in stages.

2.4.1 Beam Models

A model of an under-reinforced concrete beam, simply supported at each end, was subjected to a steadily increasing load, applied at two points, causing it to fail in bending.

For computational efficiency, a quarter of the total beam was modelled to exploit symmetry. The area of steel reinforcement was distributed to suit the mesh. No cover was modelled. The applied load was distributed over small areas to avoid numerical problems. The load was then steadily increased with time until the point of failure was reached. The load-deflection curve predicted by the LS-DYNA model, shown in Fig. 1, agrees well with the experimental results.

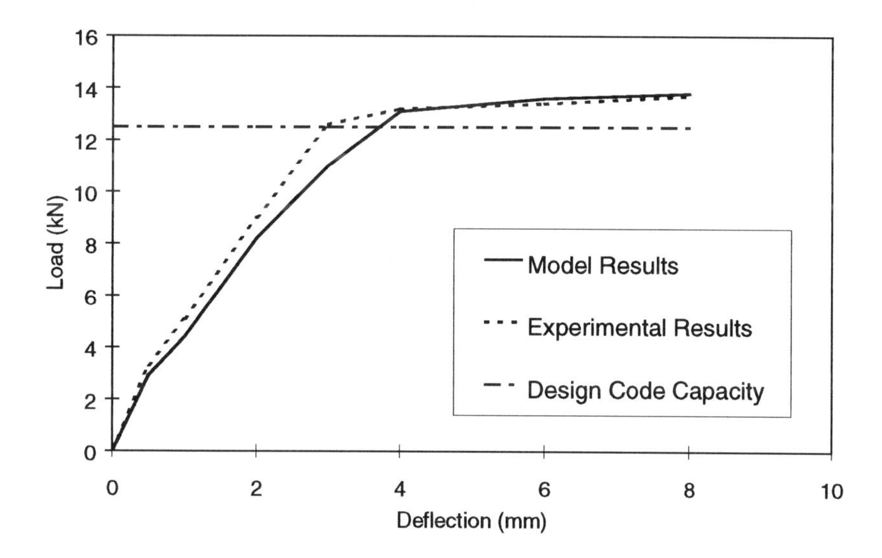

Fig. 1. Load-deflection curve for a reinforced concrete beam

Failure in shear of a reinforced concrete beam was also modelled, for comparison with an experimental test case. The beam had no shear reinforcement. Substantial tension reinforcement, consisting of four bars at the bottom face of the beam, resisted flexural failure. A vertical load was applied at the centre of the beam, and was gradually increased until shear failure occurred. Table 1 shows a comparison between the experimental failure load, the design code ultimate limit state load and the LS-DYNA model failure load.

LS-DYNA predicts a greater failure load than the design code, but remains conservative.

Table 1. Failure Loads for Reinforced Concrete Beam Failing in Shear

Test case embodiment	Failure Load (kN)
Design Code Ultimate Limit State	240
LS-DYNA Model	270
Experiment	330

2.5.2 Single column model

A model representing one half of a column in the actual support structure being assessed was constructed. The two-inch cover to the reinforcement was, initially, included. Rigid brick elements were used to model the intersecting beam sections at the top of the column. The heavy plant was modelled by attaching mass elements to the top surface of the beam section. The model is illustrated in Fig. 2 below.

One and a quarter cycles of sinusoidal displacement were imposed at the top of the column over a period of twenty seconds to suppress dynamic effects. A hysteresis curve for shear force at the column base is illustrated in Fig. 3.

A qualitative inspection of the hysteresis curve confirms that the non-linear model gives a reasonable prediction of the behaviour of the column. The hysteresis curve is steep mid-cycle, as the column 'rocks' from one side to the other. The 'rocking' effect is particularly marked in this column because it is under-reinforced with respect to bending. Bending deflection is realised by stretching steel, rather than by crushing concrete.

The second stage in the development of the single column model was to check and understand its dynamic behaviour. A vertical body acceleration was ramped up over one second, and then maintained to represent gravity. Seismic loading was superimposed, using prescribed displacement time histories in the vertical and transverse directions, for a period of ten seconds.

The maximum bending moment at the base of the column over the period was calculated. It was significantly greater than the ultimate limit state moment specified in the design code. Nevertheless, the non-linear analysis showed that the column, with its supported mass, remained stable in the seismic event.

The satisfactory performance in this instance is attributed to the low proportion of vertical reinforcement. Because the column section is under-reinforced with respect to bending, seismically induced bending moments cause concrete to crack and steel to yield in the tensile zone without significant crushing of the concrete in the compressive region.

The transverse motion is small compared with the column length, and strain in the steel reinforcement exceeds the yield strain only for very brief periods. Cracking occurs as the column rocks from side to side, as implied by the tensile stress zone in Fig. 4. Rocking tends to isolate the top of the column from horizontal seismic loads.

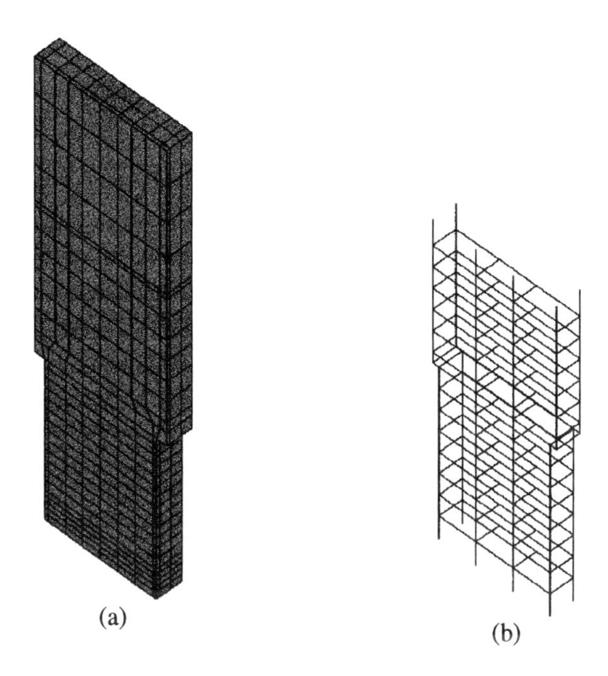

Fig. 2. Single column model (a) including cover and (b) reinforcement mesh

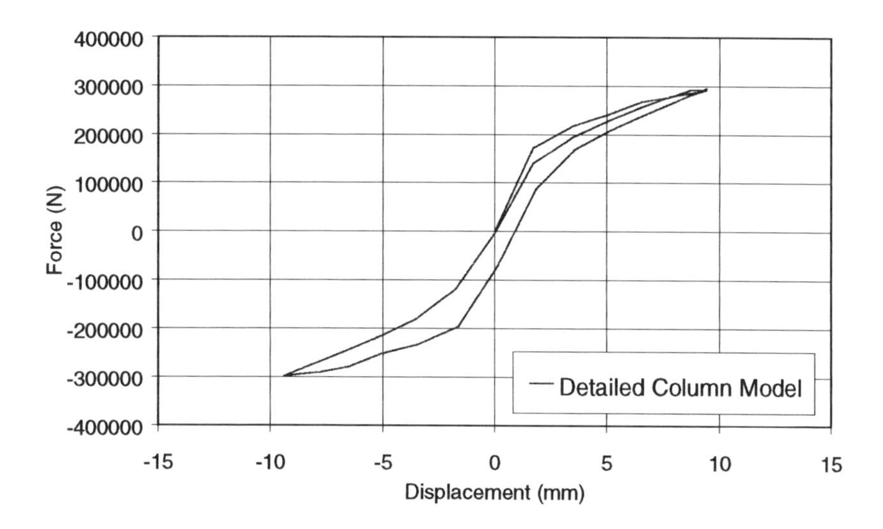

Fig. 3. Hysteresis curves extracted from single column model

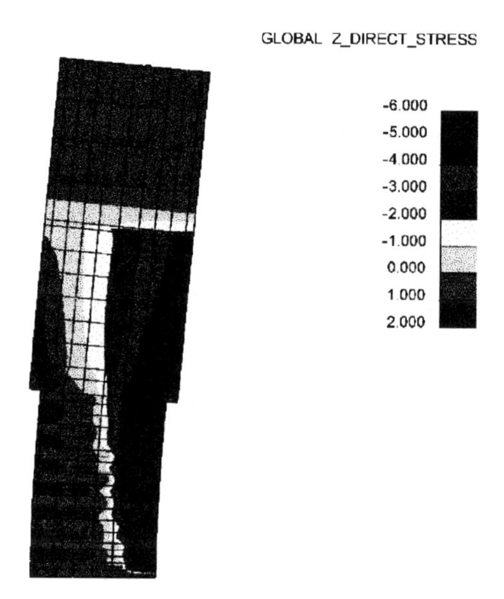

Fig. 4. Concrete stress during rocking motion, displacements magnified 100 times

The sensitivity to model construction details was investigated in a drive for modelling and computational efficiency. It was concluded that removing the concrete cover elements and judicious reduction in mesh density did not significantly degrade the accuracy of the assessment.

2.5.3 Two column model

The next step in the model development produced a complete segment of the structure. This segment comprises two columns, those parts of the beams above the columns, and a representation of the supported plant, as illustrated in Fig.5. The simplified column models were adopted. Each reinforcement bar was still explicitly modelled using fully integrated beam elements.

The sections of the beams directly above the columns were represented using a steel and concrete mesh. The beam spans were not modelled explicitly at this stage. The concrete density was increased to represent the mass of the beams extending from each column.

The model uses a combination of beam, brick, shell and mass elements, of elastic and rigid materials. The plant travels on rails. Its carriage incorporates articulating rocker pads to achieve kinematic support. The wheel rail interfaces and pads can roll, slide and ultimately separate. They were modelled using a standard contact algorithm to achieve a realistic response of the heavy plant.

The results of the analysis showed that the transverse drift of the columns was typically 0.4% of the column height, and that collapse of the plant did not occur. The structure's state at the end of the seismic event is shown in Fig.6.

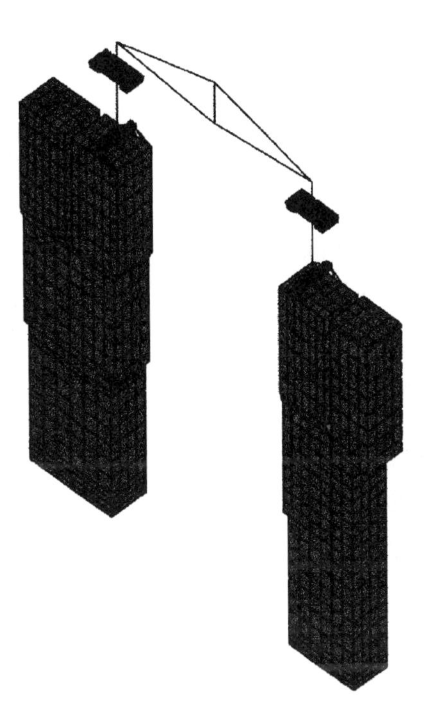

Fig. 5. Two column model including plant item (some plant item elements not shown)

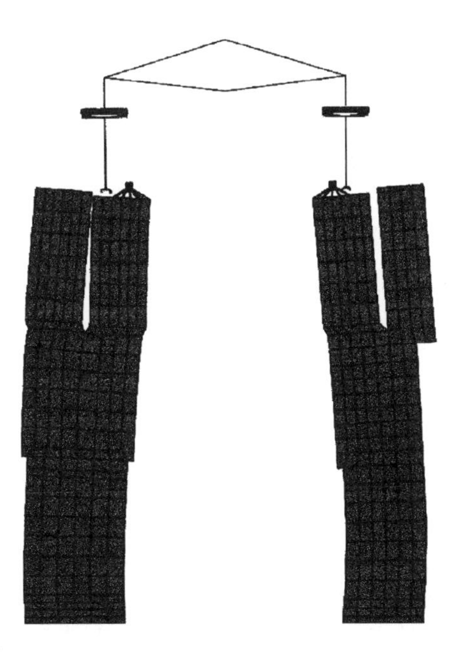

Fig. 6. Two column model after seismic event (displacements magnified 100 times)

2.5.4 Eight column model

In the final stage of the assessment, the model was expanded to include eight columns and the beams spanning between them. An area of particular interest was the central beam spans supporting the plant. Fully detailed models of the two central beam spans and the four adjacent columns were constructed. The surrounding support structure was also included but in less detail, to provide adequate boundary conditions for the central spans. The model is shown in Fig. 7.

The central beam spans were modelled using a mesh density similar to that used in the column models. The high density of bending and shear reinforcement meant that groups of bars were lumped into individual beam elements. The concrete cover to the reinforcement was not modelled.

Fig. 7. Eight column model

The surrounding beams and columns were modelled in less detail. Half of the span of each beam was modelled using reinforcement beam elements and concrete brick elements, with a mesh density half that of the central spans. The remaining half spans were modelling using a coarse, elastic brick mesh, as were the remaining four columns at the ends of the spans.

The transverse loads imposed on the columns were smaller than predicted by the previous analysis. In the eight-column model four columns support the plant, not two. The benefit of portal action provided in the longitudinal direction through the beams and surrounding columns was also clear.

The central beam spans suffered a permanent vertical deflection after the seismic event. Permanent bending of the columns under the beams caused this sag. The displaced shape of the structure can be seen in Fig. 8.

The compressive and deviatoric stresses in the central beam spans were compared before and after the seismic event. It was found that although a significant proportion of the concrete section was cracked and could not support tensile or deviatoric stresses after the earthquake, adequate compressive zones of concrete remained at the ends of the spans to transmit the shear force to the columns.

Retained shear capacity in the concrete of a cracked beam relies upon tension in the longitudinal reinforcement to sustain compression in the concrete. The peak stress in the tension reinforcement was found to be significantly below the yield value. Adequate shear capacity is therefore secure.

Fig. 8. Eight column model after seismic event (displacements magnified 100 times)

3 Conclusions

Use of non-linear finite element analysis of a reinforced concrete structure yielded significant benefits over conventional linear elastic methods. Freedom from the constraints of linear elasticity avoided the limitations of a previous force-based assessment, which had given rise to substantial conservatism.

A comparison of established test results with finite element models of structural failure has built confidence in the concrete model. The minimum mesh density required to give accurate results was established and adopted.

Exercising these models has clearly demonstrated why the structure remains stable despite certain parts straining substantially beyond code limits. Rocking at the bases of the columns effectively isolates the structure and the heavy plant from the seismic loading.

The ability of such assessment to resolve non-linear dynamics and demonstrate viable load paths through cracked concrete has avoided difficult and expensive modifications to the structure.

4 References

1. Oasys Ltd. (1997) *Oasys LS-DYNA 7.0 Environment User Manual. Version 7.0*, Oasys Ltd., London, Revision 1.

2. Kotsovos, M. D. and Pavlovic, M. N. (1995) *Structural Concrete - Finite Element Analysis for Limit-State Design*, Thomas Telford Publications, Great Britain.

DYNAMIC ANALYSIS TO DETERMINE SOURCE OF BLAST DAMAGE
Dynamic analysis for blast damage

W. G. CORLEY and R. G. OESTERLE
Construction Technology Laboratories, Inc., Skokie, Illinois, USA

Abstract
Following an internal explosion in a multi-story building, it was necessary to determine the location and source of the blast. Eyewitness reports, conditions prior to the blast, and evidence collected in initial investigations resulted in differing possibilities about the source of the blast.

Careful evaluation of damage to structural members after debris had been removed indicated more than one possible scenario. Dynamic analysis using various sources of the explosion provided conclusive evidence about the location of the explosion.
Key Words: Blast loading, damage, buildings, dynamic analysis, explosion, structure, multi-storey, sources of explosion

1 Introduction

Construction Technology Laboratories, Inc. (CTL) performed an investigation of an explosion that occurred in a low-rise building. The building was of mixed construction including reinforced concrete and structural steel. The explosion produced extensive damage to the building, particularly within the lower half of the structure. Following clean-up of the debris, the structure was demolished. The cause and location of the explosion was in question. Construction Technology Laboratories' work involved independent review of available information and the determination of cause and location of the explosion.

2 Summary of findings

Based on CTL's review of documents and structural calculations, the following findings are presented:

Abnormal Loading on Structures edited by K. S. Virdi, R. S. Matthews, J. L. Clarke and F. K. Garas.
Published in 2000 by E & FN Spon, 11 New Fetter Lane, London EC4P 4EE, UK. ISBN 0 419 25960 0

1. The origin of the explosion was in the central region of the building between the second and third floors and between the third and fourth Column Lines.
2. Origin of the explosion is supported by observed symmetry of damage from the second level down to the ground floor and up to the fourth floor. It is also supported by the directions of movement of structural elements on the north, south, east, and west perimeters of the structure.
3. Calculations of damage to beams at the second floor level support the conclusion that the origin of the explosion was between the second and third floors.

3 Objectives and scope

Objectives of the work by CTL were to evaluate available evidence and data concerning the explosion and to correlate this information with possible causes and locations of the explosion within the structure. Based on the location of the explosion and sources to get gas to that location, the actual cause was to be determined.

The objectives were accomplished within the following scope:

1. Determination of the origin of the explosion based on the analysis of documents that showed the locations of debris for the damaged structure.
2. Structural analyses for static and dynamic response of selected components to correlate observed damage of specific components with the origin of the explosion.

4 Structural analysis

To determine the origin of an explosion in a low-rise building, structural analyses for static and dynamic response of selected components were carried out. A correlation of observed damage to specific components near the origin of the explosion was used to determine where the explosion occurred. This report describes analyses of the dynamic response of a reinforced concrete beam at the second level.

4.1 Overpressure-time relationship

Two types of pressure-time relationships were used in the analyses. These pulse are shown in Fig. 1. The first pulse shown is a high pressure pulse with a short duration and represents an upper estimate of the overpressure. In contrast, the other pulse shown in Fig. 1 has a significantly lower pressure and longer duration. Both types of pulses were used in the analyses.

4.2 Reinforced concrete beam at second level

A reinforced concrete beam at the second level was severely damaged by the explosion. The diagonal direction of observed cracks indicates that the cracks resulted from shear forces. A similar beam at the first floor level and directly below the severely damaged second floor beam was not significantly damaged. This observed damage pattern indicates the explosion must have been above the first floor.

To investigate the effects of explosion of gas directly below the beam, dynamic analyses using a computer model were carried out as described in the following sections.

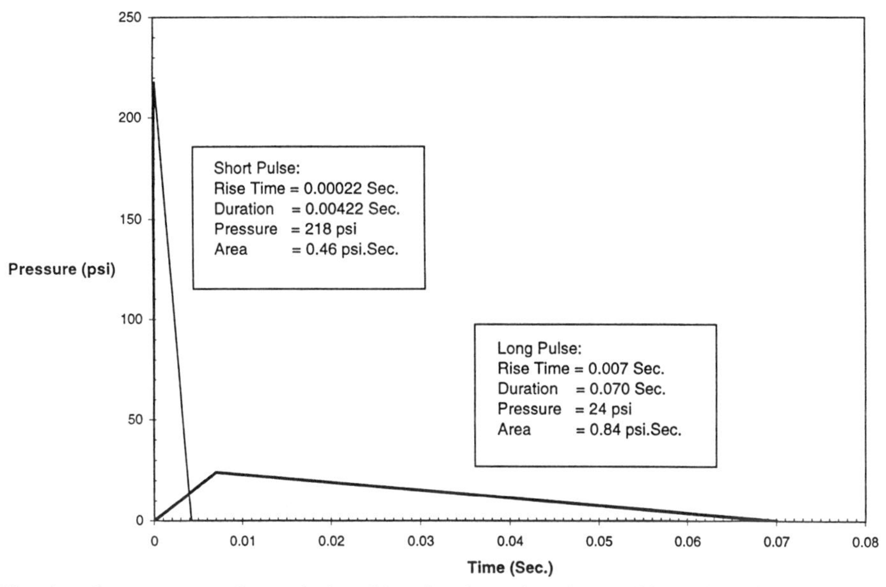

Fig. 1. Overpressure-time relationships for short duration and long duration pulses

4.3 Geometry

A two-dimensional sub-frame including the beam and the top and bottom columns at the second floor was considered in the analyses. The far end of the bottom column was fixed and the far end of the top column was fixed against rotation but free to translate. The beam was fixed for translation and rotation where it frames into a shear wall. Two-dimensional beam elements were used to model the beam and the columns. The finite element mesh is shown in Fig. 2.

4.4 Material properties

Based on tests of cores, compressive strength of the concrete was taken as 3,000 psi. This leads to a modulus of elasticity of 3,100 ksi. The damping ratio of all the vibration modes was taken as 2%.

4.5 Loading

The sub-frame was analyzed for three different load conditions. In the first load condition, dead and live loads were applied to the beam. Live load was assumed to be 4 lb/ft^2. For the second and third load conditions the bottom face of the beam was subjected to two different pulses, one pulse with high pressure of 218 psi and short duration of 0.00422 seconds, and another pulse with lower pressure of 24 psi and longer duration of 0.07 seconds. These pressure pulses are shown in Fig. 1.

4.6 Numerical results for the short duration pulse

Representative histories of mid-span vertical displacement, mid-span moment, and shear force at different locations of the beam are shown in Figs. 3 to 5. Periods and frequencies of different vibration modes are listed in Table 1.

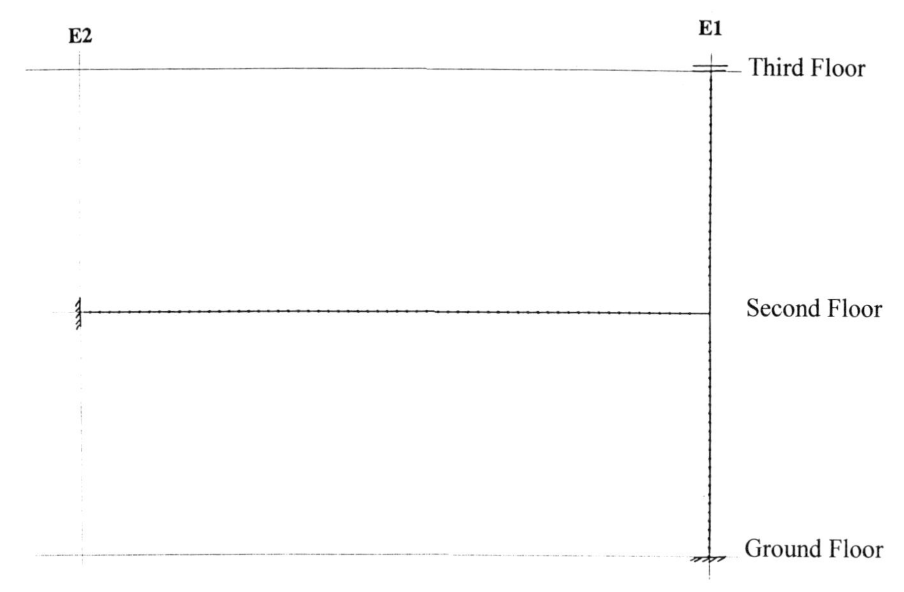

Fig. 2 Finite element model for beam E1-E2.

4.7 Numerical results for the long duration pulse

Representative time histories of mid-span vertical displacement, mid-span moment, and shear force at different locations along the beam are shown in Figs. 6 to 8.

Table 1 Modal periods and frequencies for model of beam

Mode	Period (s)	Frequency (cycles/s)	Mode	Period (s)	Frequency (cycles/s)
1	0.2582	3.87	16	0.0021	466.08
2	0.1174	8.52	17	0.0021	482.05
3	0.0876	11.42	18	0.0017	577.70
4	0.0385	25.97	19	0.0017	577.89
5	0.0173	57.88	20	0.0016	615.26
6	0.0146	68.56	21	0.0016	630.35
7	0.0075	133.61	22	0.0016	642.37
8	0.0062	160.82	23	0.0013	762.50
9	0.0045	220.38	24	0.0013	789.24
10	0.0044	227.52	25	0.0012	802.43
11	0.0034	295.23	26	0.0010	962.15
12	0.0032	316.06	27	0.0010	985.36
13	0.0031	321.87	28	0.0009	1121.46
14	0.0030	328.10	29	0.0009	1152.90
15	0.0024	414.74	30	0.0009	1153.04

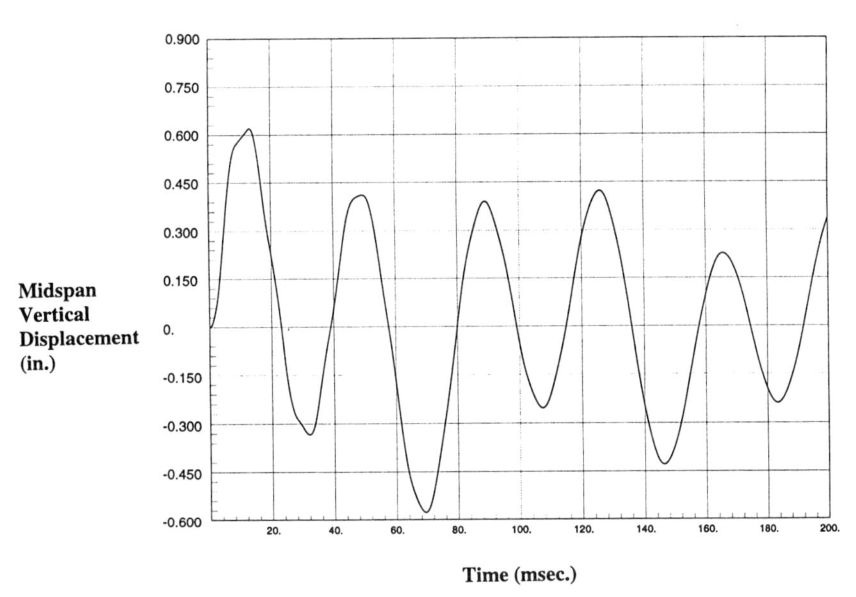

Fig. 3 Midspan vertical displacement with short duration pulse.

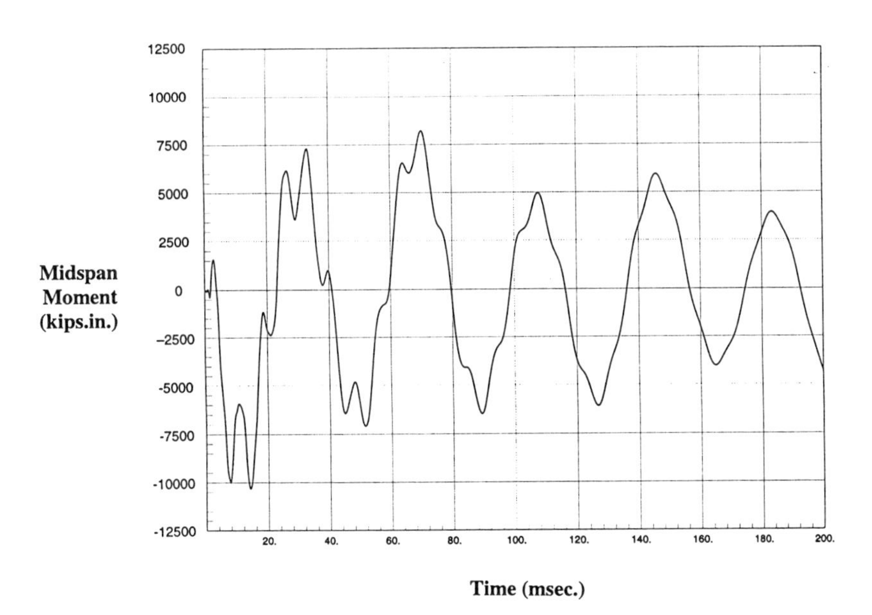

Fig. 4 Midspan moment with short duration pulse

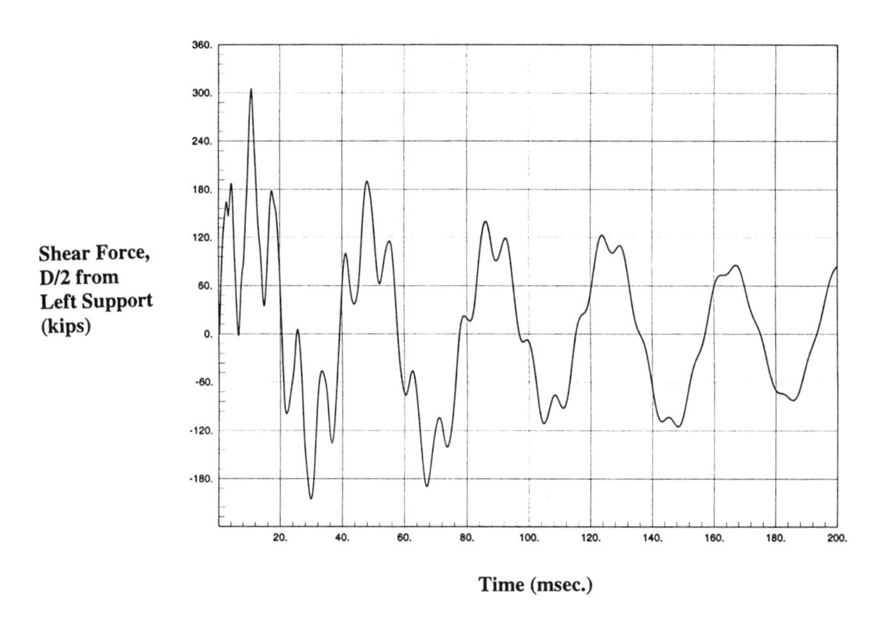

Fig. 5 Shear force, D/2 from left support for beam with short duration pulse

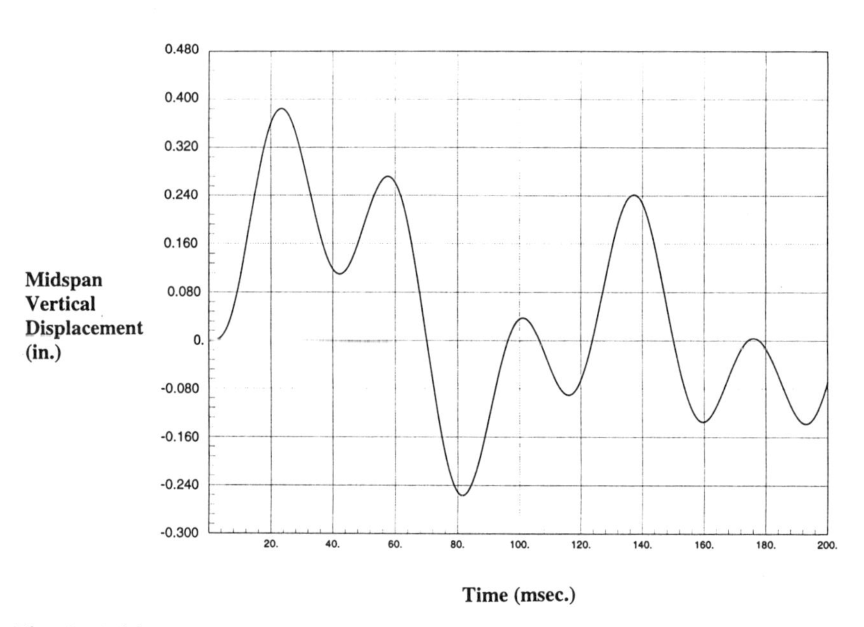

Fig. 6 Midspan vertical displacement for beam E1-E2 with long duration pulse

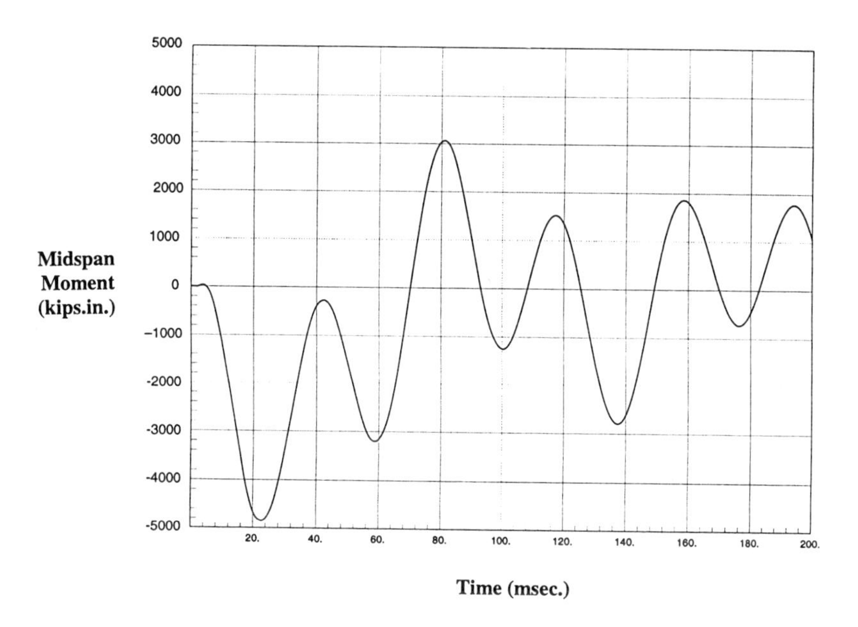

Fig. 7 Midspan moment for beam E1-E2 with long duration pulse

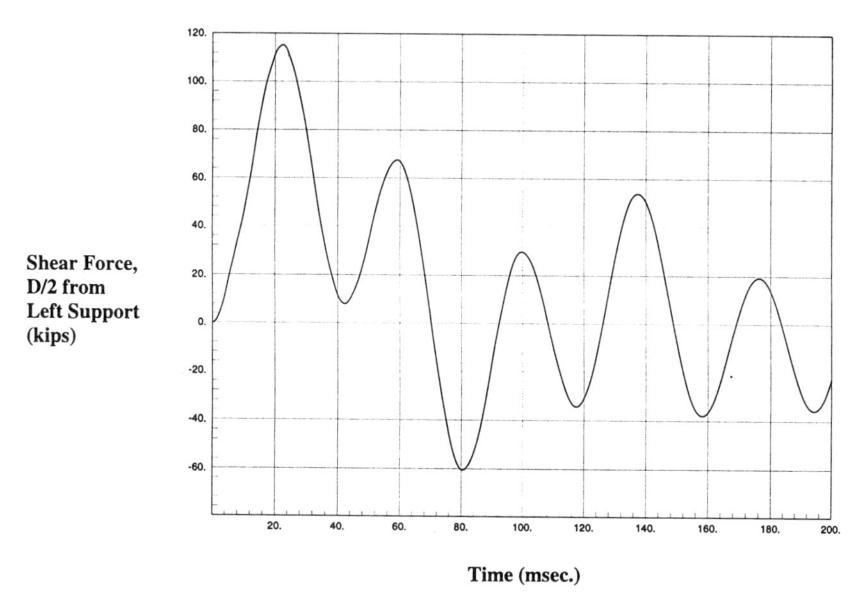

Fig. 8 Shear force, D/2 from left support for beam E1-E2 with long duration pulse

4.8 Summary of results

Mid-span vertical displacement, mid-span moment, and shear force at different locations along the beam under different load conditions are presented in Tables 2 and 3. Results of the analyses in column (6) of the tables compare the calculated shear force in the beams at selected locations to the shear force, V_c, that would be expected. However, it should also be noted that the V_c used in Tables 2 and 3 did not include the effects of tension on the beam. The explosion caused one end support to move east and the other to move west. This building movement put the beam in tension. With tension, the shear force needed to cause cracking is decreased. Therefore, the ratios somewhat less than one in Column (6) in Tables 2 and 3 still indicate cracking is expected.

Table 2. Effect of long duration pulse on beam

	(1)	(2)	(3)	(4)	(5)	(6)
	D+L Load	Static Pressure	Dynamic Pressure	(3)/(2)	(3)/(1)	(3)/Vc
Vertical displacement at midspan (in)	0.09	0.30	0.38	1.27	4.2	-
Moment at midspan (kips in)	956	-3374	-4849	1.44	5.1	-
Shear at left support (kips)	-25.6	86.7	115.3	1.33	4.5	2.0
Shear at 11 ft from left support	-11.2	39.5	55.4	1.40	4.9	1.0
Shear at 11 ft from right support	4.8	-21.9	-48.1	2.20	10.0	0.8
Shear at right support	24.5	-73.9	-105.6	1.43	4.3	1.9

Table 3. Effect of short duration pulse on beam

	(1)	(2)	(3)	(4)	(5)	(6)
	D+L Load	Static Pressure	Dynamic Pressure	(3)/(2)	(3)/(1)	(3)/Vc
Vertical displacement at midspan (in)	0.09	2.77	0.62	0.22	6.9	-
Moment at midspan (kips in)	956	-30644	-10500	0.34	11.0	-
Shear at left support (kips)	-25.6	788	310	0.39	12.1	5.5
Shear at 11 ft from left support	-11.2	359	175	0.49	15.6	3.1
Shear at 11 ft from right support	4.8	-200	-175	0.88	36.5	3.1
Shear at right support	24.5	-671	-310	0.46	12.7	5.5

5 Conclusion

Results indicate that both the short duration and long duration pulses would be expected to cause shear cracking. Visible shear cracks in the beam at the second level is consistent with an explosion of gas directly below this beam. This is consistent with gas collecting directly below the second floor and into the enclosed vacant space between the second and third floors. Lack of damage in the beam at the ground floor level indicates the explosion could not have been below the first floor.

DYNAMIC RESPONSE OF BUILDINGS AND FLOOR SLABS TO BLAST LOADING

Response to Blast Loading

P. ESPER
Arup Energy, Ove Arup & Partners, London, UK
D. HADDEN
Building Engineering, Ove Arup & Partners, London, UK

Abstract

The number of bombing attacks on modern societies has increased dramatically in recent years calling structural engineers to consider explosion loading more seriously in their design. In Mainland UK alone, four major bombing incidents took place within the last seven years. Both authors were involved in the investigation of damage and reinstatement of a number of commercial buildings that were structurally affected by these incidents, and in advising building owners and occupiers on blast protection measures. Numerical analysis as well as laboratory and on-site testing were carried out in order to investigate the dynamic response of the damaged buildings to bomb blast. The structural integrity of these buildings, including the residual strength of their floor slabs, was assessed. In addition, numerical modelling techniques to simulate the behaviour of laminated glass under blast loading are being developed.

1 Introduction

Following, what seems to be, an increasing incidence of bombing attacks on modern societies in the last decade or so, it is becoming more evident that structural engineers and architects need to consider blast loading in their design in order to protect their societies from such attacks. Four major bombing incidents took place in Mainland UK within the last seven years; the 1992 St Mary's Axe, the 1993 Bishopsgate (see Fig.1), the 1996 Docklands and Manchester bombs. These detonation devices were estimated as 450 kg, 850 kg, 500 kg and 750 kg of TNT equivalent, respectively. Both authors were involved in the investigation of damage and reinstatement of a number of commercial buildings, and in providing advice to building owners and occupiers on blast protection measures for both existing and proposed buildings. Numerical modelling as well as laboratory and on-site testing were used in the investigation of damage and assessing the dynamic response

Abnormal Loading on Structures edited by K. S. Virdi, R. S. Matthews, J. L. Clarke and F. K. Garas.
Published in 2000 by E & FN Spon, 11 New Fetter Lane, London EC4P 4EE, UK. ISBN 0 419 25960 0

of these buildings and their floor slabs to blast loading. Two buildings are discussed in this paper. The first, referred to as building (B1), was located only 11 m away from the explosive device (see Figs.2); the second, building (B2), was at a distance of 75 m away from the same bomb (see Figs.3). This paper describes the use of FE analysis, by the structural engineer, as a potential tool for blast damage assessment alongside empirical and forensic examination techniques. Correlation between the results of the FE analysis and laboratory and on-site testing is highlighted, and the reliability of the former in predicting/highlighting problematic zones and reducing the scope of the latter when assessing the extent of damage under such loading, is considered. The blast hazards to building occupants associated with conventional glazing and the development of numerical techniques to simulate the behaviour of laminated glass under blast loading, is also discussed.

2 The nature of blast loading

Home-made explosive devices, as the ones used in the above incidents, consist simply of a small quantity of military high explosive detonator (usually semtex) surrounded by a fertiliser and a fuel. The detonation of such devices results in the generation of highly pressurised hot gases which expand violently displacing the surrounding air from the volume it previously occupied. A layer of compressed air, known as the blast wave, forms in front of the expanding gas and is characterised by an instantaneous rise from ambient pressure to a peak incident pressure (see Fig.4). This pressure increase, or shock front, travels away from the blast source in the form of a hemispherical wave with a diminishing velocity. The speed of the pressure wave of most home-made explosives rates around 2,700 m/s compared to 7,500 m/s for high explosives. As the gas expands the pressure decreases until it falls, eventually, to or below the atmospheric pressure. This negative pressure is associated with the over-expansion of the gas, the result of which is a reversal of flow towards the source (i.e., suction). This negative phase has a longer duration than that of the positive phase but has a much lower peak value. As the blast waves radiate out at supersonic speeds within the confined city streets they are reflected and refracted by adjacent buildings where greater pressures may result. If the explosion is close to, or on the ground surface, a small proportion of the energy will be transmitted through the ground as seismic waves, whereas most of the energy will be absorbed by the ground surface displacing it and forming a crater (see Fig.1).

3 Structural response to blast loading

The ductility and natural period of vibration of a structure governs its response to an explosion. Ductile elements, such as steel and reinforced concrete, can absorb significant amount of strain energy, whereas brittle elements, such as timber, masonry, and monolithic glass, fail abruptly. Blast loads are typically applied to structures at rates approximately 1000 times faster than earthquake-induced loads. In the investigation of the dynamic response of a building structure to bomb blast, the following procedure is followed [Esper, 1996]: (a) the characteristics of the blast wave must be determined; (b) the natural period of response of the structure (or the structural element) must be determined; and (c) the positive phase duration of the blast wave is then compared with the natural period of response of the structure. Based on (c) above, the response of the structure can be defined as follows:

i) If the positive phase duration of the blast pressure is shorter than the natural period of vibration of the structure, the response is described as impulsive. In this case, most of the deformation of the structure will occur after the blast loading has diminished.
ii) If the positive phase duration of the blast pressure is longer than the natural period of vibration of the structure, the response is defined as quasi-static. In this case, the blast will cause the structure to deform whilst the loading is still being applied.
iii) If the positive phase duration of the blast pressure is close to the natural period of vibration of the structure, then the response of the structure is referred to as dynamic. In this case, the deformation of the structure is a function of time and the response is determined by solving the equation of motion of the structural system.

In general, a tall building will have a low frequency and thus a long period of vibration in relation to the duration of the load. Individual elements will have response times that may approach the load duration. Window panels of 1.0-1.5 m^2 for instance, have a frequency greater than 10 Hz (or less than 0.1 s response time). Floor slabs have a frequency range of 10-30 Hz (or 0.03-0.1 s response time). Rigid elements which are unable to respond dynamically within the time period of loading will under some circumstances continue to attract load until they fail in-elastically, and abruptly. Flexible elements, on the other hand, will attenuate the load by responding to the blast pressures, enabling the strain energy to be absorbed through deformation. Unlike steelwork, the rebound in concrete is small as cracking in concrete will result in internal damping. The massive nature of reinforced concrete structures performs well in resisting the impulsive load usually encountered close to the blast point.

4 Magnitudes of blast pressures

The magnitudes of the blast pressures experienced by two buildings, externally and internally, were estimated at specific points of each structure, by carrying out a 3-D simulation of the street layout and building blocks, using 'INBLAST' and 'CHAMBER' programs. Building B1 (see Fig.2), which was only 11 m away from the bomb, was a historically listed building, 7 storey high and was constructed in 1928. The floors were generally in-situ reinforced ribbed slabs with permanent hollow tile liners. On this building, the maximum external pressures experienced by the building were estimated as 1400 kN/m^2 and their duration was approximately 170 ms. The internal pressures on the floor slabs had a positive pressures range of 130 kN/m^2 south to 490 kN/m^2 north (close to the bomb), and their duration was approximately 350 ms. The estimated arrival times of these pressures were 12 ms and 140 ms, respectively. Careful examination of these pressures at various time intervals, concluded that they subjected the floor plates to highly complex transient behaviour in both upward and downward directions occurring simultaneously on individual floor panels and far exceeded their original design loads. Building B2 (see Fig.3) which was 75 m away from the bomb, was constructed in the late

1950's and consisted of a basement, ground and five upper floors. It has a concrete cased structural steel frame supporting hollow tile reinforced concrete floors. The beam column connections consisted of riveted seating cleats and site bolting of the beams to the cleats. Lateral stability was achieved through lift/staircase cores and flank walls. As the rounded corner of this building was totally exposed to the shock wave, all points on this corner would have seen very high blast pressures that reached a maximum value of 130 kN/m^2 at first floor level, and their duration was approximately 180 ms (see Fig.4). The values of the internal pressures on the ground floor ranged between 120 kN/m^2 and 170 kN/m^2, and their duration was 200 ms.

5 Investigation of damage - finite element technique

The fifth floor slab of building B1 was modelled using ANSYS (see Fig.5) in order to investigate its dynamic response to the estimated blast pressures. Two element types were used; 4-noded shell element for the reinforced concrete slab, and beam element to model the steel beams. The total number of elements in the model was 945, and the total number of nodes was 627. The column locations were considered as support points with the panel sizing varying between 7×7 m^2 to 7×4.5 m^2. Mechanical properties of concrete and steel were obtained from material testing. By carrying out a modal analysis, the natural frequencies for the first three modes of the floor plate of building B1 were determined. The fundamental natural frequency was f_1= 12.50 Hz, giving a fundamental period of vibration equal to: T_1 = 80 ms. The positive phase duration was in the range of 20-45 ms. As this was shorter than the fundamental period of the floor plate, a 3-D non-linear transient dynamic analysis was undertaken in order to determine the response of the floor slab to the internal pressures. The blast pressures were applied on the top and bottom faces of the concrete floor as pressure-time histories. Consequently, two graphs for every node of the model were obtained, one representing the deflections due to the pressures acting on the bottom face of the slab (see Fig.6), and the other representing the deflection of the floor slab due to pressures acting on the top face of the slab. Although this analysis was non-linear, superimposing the two graphs in the manner shown in Fig.6 gives a clearer picture on the way in which the slab would deflect under the two time-varying pressures applied at its top and bottom surfaces. It was demonstrated by the FE analysis that when the soffit of the floor plate saw the positive peak pressure, it took 10 ms on average to develop its maximum deflection which ranged between 16-24 mm. It was observed that a 5 ms time lag in the pressures hitting the top surface of the slab was sufficient to result in a net upward deflection of approximately 16 mm above the horizontal. This momentary uplift of the plates resulted in tension cracks to the top surface of the slab at its mid-span. This, combined with the subsequent rebound of the plate, caused crack aggravation over the supports, thus impairing the performance of the bond between the concrete and its reinforcement.

In the case of building B2, a full 3-D FE model of the whole building was generated using ANSYS (see Figs.7-9) in order to investigate the global response of the structure to the estimated blast pressures, and to ascertain if any twisting of the structural frame has resulted. Shell elements were used to model the reinforced concrete slabs, and beam element to model the steel beams and columns. The total number of elements in this model was 5232, and the total number of nodes was 3912. This building had a large window area on all elevations (see Fig.3) except for the west elevation (flank wall). In addition, it was reported by various sources [4], and confirmed by our investigation on glass panels (discussed in Section 8), that normal glass, as a brittle material, takes only 5-8 ms to break. Hence, it was more realistic not to include the glass panels in the FE model. Mechanical properties of concrete and steel used in the analysis were also obtained by carrying out material testing. Modal analysis of this building provided the natural frequencies for the first three modes of vibration with the fundamental frequency being f_1= 1.90 Hz, and the fundamental period of vibration being equal to T_1 = 527 ms. The positive phase duration was found to range between 50-80 ms. As this was shorter than the fundamental period of vibration of the building, its response was impulsive. However, the main concern, as described above, in the case of this building was its lateral stability, and a quasi-static analysis was considered sufficient to achieve this objective. The variable blast pressures were applied at different time intervals as a static load on the elements of the facades. These values were extracted from the pressure-time history graphs produced at each specific time interval considered in the analysis. This allowed to examine the maximum deflection that could have been experienced by the building at any specific time under the blast loading, while achieving a considerable saving of computer time and analysis efforts.

6 Investigation of damage – on site testing

In such events as blast damage, it is important that the general condition of the fabric is recorded (i.e. photographed and videoed) prior to its removal. This will help to pin-point local areas of over-stressing or 'hot spots' caused by a build-up of pressure internally due to the poor venting characteristics of an internal layout. Having identified areas within the structure for detailed investigation, a regime of site and laboratory testing is prepared, starting from locations of greatest visible blast damage and areas where analytical methods (i.e. computer modelling) have indicated potential over-stressing of the structure. If the damage being encountered were still severe, then the area of testing would be extended. As the findings become negligible the area of testing is curtailed. The material testing is usually undertaken by a specialist testing house, preferably one with resources to deal with most aspects of the investigation. Other surveys undertaken may include CCTV drain surveys, plumb line surveys, dimensional surveys of the building (if original drawings do not exist), level surveys, raised floor and ceiling surveys, cladding surveys, etc. The following non-destructive tests were undertaken on RC elements in the case of the two buildings discussed here:

a) Indirect method of ultrasonic pulse velocity (UPV): This was used to estimate crack depths and assess the integrity of the grouted joints between the PC floor planks.
b) Impact spectra analyser (ISA).
c) Impulse radar measurement was used to assess the integrity of basement slabs and retaining walls.
d) Crack ageing; concrete cores were taken (see Fig.10) and the following steps were carried out:

- Phenolphthalein solution to determine depth of carbonation
- Petrographic examination of thin sections

e) Residual strength assessment was carried out through crushing of concrete cores and load testing.

Tapping surveys were undertaken on all rendered elevations to identify areas of de-bonding. Detailed visual inspections of masonry elevations were carried out with all new and old crack patterns and bed joint disturbances noted. Brick and mortar samples were obtained for strength and composition testing respectively. Dye testing was used to identify surface cracking. On-site testing of brickwork supports and cladding assemblies together with laboratory testing of component parts to both ultimate and service loads was also carried out. Testing/inspection of steel frames varied from 1920's riveted beam/column connections to present day welded and bolted connections. It included confirmation of material grade, visual inspection of bolts and rivets, magnetic particle inspection to detect surface tearing or cracking of joint assemblies, and ultrasonic inspection of butt welds and rivets. In those locations where crack like features were observed from the ultrasonic inspection of the welded joints, acetate replicas and metallurgical sectioning to allow macro and micro examination were undertaken. Bolts were tested for tightness on site and where bolts were found to be loose they were replaced (see Fig.11).

7 Correlation between damage, testing and finite element analysis results

In the case of building B1, the FE analysis suggested excessive upward floor slab displacements. Crack ageing and floor plate load tests corroborated these findings. Ultrasonic inspection of the riveted steelwork connections and MPI of the plated assembly confirmed that no over-stressing had occurred. Load testing of selected floor panels was undertaken in order to check their structural adequacy to carry the original design loads. The upper floor slabs had failed, and consequently, were replaced. Although building B2, was 75 m from the blast epicentre, damage to its fabric was nevertheless extensive. Diagonal shear cracks were observed in the flank wall enclosing the staircase at various levels above ground floor (see Fig.9). From these observations and the results of the plumb survey, it was concluded that the structure had undergone rotational movement. Also, the FE analysis results confirmed the deflected shape observed and the calculated maximum stresses coincided with the above-mentioned shear cracking. Design checks were undertaken in order to assess the structural adequacy of the frame to carry the additional forces and moments resulting from the blast induced displacements. Cracking was also observed in the basement retaining walls and petrographic inspection of the cores confirmed this to be of a recent origin. The top two floor plates replaced and vertical wind bracing added within the existing cores.

8 Glazing

One of the greatest source of injuries and internal disruption from an external bomb explosion is the fragmentation of conventional annealed glass in windows which, when it breaks, does so in the form of dagger-like shards. These shards are then thrown as hazardous projectiles either deep into the building, or into the surrounding exterior areas. Toughened glass will shatter at higher load than annealed glass and forms dice-shaped particles rather than elongated shards. Although less hazardous than annealed glass shards, these toughened "dice", travelling at high velocity, can still cause injuries deep inside the building. The most effective protection to reduce this hazard is to use laminated glass anchored to its supporting frame in deep rebates (typically 25-30mm deep). When broken, the laminated glass remains stuck to the interlayer material rather than forming shards. By securely bonding the pane to its frame by means of a structural sealant, the interlayer can act as a ductile membrane which bulges inward. By remaining intact the membrane prevents blast pressures and debris entering the building. If the membrane is stretched beyond the limit of its ductility, it will tear at mid-pane rather than at the perimeter so that the pieces of the torn laminate remain attached to their frame. The effectiveness of laminated glass used in this way, in combination with appropriate frames and fixings to the structure, is well proven in both tests and actual terrorist bomb explosions. Empirically derived data exists for a range of glass types and thickness in the form of iso-damage curves plotted against pressure and impulse [7]. From such charts the façade designer can determine the likely level of damage that a particular glazing configuration will sustain under a given blast event. Analytical techniques to model the behaviour of laminated glass under transient loads are not yet well established. By developing suitable numerical methods, and validating them against the empirical data referred to above, it should be possible to design glazing systems that can meet the demands for natural light and transparency required in many current buildings, while achieving at the same time a worthwhile level of blast resistance. FE models, which combine elements to reliably simulate the post-crack behaviour of laminated glass and their metal framing, would also allow full benefit to be taken of the ductility in such systems so as to minimise the blast loads imparted by the glass to the frames and by the frames to the primary structure of the building. Preliminary work has been undertaken using LS-DYNA to develop computationally efficient methods of modelling single laminated glass windows under a representative blast pressure-time history (see Figs.12&13). While further investigation and testing is required to confirm the high strain rate properties of the polyvinyl butyral interlayer, the analytical results to date show promise in terms of the predicted reactions on frames, deformations, glass fracture patterns, and energy balance versus time.

9 Discussion

The main concern when undertaking bomb damaged investigation of a structure is to identify hidden damage. The fracturing of the ashlar within its depth on building B1, for example, only became apparent during the dismantling of the elevation. Given the unpredictability of blast effects on buildings, it was found more cost and time effective to implement methods, such as FEA, that highlighted areas of hidden damage prior to undertaking extensive opening up of the structure. This was the case for both building B1 and building B2. Reinforced concrete and steel-framed buildings are able to survive better the effects of blast loading than load-bearing masonry buildings. For example, a terrace of Victorian masonry buildings situated next to building B2 sustained such blast damage that they had to be totally demolished. Also, good detailing of beam-column connections in both concrete and steel structures will greatly enhance their performance under dynamic loads. Inspection of steelwork connections indicated that beams having both bolted or riveted top and bottom flange cleats performed much better than welded connections with flexible end plates or butt welded portal frames. Structures that are unsymmetrical in plan are vulnerable to damage when subjected to dynamic loading due to torsional effects. For example, the lack of symmetry in the plan layout of building B2 caused twisting along its height from the blast pressures and resulted in a permanent deformation. Of all the buildings investigated by the authors, no structural failures of the foundations were encountered. However, new micro-cracking within concrete retaining walls up to 75 m from the blast epicentre and cracking to asphalt tanking were observed.

10 Recommendations and protection measures

It is generally not possible to design 'soft target buildings' to be bomb proof. Because of this, a threat assessment should be undertaken to assess what level of protection the building is expected to achieve against a potential threat. This should be based on limiting the structural damage to an acceptable level and minimising injuries to personnel. However, it should be recognised that there is a level of safety where the cost of protection with respect to the cost of loss is optimised. When undertaking a risk assessment of a new construction or an existing building, the following points should be considered [3]:

a) Stand-off distance; in the case of an external threat (vehicular bomb) the distance between the vehicle and the building should be considered and maximised, if possible.
b) An open ground foyer providing parking beneath the first floor should be avoided as this will reflect and magnify the incident blast pressures. Similarly, underground car parks and access roads should be avoided, because of the tremendous build up of gas pressure, which generally vents through riser and lift shafts causing extensive secondary damage throughout the upper floors.
c) Glazed atriums pose a particularly high risk to personnel from injuries caused by flying glass fragments, as do glazed curtain walling and roofs. This risk, however, can be mitigated by the use of laminated glass, particularly if mounted on blast enhanced framing.
d) Shelter areas within the building should be identified to relocate staff in the event of a police warning. It is now recognised that occupants are less at risk from flying debris if kept within the building rather than being evacuated into the streets in the event of a large vehicle bomb. Also, designated safe routes should be provided within the building to access these shelter areas.
e) As the lower floors are at most risk from a terrorist attack, it is important that their construction be more substantial than the upper floors. Reinforced concrete slabs and columns or reinforced concrete encased steel columns are preferable.
f) The detailing of steel and concrete frames can greatly influence the manner in which a structure responds to blast loading. In particular close attention should be paid to the connections [1], [5].
g) Outside the blast affected zone, secondary damage can be greatly reduced if consideration is given to glazing protection.

11 Acknowledgement

The LS-DYNA FE analysis of the window glass panel response to blast loading, which is reported in Section 8, was carried out by the advanced technology group at Ove Arup & Partners. The main analysis work, however, of the dynamic structural response of both buildings B1 and B2 discussed in this paper, was carried out by the first author using ANSYS.

12 References

1. Esper, P. (1996) Non-linear Transient and Quasi-static Analyses of the Dynamic Response of Buildings to Blast Loading, ANSYS 7th International Conference and Exhibition on Finite Element Modelling and Analysis, Pittsburgh, USA, May.
2. Esper, P. (1996) Dynamic Response of Steel Building Frames and Connections to Bomb Blast Loading, 1st International Conference on Impulsive Analysis, Osaka, Japan, November.
3. Esper, P. and Keane, W. (1996) Structural Integrity Assessment and Repair of Bomb Damaged Buildings, AEA 3rd International Conference on Engineering Structural Integrity Assessment, Churchill College, Cambridge.
4. Mays, G. C., and Smith, P.D. (1995) Blast Effects on Buildings, London.
5. Rhodes, P. S. (1974) The Structural Assessment of Buildings Subjected to Bomb Damage, The Structural Engineer, volume 52, London, September.
6. The Institution of Structural Engineers. (1995) The Structural Engineer's Response to Explosion Damage, London.
7. The Steel Construction Institute. (1999) Protection of Buildings against Explosions, Ascot.

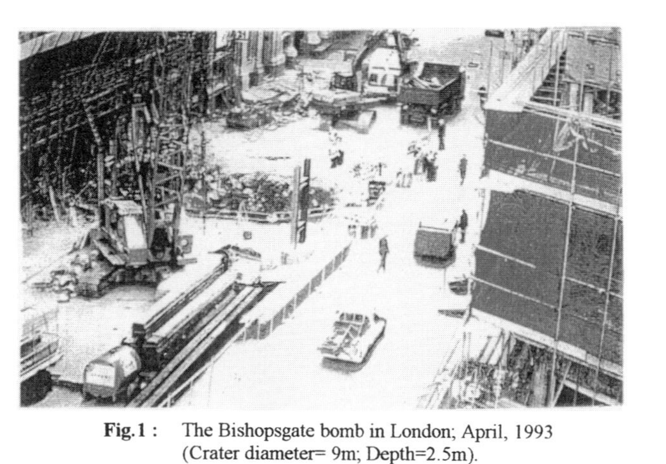

Fig.1 : The Bishopsgate bomb in London; April, 1993 (Crater diameter= 9m; Depth=2.5m).

Fig.2 : Building B1.

Fig.3 : Building B2.

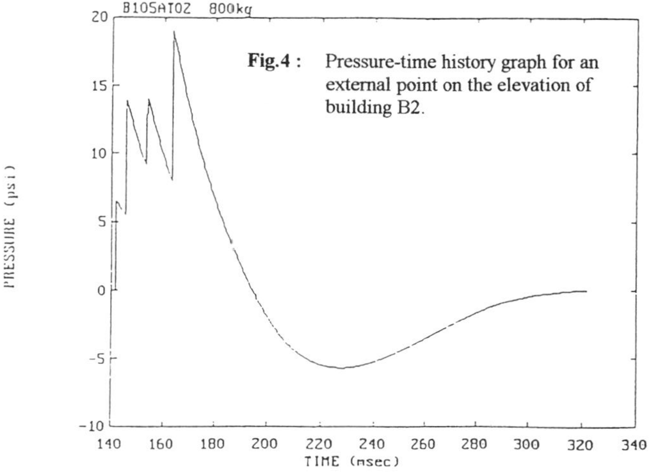

Fig.4 : Pressure-time history graph for an external point on the elevation of building B2.

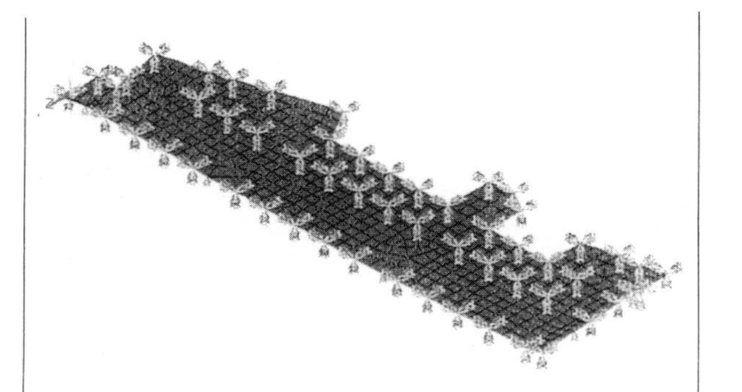

Fig.5 : FE model (showing boundary conditions) of a typical floor plate of building B1.

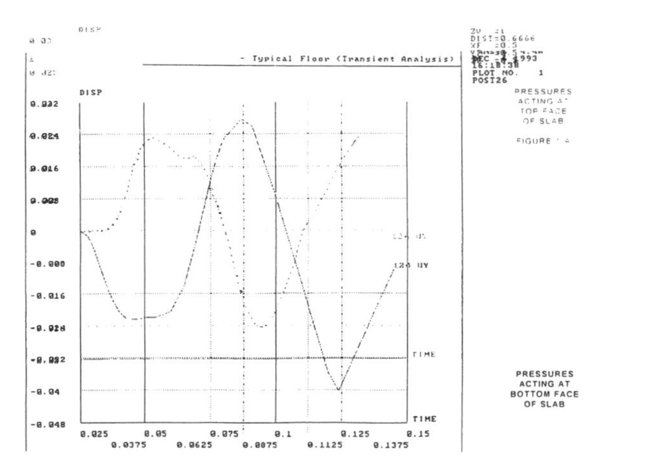

Fig.6 : Displacement-time history graph resulting from the non-linear transient dynamic analysis on the floor plate of building B1, shown in Fig.5.

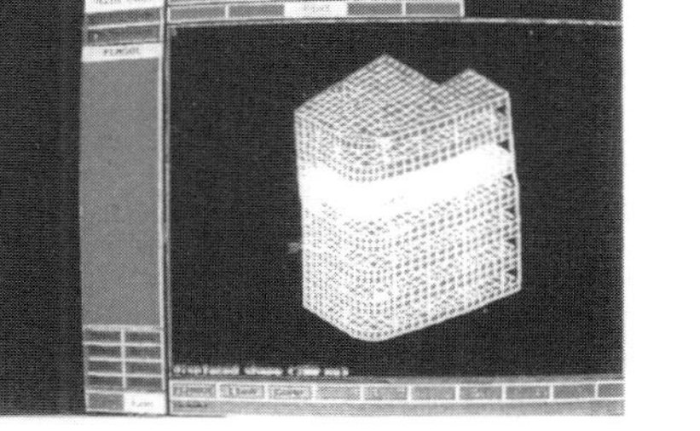

Fig.7 : 3-D view of the deformed shape of building B2, at t = 200 ms.

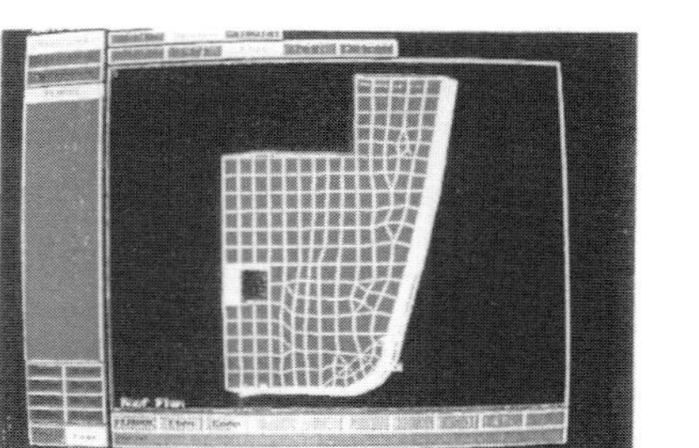

Fig.8 : Top view of the deformed shape of building B2, at t = 200 ms indicating that the building has twisted anti-clockwise.

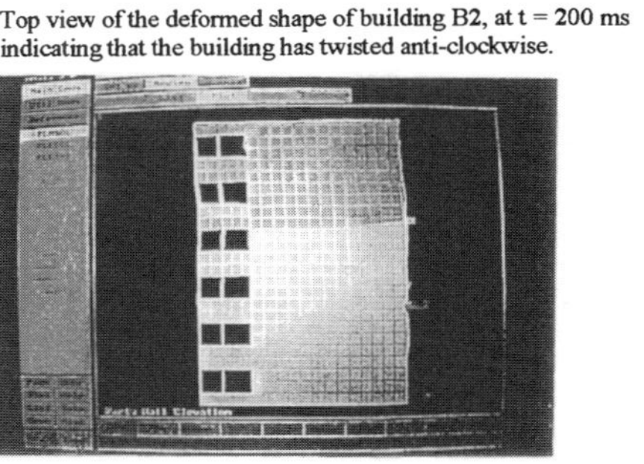

Fig.9 : Deformed shape of party wall elevation of building B2 showing shearing of wall at two levels.

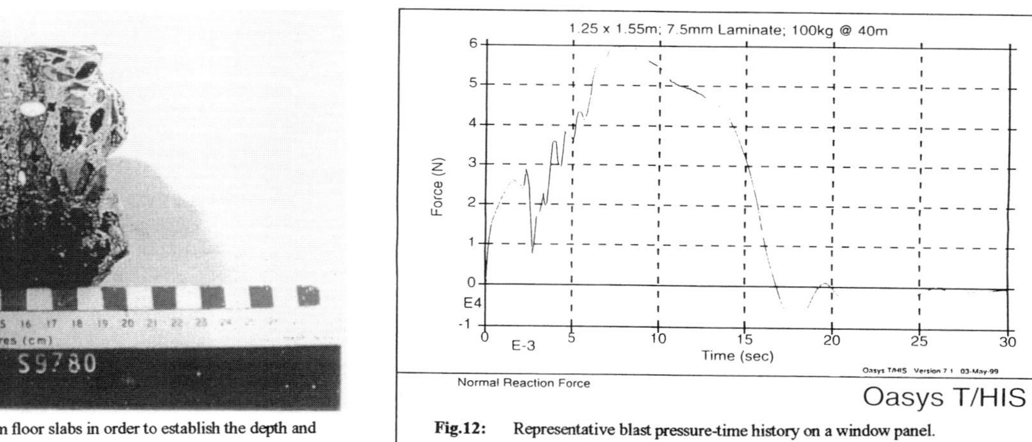

Fig.10: Concrete cores were taken from floor slabs in order to establish the depth and age of cracks and micro-cracks.

Fig.11: Steel Connections were opened up for visual and ultrasonic inspection.

Fig.12: Representative blast pressure-time history on a window panel.

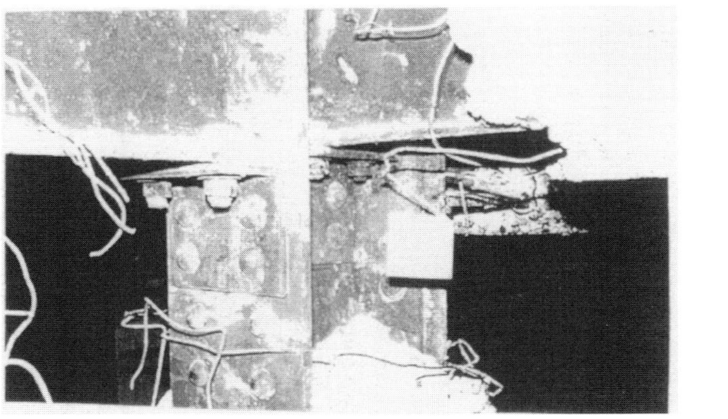

Fig.13: FE Model of a single laminated glass window panel.

MODELLING PROGRESSIVE COLLAPSE OF STRUCTURES AS A RESULT OF EXPLOSION

Progressive collapse due to explosions

J. R. GILMOUR and K. S. VIRDI
Department of Civil Engineering, City University, London, UK

Abstract
A study of progressive collapse of structures exposed to explosions is made and two examples of the phenomenon described. Attention is drawn to potential sources of abnormal loads that should be examined when designing for progressive collapse performance. A plane frame computer program is described for the collapse analysis of reinforced concrete frames using finite elements and a quasi-static approach. The paper concludes by discussing the on-going development of a 3-dimensional quasi-static non-linear finite element computer program for more accurate analysis.
Keywords: Finite element method, progressive collapse, reinforced concrete structures, structural analysis.

1 Introduction

1.1 General

Improvements in structural analysis and knowledge of materials over the last 100 years or so have led engineers to build structures that are more efficient in performance than in the past. This leads increasingly to stretching constituent materials to the limit of their operational envelope. The result is that modern structures tend to lack the reserve of strength that was inherent in older structures engineered by empirical knowledge and instinct. Thought, therefore, must be given as to how such structures will perform when subjected to abnormal loads.

Progressive collapse occurs when a structure has its loading pattern or boundary conditions changed such that elements within the structure are loaded beyond their capacity and fail. The residual structure is forced to seek alternative paths for the applied loads. As a result other elements may fail causing further load redistribution. The process will continue until the structure can find equilibrium either by shedding load as a by-product of elements failing or by finding stable alternative load paths.

Abnormal Loading on Structures edited by K. S. Virdi, R. S. Matthews, J. L. Clarke and F. K. Garas.
Published in 2000 by E & FN Spon, 11 New Fetter Lane, London EC4P 4EE, UK. ISBN 0 419 25960 0

1.2 Case Studies

1.2.1 Ronan Point collapse

Perhaps the most dramatic example of progressive collapse occurred in 1968 when an internal gas explosion seriously damaged the Ronan Point residential apartment block in London, UK. According to a Ministry of Housing report [1] into the incident the collapse and initiating explosion occurred as detailed in the following paragraphs.

The structure was a 22-storey tower seated on a heavily reinforced concrete platform that transferred the structure loads to the supporting piles. The structure above the platform was based on the Larsen Nielsen precast system. The floor plans were identical for each floor, a central stiff core providing lateral restraint to the unreinforced concrete wall units that composed the structure's vertical load path. The slab units used were lightly reinforced and at the extreme corners of the structure the wall units had no direct connection to the stiff core. The exterior wall panels were attached to the structure at their top face by a longitudinal bar threaded through the overlapping lifting eyes of the top face of the wall panel and the floor slab of the level above made good by encasement in grout. At the bottom face the wall panels were restrained by friction. The friction resistance generated as a function of the vertical load passing through the wall panel was considered adequate to withstand the design wind pressures in both directions. The wall units themselves were 178 mm thick plain concrete. This was arranged so that the most heavily loaded units at the first storey level adjoining the RC platform could bear the loads from the upper levels of the structure, in order to achieve the required one hour fire resistance.

The explosion occurred on the 18th floor as a result of a build up of gas from a faulty connection to a domestic cooker. The explosion blew out the non-load bearing walls of the kitchen and living room and the external load bearing wall panels of the living room and bedroom. The loss of the external wall panels removed support for the floor units on level 19. These units subsequently failed. The loss of support to floor slabs progressed upwards to the roof, the debris load from the floor slabs falling on levels 17 through 1 caused each level to fail progressively as the lightly reinforced floor units failed in a brittle manner.

The owner of the flat, in which the explosion took place, was not among the 5 people who died, all from injuries associated with the collapse and not the blast, and was able to give a detailed statement of the condition of articles in the flat prior to the explosion. Investigators then carried out tests to duplicate the level of deformation found in certain metallic objects, most notably the steel door covering the electric meter, in order to estimate the overpressure generated by the blast. It was concluded that an overpressure of 0.02 - 0.08 N/mm^2 was generated in the various paths taken by the blast. A further analysis of the structure showed that at the time of design the figures used for wind speed were optimistically low given the location of the block, near the Thames estuary, and the design life of the structure, which was 60 years. The 1 in 60 year gust for a structure 61 m high was 169 kph compared to the 95-113 kph figure that was used to calculate the required lateral resistance of the structure. Ronan Point would have suffered structural damage during winds higher than 113 kph as suction pressures would have exceeded the frictional resistance of the upper storey wall panel joints. These joints were pivotal in the collapse, as the overpressure from the blast exceeded the frictional resistance of the joint causing the panel to move outwards at its base. The outward movement at the base could only be resisted by the joint at the top of the panel, between the panel and the upper level floor units. This joint was not capable of offering any substantial rotational resistance allowing the base of the wall unit to continue to move outwards under the action of the blast pressure. It was only necessary for the blast pressure to act long enough to move the panel clear of the vertical line of the structure before the initiating mechanism for the progressive collapse was formed.

Ronan Point complied with all the prevailing local building regulations and was found to contain no notable workmanship defects. Until that time building regulations both in the UK and USA were concerned with individual member performance and gave little consideration to the whole structure. Ronan Point was a precast large panel and slab structure with unreinforced walls and as such was stable only while continuity existed in the vertical load path. As soon as the vertical load path was disrupted and joints were expected to carry moments as the upper floors cantilevered, a disproportionate collapse was inevitable, since the system of construction was analogous to a house of cards, i.e. had no demonstrable ability to redistribute loads following a local failure. Following the collapse much work was done by UK code writing committees that ultimately resulted in clauses being included in the 1975 UK Building Regulations to guard against disproportionate collapse. The clauses cover horizontal and vertical continuity, horizontal loading and ductility. For structures greater than 5 stories, where ties do not reach the desired minimum, any single vertical structural member must be able to be rendered incapable without causing significant collapse. Where any vertical element may not be removed, it and its connections must be able to withstand a specified overpressure applied in any direction [2].

1.2.2 Alfred. P. Murrah Building, Oklahoma City, USA

On 19th April 1995 a fertiliser-based bomb exploded in a truck parked adjacent to the Murrah federal building in Oklahoma City. The blast damaged a number of buildings in a 1km radius from the seat of the explosion. The damage ranged from failure of glazing, to partial or total collapse of a number of structures, in the immediate vicinity of the blast. All 268 fatalities occurred in the bomber's intended target, the Murrah Building. A medical examiner's report found that 80% of fatalities were as a result of crush injuries caused by falling debris. The combined Federal Emergency Management Agency and American Society of Civil Engineers report [3] into the incident considered the performance of the structure and how it could have performed better to have minimised casualties.

The Murrah building was composed of a nine-storey tower, a single storey annex and an underground parking garage. The building was built as an in situ reinforced concrete ordinary moment frame, designed in the early 70s and completed in 1976. The structure had a discontinuity in the vertical load path at the second storey level. This was as a result of exterior column for the ground and first level, at a spacing of 12.2 m, being located at twice the spacing for the upper storeys. The loads from the columns that only existed from the second storey upwards, were transferred to the foundations via a transfer girder located at the second storey level that spanned all four sides of the structure and which was supported by the lower storey columns. The blast caused one of the lower storey columns, G20, to fail completely (Fig. 1).

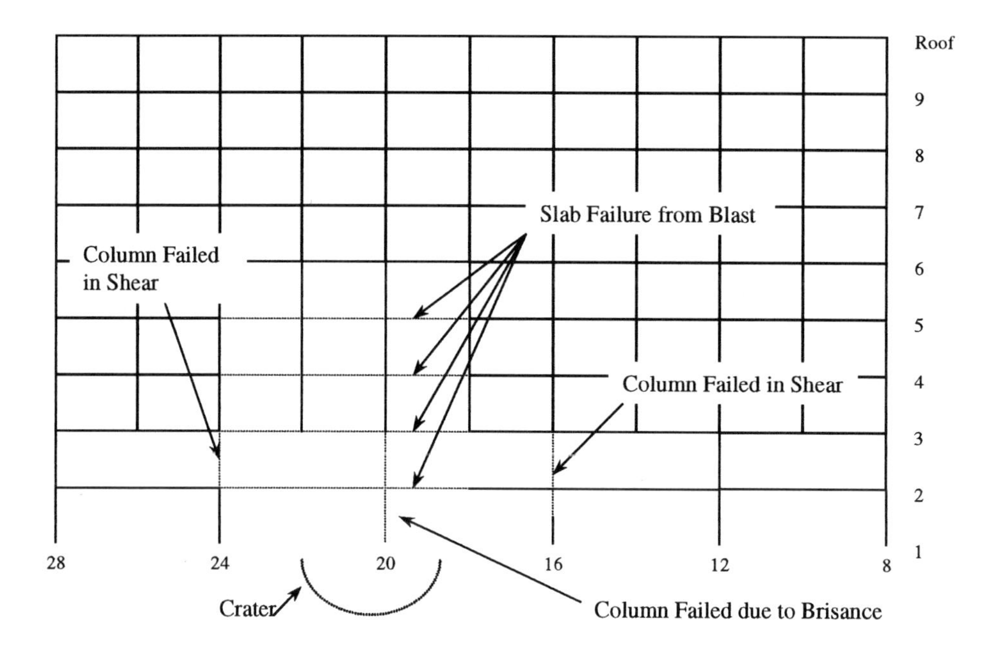

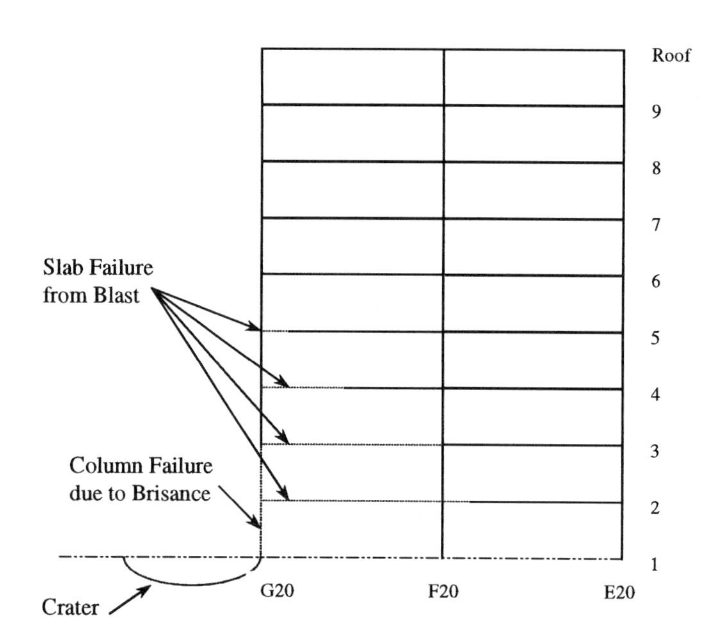

Fig. 1. Elevation and cross-section of the blast response of the Murrah building [3].

This was assumed since no trace of this column was found in the rubble and is consistent with military research with regard to the explosive yield and proximity to the structure, 1814 kg (TNT equivalent) at 4.75 m estimated maximum overpressure 68 N/mm^2 at the column location. The columns, G24 and G16, on either side of the shattered column were also heavily loaded by the blast; respective peak overpressures were calculated to be in the region of 9.65 N/mm^2 and 4.42 N/mm^2. These two columns were analysed using a single degree of freedom model. Given the triangular blast loading and column cross section details the single degree of freedom (SDOF) analysis indicated that lateral deflection caused by the blast would create shear forces in excess of the shear capacity of the section. Inspection of the column remains after the collapse showed that these columns had failed in a manner that was indicative of a shear type failure given the prevailing boundary conditions.

The loss of support to the transfer girder caused it to fail, resulting in virtually the entire elevation failing. Blast attributable damage was limited to the removal of the column at ground level by brisance, shear failure of the columns on either side, and floor slab failures up to the sixth storey level due to differential top and bottom face blast overpressures exceeding the ultimate moments of the slab (Fig. 1). The remaining, and major, proportion of the collapse was due to the progressive failure of the structure due to a lack of available alternative load paths. With reference to the column labels used in Fig. 1, the progressive collapse ran from ground through to roof level on column lines G28 - G8 and F24 - F20. Ignoring the vertical column arrangement, the main reason for the lack of alternative load paths was because of shortcomings in structural detailing. It was noted that the detailing as-built was, in fact, in excess of that required by the prevailing codes at the time of construction. The discontinuous reinforcement in the beam elements meant that sections were not capable of dealing with stress reversal or the removal of columns. Lack of spiral reinforcement in columns made it impossible for the element to retain its material once it had shattered.

As was the case with Ronan Point, the Murrah Building complied with the local building codes in force at the time of construction and with required engineering by the client. The client, U.S. Government, did not require blast resistance to be taken into account during the design. It was fortunate that the lateral strength and stability required for wind loads, provided sufficient lateral stiffness for the structure to withstand whole structure lateral and torsional loads, with little distress. It was concluded in the FEMA report that the structure was not capable of containing structural failure if a column supporting the transfer girder was removed. The conclusion went further in saying that had the structure been built using the rules governing special moment frames, then the additional toughness and ductility that would have resulted, whilst still exhibiting plastic deformations, would have kept the majority of elements in place and the building sufficiently erect for the evacuation of occupants.

1.3 Summary

It is now clear that abnormal loading must be taken into account when designing structures. Abnormal load events come from a number of sources: gas explosion, confined dust or vapour conflagration, machine malfunction, bombs, or projectile impact. Given the sensitivity of modern business to disruption either of staff or information, knowledge of how a structure will perform under a particular set of conditions is of prime importance when calculating company losses due to unforeseen occurrences. It is therefore desirable that engineers should have at their disposal tools that enable them to analyse structures cost-effectively with regard to performance under abnormal loading conditions without demanding full familiarity with the complexities of the problem. To date no single tool exists, to the knowledge of the authors, that can be used to perform a progressive collapse analysis. To that end, the Structures Research Centre at City University has developed a program for the analysis of plane reinforced concrete frames and is currently engaged in developing a computer program capable of analysing both steel and concrete 3-dimensional frames with regard to progressive collapse performance.

2 Plane Frame Analysis

The reinforced concrete plane frame solution developed at City University by Virdi and Beshara [4] follows a quasi-static approach in analysing structural behaviour. The benefits of a quasi-static approach as opposed to a full dynamic analysis are twofold. Firstly, the problem formulation is the same as for a regular static analysis, and secondly the lack of dynamic terms in the equations, means that the computer time required to obtain a solution is very much reduced. The drawback of using such an approach when applied to a structure exposed to true dynamic loading, i.e. those members local to the explosion or impact, have been highlighted by Jones [5] and Pretlove *et al* [6]. Both workers point out that the quasi-static approach is non-conservative because it ignores the effects of energy released in the system as members fail. The energy can cause transient loads and displacements greater than those apparent from the static case and hence the quasi-static approach may miss some potential member failures or elect to remove members in the wrong order. Jones gave an example of an impact on a beam; dynamically the response would be governed by the beam's response to transverse shear, with transverse shear failure a possibility, whereas in static analysis transverse shear effects would not be a critical parameter.

The computer program *PROCSIE* developed at City University by Jeyarupalingam and Virdi [7] and Gilmour [8] takes a quasi-static non-linear finite element approach to modelling structural behaviour. The three main facets of the program are:

- local damage analysis
- alternative load path analysis
- debris load analysis.

Local damage assessment is carried out using a single degree of freedom approach suggested by Biggs [9]. Other more rigorous techniques were examined including Dyna-3D and FEABRS (a program developed by Beshara [10] for blast analysis of reinforced concrete beams). A parametric study carried out by Iyengunmwena [11] showed that Biggs' method was accurate to within 8% in most cases.

In the proposed method, a member is selected by the user to be subject to the abnormal load. It is then analysed by Biggs' method, and end moments and shears are determined. At the instant immediately prior to failure the frame will be reacting equal and opposite shears and moments on the member.

In order to obtain the quasi-static response of the rest of the structure, the frame is analysed with the member removed and the end moments and shears applied initially at 50% of their magnitude. Assuming equilibrium can be found then the abnormal load factor is incremented until it reaches 100%. At each iteration, the program monitors the strain at specific cross-sections located in all members checking for additional member failures as a result of the applied abnormal end reactions and redistributed loads.

Alternative load path analysis refers to the removal of elements from the frame that have failed the strain limit criteria at the cross-sections along the member. The failure criterion used was that strains greater than concrete crushing strain when detected at both ends of a member implied failure of the complete member. Such a member is then removed at the next iteration. The frame is re-analysed and loads are redistributed by the user to the surrounding frame elements as appropriate. This continues until the abnormal load factor reaches unity, and no further members breach the strain failure criterion.

Debris load analysis is carried out by the user when a vertical load carrying member is removed, e.g. a beam fails and it is necessary to know where to put the beam's imposed and dead load. The debris loads are factored up by 1.25 in order to take into account the effects of impact from the falling debris. The load vector entry is updated for the member to which the debris has been directed and the analysis continues until a steady state is reached.

A number of assumptions have been made in order to allow for plane frame modelling to take place:

- the cross-section is subjected to uniaxial or biaxial bending moment and axial force shear stresses are neglected
- plane sections remain plane before and after bending
- the frame is made up of beam-column finite elements
- each member is considered to be shallow with respect to its length and so can be represented entirely by its mid-height longitudinal axis
- out of plane displacements are neglected
- loads are applied only in the plane of the frame.

The finite element scheme adopted is relatively straightforward. The frame members are modelled by evaluating the displacement field at a number of stations along the member length using Gauss integration techniques. At each of the Gauss points along the member the entire cross-section is evaluated and a biaxial grid of Gauss points is used to integrate over the section to obtain the required stress resultants. The value of Young's modulus at each Gauss point on the cross section is monitored and from an assumed strain displacement relationship the stress at each Gauss point can be evaluated using the selected material relationship.

2.1 Plane Frame Solution Algorithm

The direct method is employed for the resolution of the equation of equilibrium:

$$[K_s]\{Q\}=\{F\} \tag{1}$$

where, $[K_s]$ is the global secant stiffness matrix and is determined by iteration as initially the internal forces and displacements are unknown. The following has been employed as a solution procedure:

1. Form the elastic stiffness matrix for each element assuming no axial force and elastic material response.
2. Form the transformation matrix from the initial un-deformed geometry and form the global stiffness matrix.
3. Assemble the applied load vector $\{F\}$.
4. Solve the equations for the unknown degrees of freedom,

$$[K_s]_{i-1}\{Q\}_i=\{F\} \tag{2}$$

   note that K lags behind the current iteration i.
5. Using $\{Q_i\}$ obtain updated material state, strains, stresses and rigidity matrices.
6. Re-form the element stiffness matrices, transformation matrices and global stiffness matrix $[K_s]$.
7. Using $\{Q_I\}$ and $[K_s]$ calculate the member forces at each node

$$\{F\}_i=[K_s]_i\{Q\}_i \tag{3}$$

8. Compare applied load vector $\{F\}$ and calculated $\{F_i\}$. Continue to iterate from step 4 to step 7 until the external and internal forces converge.

The above procedure forms the basis of the computer program *PROCSIE* [7] [8].

3 Three Dimensional Analysis

In order to increase the scope of the solution procedure, a 3-dimensional non-linear quasi-static finite element code is currently being developed. Whilst it is acknowledged that blast loads are dynamic and that areas of the structure will respond in a dynamic manner, it is the intention to continue using the quasi-static method. The justification for this is found in part in the Oklahoma City bombing report [3] and the successful use of plane frame analysis, quasi-static methodology and empirical knowledge of the performance of composite members under explosive loads to model the most likely collapse mechanisms. The computer solution under development is being written in a modular form in order to facilitate a combined dynamic quasi-static analysis to be included at some future date. Such an analysis would involve the decomposition of the finite element domain into dynamic and quasi-static regions. From a finite element perspective, the problem is analogous to a fluid-structure interaction problem, i.e. interface nodes passing data between analyses being carried out on separate meshes and at differing time steps.

The quasi-static analysis currently under development uses a total Lagrangian approach and 3-dimensional curved beam elements developed by Surana and Sorem [12]. The elements are well suited to the modelling of the potentially large rotations that may be obtained during a progressive collapse analysis, and allow the reference axis to exist outside of the beam-column cross-section, thereby making it easy to incorporate shell elements to model slabs. This makes the beam reference axis follow the mid-depth position of the slab, thereby removing the need to use linking elements to couple beams to slabs.

The beam-column 1-dimensional elements are formed by degenerating 3-dimensional isoparametric hexahedron elements. The problems associated with large rotations and the non-linear functions used in the element derivation are fully explained in [12]. As was the case in the earlier plane frame work, reinforced concrete sections were modelled by a grid of Gauss points. Each Gauss point has its own material parameters and so the incorporation of reinforcing bars is made very easy as the bar area and Cartesian coordinates are transformed into natural space and incorporated in to the integration scheme for the assessment of the strain-displacement matrix. The solution proceeds as per a typical textbook non-linear scheme [13] and solves the global equation:

$$\delta W = \int_V \sigma_z \delta\varepsilon_z + \tau_{xz}\delta\gamma_{xz} + \tau_{yz}\delta\gamma_{yz} dv - \langle Q \rangle \{\delta q\} = 0 \tag{4}$$

where, the symbols have their usual meanings. In the program, Eqn. (4) is inserted within a load control loop that increments the hazard load applied to selected members and secondary loads caused by the failure of elements subjected to the primary hazard load and any subsequent load redistribution that may need to take place. The quasi-static analysis finishes once all of the load factors have reached unity or when the structure becomes statically unstable.

The key to modelling the progressive collapse phenomena correctly is the application of failure criteria. The term *failure* is itself rather misleading as it is generally applied to a material when the material stops acting in the manner in which it was expected to act at design. In fact the material, or rather the member composed of that material, may still be capable of influencing the distribution of forces in its locality. This is particularly the case for reinforced concrete sections that have begun to crack and have lost the majority of their tensile and flexural load-carrying capacity but can still bear compressive forces and some tensile force until ductile failure of the reinforcement occurs.

It should also be remembered that concrete is often used in high risk structures due to its ability to absorb energy during fracture. Therefore, some energy-wasting function should be incorporated in to the concrete model. The form that such a function should take is as yet undecided although its justification follows along the lines of the smeared crack approach, i.e. specific crack paths are not important but accounting for crack energy is. The smeared crack approach cannot be employed 'as is' since insufficient strain data is available along the length of the member as a result of the one-dimensional elements used to represent the beam-columns. The use of such elements is necessary in order to allow for the analysis of the whole structure within a reasonable time on a PC or single processor workstation.

4 Conclusion

The plane frame analysis, although not fully automatic, gives a good graphic visualisation of the phenomenon of progressive collapse. The 3-dimensional analysis under development represents the physical problem in a more satisfactory manner. The formulation retains a number of higher order terms in order to model structural behaviour more accurately when large rotations are present. Whilst the theoretical shortcomings of the quasi-static approach are acknowledged, the work by Biggs and the FEMA analysis of the Oklahoma City bombing demonstrate that it is a credible method.

5 References

1. Ministry of Housing & Local Government (1968), *Report of the Inquiry into the collapse of the flats at Ronan Point, Canning Town,* HMSO August 1968
2. Popoff, A. (1975), Design against Progressive Collapse, *PCI Journal*, March-April 1975 pp45-57
3. FEMA, ASCE (1996), *The Oklahoma City Bombing: Improving building performance through multi-hazard mitigation,* FEMA Report 277, Washington, USA
4. Virdi, K.S. Beshara, F.B.A. (1992), Numerical Simulation of Building Decommission as a Progressive Collapse Process, *Decommissioning & Demolition Proceedings 3rd Int. Conference*, UMIST, UK, March 25-26 1992
5. Jones, N. (1995), Quasi-static Analysis of Structural Impact Damage, *Journal of Construction Steel Research* 33 1995 pp151-177
6. Pretlove, A. Ramsden, M. Atkins, A. (1991), Dynamic Effects of Progressive Failure of Structures, *International Journal of Impact Engineering*, v11 n4 pp539-541, 1991
7. Jeyarupalingam, N. and Virdi, K S. (1993), Progressive Collapse of High Rise RC Structures. SERC Research Grant Final Report GR/G39587, Structures Research Centre, City University, London, UK.
8. Gilmour, J. R. (1999), Numerical Modelling of the Progressive Collapse of Structures as a result of Impact or Explosion. PhD Thesis to be submitted, City University, London, UK .
9. Biggs (1964), *Introduction to Structural Dynamics*, McGraw-Hill 1964
10 Beshara, F.B.A. (1991), Non-linear Finite Element Analysis of Reinforced Concrete Structures Subjected to Blast Loading. PhD Thesis, City University, London, 1991.
11 Iyengunmwena, M. (1991), Evaluation of End Reactions of RC Structures Under Dynamic Loading Conditions. MSc Dissertation, City University, London, 1991.
12 Surana, K. and Sorem, R. (1989), Geometrically non-linear formulation for three dimensional curved beam elements with large rotations. *International Journal for Numerical Methods in Engineering*, Vol.28, pp 43-73.
13 Zienkiewicz, O.C. (1991), *The Finite Element Method: Fourth Edition, Vol. 2: Solids and Fluid Mechanics Dynamics and Non-linearity*, McGraw-Hill 1991.

TUNNEL SHELLS AND FIRE PROTECTION - TESTS AND ANALYSES

Fire protection for tunnel shells

E. RICHTER and K. PALIGA
Institute for Building Materials, Concrete Structures and Fire Protection (iBMB) of the Technical University of Braunschweig, Braunschweig, Germany

Abstract

The present paper reports on experiments and calculations made with the aim of optimising the fire-engineered design of single-shell tunnel lining systems. Preliminary tests performed in this connection on small-scale specimen were aimed to show the effect polypropylene-fibre inclusion has on the mechanical high-temperature properties of normal weight concrete. These tests supplied data on the composition of reinforced-concrete tubbings that were to be subjected to full-scale tests in order to study their structural response and deformation behaviour in a typical tunnel fire. It was demonstrated that, a certain installation input provided, it is possible to realistically portray the conditions in a tunnel and to predict these conditions with satisfactory precision by means of FE calculations. This is true both for the service and the fire loads encountered in a tunnel. Plastic fibres embedded in the concrete allow destructive spalling to be reduced substantially.
Keywords: FE analyses, fire, fire protection, full scale test, material behaviour, polypropylene-fibre, tunnel shells.

1 Introduction

Where tunnelling proceeds on the basis of a high degree of mechanisation, drilling and cutting machinery can also in unconsolidated soil successfully be employed for single-shell tunnelling procedures using reinforced-concrete tubbings. Single-shell tunnel lining with reinforced-concrete tubbings, however, implies that new fire-engineering requirements have to be met that are not always easily complied with.

Tunnels have to be conceived such that a fire cannot produce damage that jeopardises the load-carrying capacity or the structural integrity of the tunnel as a

Abnormal Loading on Structures edited by K. S. Virdi, R. S. Matthews, J. L. Clarke and F. K. Garas.
Published in 2000 by E & FN Spon, 11 New Fetter Lane, London EC4P 4EE, UK. ISBN 0 419 25960 0

whole or of any of its major elements. Fires must not be allowed to impair the serviceability of a tunnel (e.g. impermeability of underwater tunnels), and neither must the structure suffer from any inadmissible deformations. Rehabilitation measures taken must ensure the lowest possible input in terms of time, equipment and cost [1].

Fires occurring in road tunnels are expected to give rise to maximum temperatures of the order of 1200°C within 5 minutes. Local peak temperatures of 1300°C will only be reached in fires involving vehicles with combustible liquids or plastics representing a high calorific potential. Temperatures tend to reach their maximum during the first few minutes of a fire. Depending on such factors as fire load, ventilation conditions, commencement of fire-fighting measures and specific local conditions, they remain at this level for about 30 to 120 minutes before the decaying phase sets in.

Reinforced-concrete tubbings used in road tunnels generally are 0.30 m to 0.50 m thick. With such a thickness, spalling of concrete, a phenomenon observed during the first 30 minutes of the fully developed fire at the fire-exposed tubbing surface, can have devastating effects [2]-[5]. To prevent the concrete from spalling and the tubbing surface from heating too rapidly, the tubbing surface has to be provided with a plaster or similar means of protection [6]. Another measure designed to prevent spalling is an admixture of polypropylene fibres in the concrete [7][8], which in the event of a fire helps to relieve the steam pressure in the near-surface concrete layers.

The present paper reports on experiments and calculations made with the aim of optimising the fire-engineered design of single-shell tunnel lining systems. Preliminary tests performed in this connection on small-scale specimen were aimed to show the effect inclusion of polypropylene-fibre has on the mechanical high-temperature properties of normal-weight concrete. These tests supplied data on the concrete composition of reinforced-concrete tubbings that were to be subjected to full-scale fire tests in one of the iBMB furnaces in order to study their structural response and deformation behaviour in a typical tunnel fire.

2 Preliminary testing

2.1 Tests on small-scale specimen

The tests conducted on small-scale specimen were designed to supply data for the temperature-related strength and deformation characteristics of concrete comprising polypropylene fibres. For the tests, concrete cylinders 300 mm long and 80 mm in diameter were produced from concrete varying in composition, i.e. in respect of the PP-fibre content (1.5 and 2.0 kg/m³) and the type of aggregate used (sand / gravel and sand / granite chippings).

Analysis of test results reveals a highly similar thermo-mechanical response of the different types of concrete tested, as compared with quartzitic normal-weight concrete [9]. Fig. 1 contrasts readings obtained in the test series using 2 kg of PP-fibres per cubic metre of concrete with the temperature-related stress-strain relationships of normal-weight concrete. As both with and without PP-fibres the thermo-mechanical material response of the concrete proves to be similar, theoretical/numerical assessment intended to accompany the tubbing fire tests can start from the temperature-related stress-strain relationships known for normal-weight concrete.

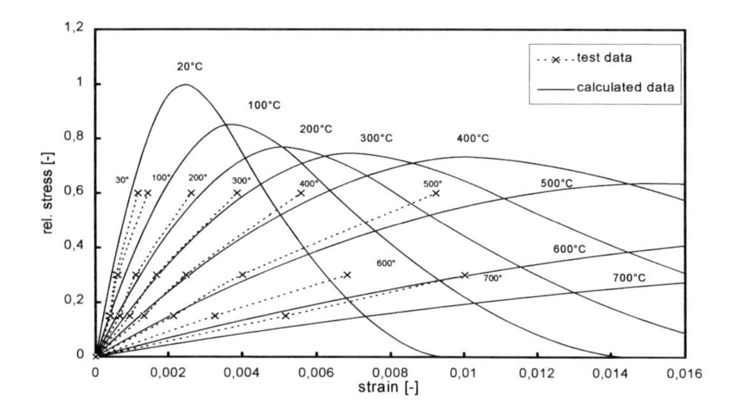

Fig. 1. Readings for concrete with 2 kg PP-fibres per cubic metre of concrete *vs.* calculated data for temperature-related stress-strain relationships of normal-weight concrete

2.2 Spalling tests

The spalling tests were targeted to supplement data available for the spalling behaviour [2] of normal-weight concrete by new data for concrete including PP-fibres. The specimen produced for this purpose had dimensions of 45×80×20 cm (W×H×D). They were loaded centrally in a purpose-made test frame, and one side face (W×H) was exposed to temperatures typically expected in tunnel structures. The furnace was heated at approx. 200 K/min for a period of six minutes, the maximum temperature of 1200°C being maintained at a constant level for 60 minutes. Once the burners had been stopped, the fire test was terminated after a 90 minute decaying phase.

The first few minutes of the fire were in all the tests accompanied by brief and sharp cracking noises with concrete sections about 2 to 3 cm in diameter coming off the fire-exposed surface. Such spalling continued for about 5 to 10 min, following which no further changes were observed at the concrete surface. The cavities recorded in the fire-exposed concrete surface after the fire tests were measured to be up to 3 or 4 cm deep.

On the basis of the spalling tests and calculations conducted in parallel for the heating rate, the tubbing concrete cover was decided to have a thickness of 8 cm. This was to ensure that the reinforcement temperature would not exceed a mean value of 300°C during the fire tests.

3 Tubbing tests

3.1 Basic principles

The test was conceived such that it reflected the geometrical and static restraint conditions of the real tunnel in a realistic manner. Great care was in particular taken in that the tubbing dimensions, jointing pattern, concrete composition, internal forces and strains expected to result from service load application, and bedding of the surrounding soil adequately portrayed the actual conditions.

With a thickness of 40 cm, the test tubbings remained by 5 cm below the thickness of the real tunnel tubbings. The same applies to the radius which at 3.92 m was smaller by 1.03 m. Calculations demonstrated that the temperature distribution to be expected for the thinner test tubbings deviates only slightly from the temperatures encountered in connection with the full thickness. This can be explained by the fact that the temperature influence is effective down to a depth of only about 30 cm. The tubbing joints in the tests corresponded in their main details to those of the real structure.

Cold design for the structure is based on a constant distributed normal-force of 4600 kN/m and the moment distribution reaching its maximum at -70 kN/m. The load condition "compressive stress - bottom face" became decisive at -12.3 N/mm² on the bottom face and at -8.1 N/mm² on the top face.

3.2 Test set-up

The basic principles relevant to the structure could largely be realised with the test set-up shown in Figs. 2 and 3. The tubbings were installed so as to form a vault above the furnace, providing a rise of 60 cm. The floor area covered was 4.94 m×3.22 m. Two tubbing rings, each 1.5 m wide, were placed next to each other. Each tubbing ring was composed of a complete together with a halved tubbing, thus allowing longitudinal and transverse joints to be examined in the fire test.

Steel cables (57 mm in diameter), running along the upper face of the tubbings, served for load application and were tensioned to restrain the tubbing rings during the test.

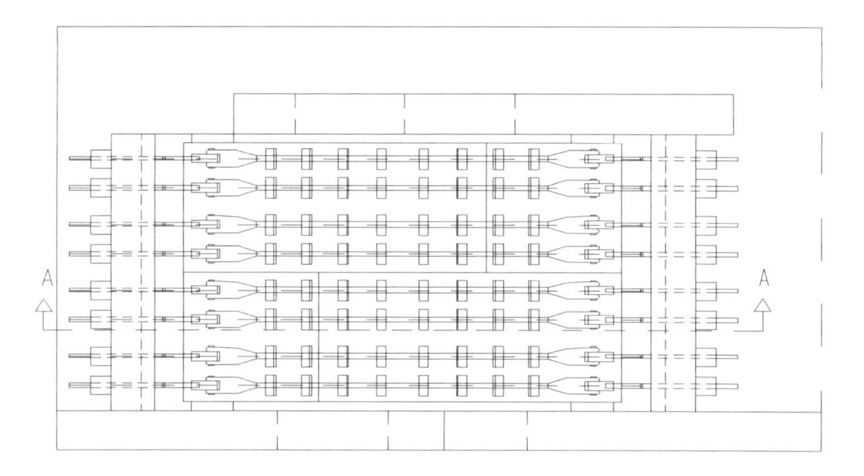

Fig. 2. Test set-up - Top view

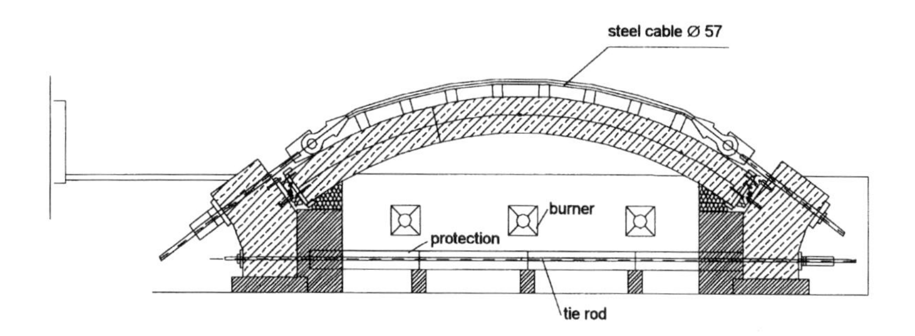

Fig. 3. Test set-up - Section A-A

The tensile forces produced by the steel cables and the tie rods are in equilibrium with the tubbing forces at support. The force direction of the resultant R produced by the two tensile forces at the support, however, does not coincide with the direction of the tubbing gravity axis (Fig. 4), which is why compressive stress is produced on the bottom face.

The tie rods passing through the furnace were to keep the installation in balance. As failure could by no means be accepted, the principle requirement here was the insulation during fire exposure.

The test set-up described here allows a representative sub-system to be analysed for a single-shell tunnel lining system using tubbings, which realistically reflects both the loading and restraint conditions of the complex load-carrying structure in the event of a fire. The static stability of tunnel shells can thus be verified with adequate accuracy.

Fig. 4. Principle of force application

3.3 Test procedure

For the first fire test, the fire exposure corresponded to that of the modified hydrocarbon curve; for the second test, the extended hydrocarbon curve was used (Fig. 5). The calorific value under the modified hydrocarbon curve compares with a fire exposure according to the standard ISO temperature-time curve for a period of 120 minutes; the calorific value of the extended hydrocarbon curve corresponds to the so-called Rijkswaterstaat curve. For the second test, half the longitudinal and transverse joints were sealed with mineral wool.

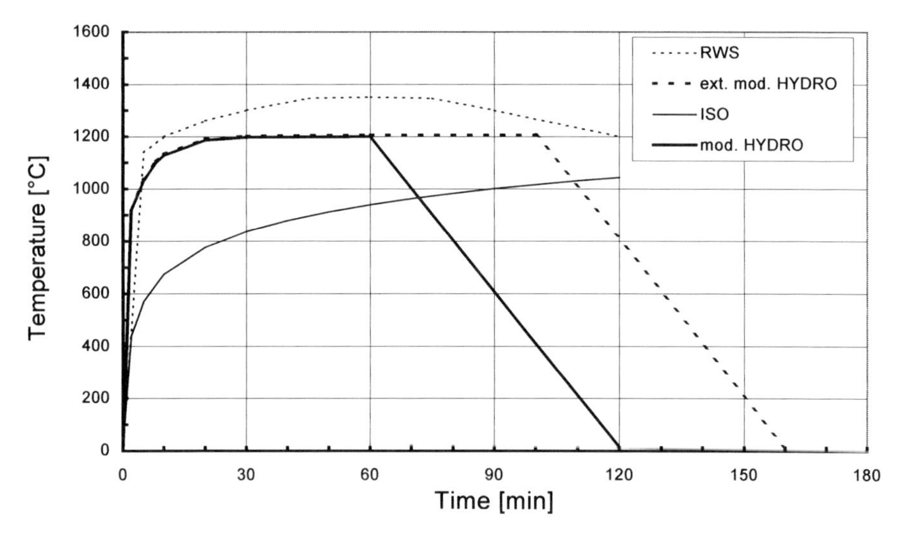

Fig. 5. Furnace temperature-time curves

So far, a total of three fire tests have been performed for different restraint conditions. The test parameters are shown in Table 1.

Table 1. Fire test parameters

Test Ref. No.	Joint insulation	Fire-protection insulation	Fire exposure	
			mod. HYDRO curve	extended mod. HYDRO curve
II	no	no	yes	no
I	yes	no	no	yes
III	no	16 mm	no	yes

The spalling and cracking behaviour, and the emergence of moisture on the unexposed tubbing face was checked visually, and the observations made were logged. The extent of concrete spalling was recorded after the fire tests.

3.4 Test results

Test III with 16 mm fire protection layer showed no spalling at all, so only the results of tests I and II are reported here.

Plastic fibres in the concrete apparently tend to limit the risk of spalling, which primarily takes place during the first 10 to 15 minutes, i.e. before the fibres reach their melting point. Fibre inclusion has the effect of reducing the tensile stress produced by friction at the pore walls within the temperature-exposed concrete sections as the water vapour escapes [2]. With concrete that has no plastic fibres embedded in it, spalling may very well extend over the entire height of the cross-section.

Furnace temperatures

Fig. 6 compares the given furnace temperatures with those measured during the second fire test.

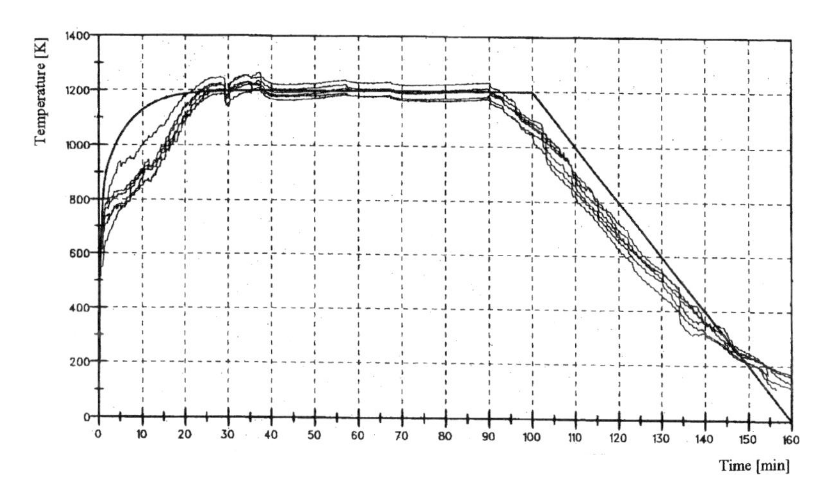

Fig. 6. Given and measured furnace temperatures

Spalling

Spalling on the fire-exposed tubbing face was found to occur between the 1st and 15th minute of fire exposure. In both tests, the concrete spalled down to depths of 4 to 5 cm on average; individual sections revealed depths of up to 10 cm so that at these points the reinforcement was exposed (concrete cover = 8 cm). Special mention should be made of the spalling pattern. With all the tubbings, maximum spalling was concentrated in the centre within an oval-shaped area, while tubbing edges were found to have remained unaffected (Fig. 7). The tubbing halves were affected to a much lesser degree than the whole tubbings. It was also noted that spalled surfaces almost always passed through cracked or broken aggregate particles.

Plastic fibres in the concrete apparently tend to limit the risk of spalling, which primarily takes place during the first 10 to 15 minutes, i.e. before the fibres reach their melting point. Fibre inclusion has the effect of reducing the tensile stress produced by friction at the pore walls within the temperature-exposed concrete sections as the water vapour escapes [2]. With concrete that has no plastic fibres embedded in it, spalling may very well extend over the entire height of the cross-section.

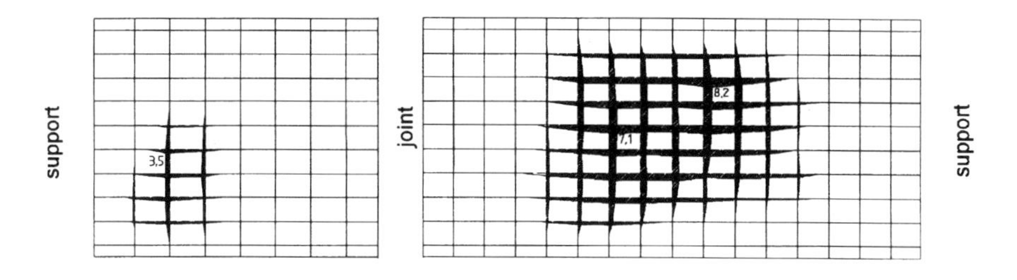

Fig. 7. Spalling pattern of a tubbing ring

The distinct spalling pattern is to a certain extent also attributable to the extremely rapid heating rate and the very high temperature level of 1200°C, conditions which are much more stringent than the ISO-834 temperature-time conditions in accordance with DIN 4102, Part 2. Aggregate cracking suggests that the high temperatures favoured physical and chemical changes in, and ultimately destruction of, these aggregates, a phenomenon on which plastic fibres have no influence. The oval spalling pattern may have been caused by restrained thermal strain. A more detailed review of possible causes will, however, require further investigation into fundamental principles.

Member temperatures

Temperatures across the concrete and in the reinforcement were found to have been influenced considerably by the extent to which the concrete spalled. In areas of minor spalling depths (approx. 2 cm), temperatures reached approx. 450°C at a depth of 4 cm and approx. 250°C at the reinforcement at a depth of 8 cm (Fig. 8).

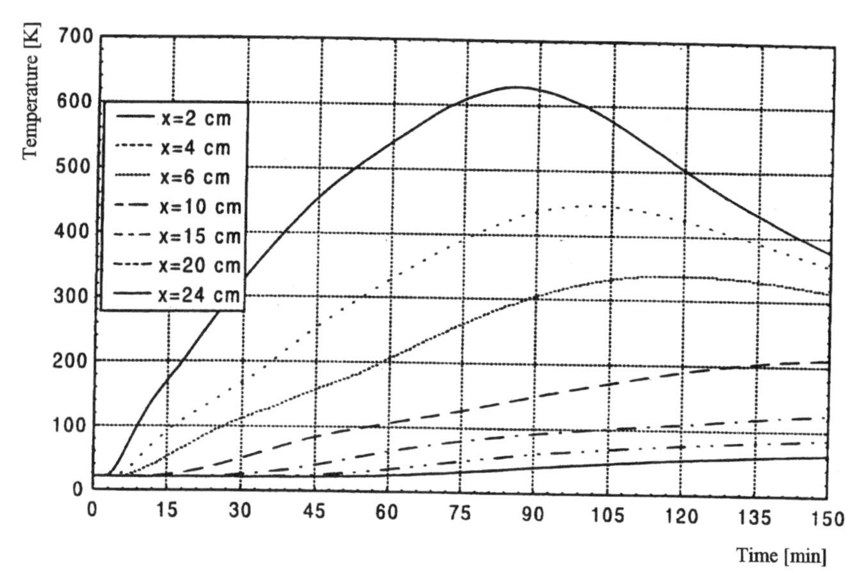

Fig. 8. Temperature distribution in the concrete - areas affected by minor spalling

Deformations
Vertical deformations were first observed to be directed downwards, i.e. towards the flames. Having reached a maximum in the 120th minute of the fire they diminished. In the quarter points, maximum deformations were measured to be as much as 11 cm to 13 cm. After about 10 hours, the deformation in the quarter points had reverted to zero, while in the crown deformations of 21 mm were measured, and a 3 mm permanent deformation directed upwards was recorded.

The angular deformation of joints along the longitudinal tunnel axis was approximately 0.35° as a maximum. As a result of this angular deformation, the joints were closed on the fire-exposed face, while they had opened on the top face. In the transverse joints, the angular deformation was more marked at almost 1° in the quarter points and 0.35° in the crown in the other direction. Major angular deformations may have the effect of the joints opening excessively at their upper ends so that the function of the sealing strip will be lost.

Forces in steel cables and tie rods
The steel cable forces proved to increase by about 5% of their prestressing force during the first phase of the test, and declined towards the end of fire exposure.

Cracking
Around the joints, a concrete layer approximately 10 cm thick revealed shear cracks at the fire-exposed face.

4 Summary

In areas affected to a lesser extent by spalling, the reinforcement temperatures remained below 300°C. Deformations were recorded to be a maximum of 21 mm, the permanent deformation after cooling being approximately 3 mm.

It was demonstrated in the full-scale tests that - a certain installation input provided - it is possible to realistically portray the conditions in a tunnel and to predict these conditions with satisfactory precision by means of FE calculations. This is true both for the service and the fire loads encountered in a tunnel. Plastic fibres embedded in the concrete allow destructive spalling to be reduced substantially. Any additional fire-proofing insulation can thus often be dispensed with altogether. For more reliable results, additional experiments have to be made to optimise the fibre content, the type of aggregate and cement used, etc. Findings available from extensive spalling tests conducted under ISO-834 temperature-time conditions can unfortunately not be referred to for tunnel fires, as the latter typically imply both higher heating rates and higher peak temperatures.

The tests made also suggest the need for additional experimentation as regards residual load-carrying capacity and possible rehabilitation measures following a tunnel fire so as to arrive at a structure that economically can be regarded as an optimum solution.

5 References

1. Kordina, K. (1982) Brandentwicklung in unterirdischen Verkehrsanlagen und Beanspruchung der Umschließungsbauteile. Beitrag in: 3. öffentliches Forschungskolloquium am 20.09.1982 an der Universität Karlsruhe. Sonderforschungsbereich 148 "Brandverhalten von Bauteilen", TU Braunschweig, 9/1982.
2. Meyer-Ottens, C. (1975) Zur Frage der Abplatzungen an Bauteilen aus Beton bei Brandbeanspruchung. Deutscher Ausschuß für Stahlbeton, Heft 248, Verlag W. Ernst & Sohn, Berlin.
3. Meyer-Ottens, C. (1974) Verhalten von Betonbauteilen im Brandfall: Abplatzungen von Bauteilen aus Normalbeton bei Brandbeanspruchung - Ursachen und Verhinderungsmaßnahmen. Sonderdruck aus beton, Herstellung und Verwendung 24, Heft 4, S. 133/136 und Heft 5, S. 175/178.
4. Gutachtliche Stellungnahme Nr. I 69 2935 vom 2.5.1969 zum Brandschaden des Autobahntunnels K 12 der Bundesautobahn, südliche Umgehung Hamburg. Institut für Baustoffkunde und Stahlbetonbau der TU Braunschweig (unveröffentlicht).
5. Kordina, K. (1994) Laborbrandversuche Fahrnau. Schlußbericht. FuE-Vorhaben des Bundesministeriums für Forschung und Technologie:RGB 8817/9 P 145/06/85. Studiengesellschaft für Stahlanwendung, Düsseldorf und Institut für Baustoffe, Massivbau und Brandschutz der TU Braunschweig. Dezember.
6. Kordina, K.(1981) Meyer-Ottens, C.: Beton-Brandschutz-Handbuch. Beton-Verlag GmbH, Düsseldorf.
7. Gutachtliche Stellungnahme Nr. G 94 099b vom 18.1.1995 zum Brandverhalten von Wänden aus hochfestem Beton der Festigkeitsklasse B 65 mit und ohne Faserzusatz beim Bauvorhaben Forum in Frankfurt. Univ.-Prof. Dr.-Ing. Dietmar Hosser, Institut für Baustoffe, Massivbau und Brandschutz der TU Braunschweig (unveröffentlicht).
8. Diederichs, U.; Jumppanen, U.-M.; Morita, T.; Nause, P.; Schneider, U. (1994) Zum Abplatzverhalten von Stützen aus hochfestem Normalbeton unter Brandbeanspruchung. Beitrag in: Forschungsarbeiten 1990 - 1994. Institut für Baustoffe, Massivbau und Brandschutz der TU Braunschweig, Heft 109.
9. Richter, E.(1987) Spannungs-Dehnungs-Beziehungen zur Berechnung des Trag- und Verformungsverhaltens von Konstruktionsbauteilen unter Feuerangriff. SFB 148, "Brandverhalten von Bauteilen": 84/86, TU Braunschweig, 12/1987.

FIRE INDUCED PROGRESSIVE COLLAPSE ANALYSIS OF SAFETY CRITICAL STRUCTURES

Optimisation of fire protection

C.P. ROGERS and S. MEDONOS
CREA Consultants Limited, Buxton, UK

Abstract

This paper examines the technologies involved in Fire Induced Progressive Collapse Analysis (FIPCA), the derivation of the fire loading, the time-dependent temperature distribution and the structural collapse analysis. In developing the techniques, some areas of concern in using the technology have been identified and these are discussed. The paper concludes that scenario based fire definition and holistic (whole structure) modelling provide the most realistic solution, although the techniques do pose problems if not applied with care.
Keywords: Fire protection, optimisation , progressive collapse.

1 Introduction

Fire Induced Progressive Collapse Analysis (FIPCA) has evolved into a valuable tool in the cost-effective provision of fire protection. The provision of fire protection whilst valuable in its function can be expensive. Very few fire protection schemes are maintenance free, this is obviously true for the active systems such as water spray and quenching gasses. However, less evident is the cost of maintaining passive systems such as intumescent coatings and concrete casings.

In some cases, the very provision of fire protection has cost implications in that the structure has to be designed with additional load-bearing capacity to allow it to carry the weight of the fire protection system. This can be a major disadvantage to weight critical structures such as offshore oil and gas production facilities. FIPCA has become an almost standard tool for the optimisation of fire protection on offshore structures, with the

Abnormal Loading on Structures edited by K. S. Virdi, R. S. Matthews, J. L. Clarke and F. K. Garas.
Published in 2000 by E & FN Spon, 11 New Fetter Lane, London EC4P 4EE, UK. ISBN 0 419 25960 0

increasing use of Floating Production Storage and Offloading (FPSOs) systems for deep-water production. This is set to increase.

2 Fire protection of structures

Fire remains one of the most potentially disastrous hazards threatening structures. Where a structure envelopes, or is sited close to hazardous processes or significant quantities of stored flammable inventories; structural fire protection is provided for several reasons. For example:

- Protection of occupants to allow a safe means of escape or to provide a safe refuge whilst rescue is effected;
- Protection of plant items to prevent escalation;
- Protection of neighbouring structures and plant; and
- Protection of assets to minimise financial loss.

The order and importance of the reasons for applying fire protection will vary depending upon the level of occupancy, training of occupants and function of the structure or process.

Fire protection can be provided as an active system, a passive system or a combination of both. Active systems are provided for several reasons: fire fighting; fire suppression; short term protection (e.g. water curtains and deluge applied to plant items); and fire prevention. Active systems have their drawbacks: reliability and being one-time systems are two primary concerns.

Passive fire protection (PFP) is attached to or closely surrounds the protected structure or plant. Passive systems have the advantage that they are always in place and available. Their major drawbacks include initial cost, their size and location, and maintenance. In the case of firewalls in environments where volatile liquids and combustible gasses are present, the firewall can increase the confinement; this increases the likely peak overpressure from any ensuing explosion and can therefore create one problem, whist solving another.

There are many cost and accident mitigation issues surrounding the provision of fire protection, these have lead the drive to develop reliable design optimisation methodologies.

1.1 Design stages

There are three design stages to developing a fire protection design:

1. The development of the fire loading;
2. The identification of the resulting material temperatures; and
3. The assessment of the structural response to the temperature rises.

Closely coupled with these phases are the designation of escape routes and temporary accommodation, and the decisions surrounding the preferred final state of the assets.

1.2 Development of the fire loading

Fire loading is an important aspect of the determination of fire protection. Flame intensity, duration, luminosity or presence of soot and its spatial spread are all important for the determination of how severe the fire is, and thus dictates the extent of the protection required. It should be realised that the fire parameters vary with time. Celulosic fires need more time from ignition to a fully developed fire; a high-pressure gas jet fire achieves its full strength practically instantaneously. The flame intensity depends on the characteristics of the fuel. The duration of a fire and the flame length depend on the rate of fuel depletion. For a fire that engulfs or impinges an object, the heat is transferred from the flame to the object by the combined mechanisms of radiation and convection. The convection contribution is negligible when the flame does not touch the object.

For fires in the open, only a part of the heat energy in a flame is transferred to the object, most of it is lost to the atmosphere. In case of fires within enclosures, the heat energy reflects from the walls back into the flame and onto exposed objects, therefore, the heat received by the object increases. The properties of heat transfer from a flame to an object are described by emissivity for the radiative part and by a heat transfer coefficient for the convective part. Flame emissivity is strongly dependent on the luminosity of the flame; i.e. less heat is radiated from obscured flames with a high soot content. Absorptivity is the term often used for the emissivity of the object receiving the heat. Dark coloured objects with a rough surface absorb more heat than light coloured objects with polished surface. Fire loading may be described by simplistic or complex methods. Simplistic methods tend to be semi-empirical whilst complex methods are based on Computational Fluid Dynamics (CFD) techniques.

The likelihood of a release of flammable fluid, its probability of ignition, fire and the exposure of an object can be determined using Event Tree calculations. Since a probability can be assigned for any particular fire occurring, it can be treated as any other hazard load and be assessed as to its credibility. A fire that has a probability of occurring of 0.5×10^{-4} per year could be ignored as being less that the 10^{-4} threshold. Thus, the first pre-condition for fire-protection design is understanding the facility being designed.

2 Fire protection

2.1 Material temperature determination

The calculation of the temperatures achieved by the structure enables the basic design of fire protection. The analysis of temperature rise in materials is well understood and is implemented in many well-featured structural analysis programs. The thermal analysis used in fire protection analysis ranges from the steady state response of single elements and structures to the time-domain response.

2.2 Basic fire protection

The simplest method used to decide on the application of fire protection is the basic consideration that a flammable inventory exists, it is possible for the inventory to leak and there is a likely ignition source. The result will then be to apply fire protection, sufficient to resist the type of fire being considered for a given time period. Often for residential or commercial premises, this will be two hours exposure to a cellulose-based fire. In general, this is sufficient because the fire resistance can be achieved by means of construction materials and/or architectural finishes. However, where the likely fire is more severe, hydrocarbon fires in refineries for instance, the required fire protection is more significant. The basic event chain:

Hydrocarbon source ⇒ leak ⇒ ignition source ⇒ fire ⇒ fire protection;

can lead to expensive and unnecessary protection. This simple consideration has lead to the development of the optimisation techniques.

2.3 Target temperature rule

The first and still possibly the most common means of optimising fire protection is to restrict the temperature of the structure to some reasonable upper limit. For carbon steel 400°C has been used for many years, although recently there have been cases where higher temperatures have been applied. The logic behind the choice of 400°C is most likely based on the change in steel properties as the temperature rises. There are several published works presenting the change in steel properties with temperature. Examples of the variation of elastic modulus and yield strength with temperature are presented within References [1-3]. In all cases, the properties have been measured up to 400°C and in some cases, they have been determined up to near melting point of the material. A large part of the work carried out on materials at high temperature has been driven by the need to contain process materials during accidental operational temperature excursions, with the civil nuclear industry leading the way. The maximum temperatures generally encountered are therefore of the order of 600°C. Since Piper Alpha and other significant fires in the 1980's more research has been carried out into the properties of materials at significantly higher temperature, with steel properties being investigated at temperatures of up to 1000°C.

The protection of steel to 400°C stems from the knowledge that in the worst case the steel properties at 400°C are no worse than 60% of their value at 20°C. The importance of this observation comes from the use of limit-state design codes where the design loads are factored. In general, in limit state design the load factors are not applied when considering accidental or extreme hazard loads. Therefore, if the dead load factor for normal design is 1.4 and the live load factor is 1.6, then by inspection the minimum reserve is $1/1.4 = 0.71$. That is if the steel retains at least 71% of its original strength then it will still be able to carry its design load at 400°C. Obviously, if the load is entirely dead load, then 400°C can be a little high. In most cases, the structure will be capable of carrying more live load than dead load and the live load may not be a permanent load.

The target temperature rule therefore requires the designer to calculate the material temperature due to the fire. If the calculated temperature exceeds the target, then protection is provided to limit the structure to the target temperature. Two methods are commonly applied here, protecting simply because an inventory exists, and calculating the maximum temperature due to the fire using steady state thermal analysis. Both of these result in a protect/don't protect decision. The optimisation applied here is that only areas likely to be subjected to a significant fire are protected.

2.4 Fire duration studies

A given inventory of material can only burn until that inventory is depleted, although some inventories can be effectively infinite. One such example is the presence of long large bore pipelines carrying hydrocarbons. Whilst it is noted that many of these installations are provided with isolation valves, they can still present large volumes of material between the valve and the leak. It is increasingly common to use time-domain thermal analysis to derive the temperature rise history of the material. This is often carried out using single member models since these can be coded in spreadsheets. The net result is that a protect/don't protect decision is made based on the temperature at a specified time. One advantage of this methodology is that if the inventory burns for say only 15 minutes, then a 15-minute protection will also comply as a one-hour or two-hour protection.

The drawback of the single element approach is that it does not consider all aspects of the thermal response of the structure. Most structures and plant items have areas with large thermal masses. As an example, connections in structures usually consist not only of the members framing into the joint, but also the connection plates or even castings. A thermal analysis of the whole structure near the fire will therefore allow the actual temperature distribution to be identified. Such an analysis will include the heat flow caused by conduction into the heat sinks. This technology gives a more rigorous analysis of the temperature distribution, and hence the fire-protection can be optimised to protect only those members that reach critical temperatures before critical time.

2.5 Fire scenario studies

The time-domain thermal analysis leads to the highest level of fire study, the fire scenario analysis. If a fire source study has presented inventories and directions, then it is possible to apply the directional data to a model of the structure. Thus, the local temperature time distributions can be established. Various fire models are now available from those that simply derive flame temperatures and fluxes, to phenomenological models that include flame buoyancy to CFD that can derive complex flame data. CFD models can represent cases ranging from the flame shape in the free field to the flame flow around obstacles. However the flame data is derived, scenario based flame analysis presents a good methodology to optimise fire protection. The basic decision using these methods may still be based on protection to a given temperature at a given time.

3 Modelling the fire protection

Using thermal analysis and in particular time-domain analysis, it is possible to consider the effect of fire protection on the structure. Combining fire scenario information with the fire-protection layout allows fire protection schemes to be designed and ranked in terms of cost and degree of protection. This is the highest level of optimisation possible without resorting to structural analysis.

Fire protection modelling includes the effect of intumescent and other passive coatings, the shielding effects of walls and floors (with time dependence if necessary) and the effects of radiation barriers. These effects are modelled by adjusting various heat transfer properties to mimic the presence of barriers. Active fire protection is modelled by adjusting the heat load provided from the fire. These advanced thermal models provide a significant insight into the likely structural performance of the structure in fire.

4 Structural analysis

Materials, in particular steel, retain strength and rigidity up to moderately high temperatures. By inspection, if the stress state of a member is known, then it is possible to fix a temperature in excess of 400°C at which it may be deemed that the member is still capable of carrying its design load. As with the thermal analysis discussed above there are various structural analysis types with increasing degrees of complexity. The types of analysis reflect the types of thermal analysis and it is not uncommon to mix the two. Analysis can be carried out on a single member or on a 3D representation of the part of the structure subjected to fire or on the structure as a whole. The 3D modelling is often extended to allow for member loss and force redistribution. In its ultimate state, the 3D analysis is run as a non-linear analysis with material plasticity, large deflection and temperature dependence. It is this high level structural analysis combined with equivalent time-domain thermal models that has been named Fire Induced Progressive Collapse Analysis or FIPCA. FIPCA is a very powerful tool in the design optimisation of fire protection; however, it has to be used with care as it is possible to be misled into thinking that a safe optimum has been reached.

4.1 Single element analysis

Single element analysis is the simplest form of structural consideration and is generally a hand analysis. This method calculates the stresses and deflections within the element using the temperature modified material elastic modulus and yield strength. Two outcomes are generally taken from this analysis: (i) confidence that the member will survive the fire as it retains sufficient structural competence or (ii) the knowledge that the member will fail. If the second outcome is the case, then it is common practice to carry out a member removal analysis. These analyses are elastic analyses with the failed member(s) removed. If the structure still stands, then the member loss is deemed acceptable.

The single element analysis is, however, only valid if the end conditions are reasonably modelled *and* the thermal expansion is properly accounted for. Thermal expansion can only be ignored if it can be shown that one end of the member has the freedom to expand. By inspection thermal expansion induces additional axial load and if the member has moment carrying ends, additional moment. The basic premise of using the design load without load factors and code checking using reduced elastic modulus and yield strength is not necessarily the clear cut conservative consideration presented in the early sections of this paper. It is also possible to demonstrate that the simple member removal exercise is not necessarily a conservative examination of the global structural response.

4.2 3D frame analysis

Over the past ten years, various groups have developed full 3D frame analysis methodologies that consider the effect of the thermal loading on entire structures or significant part structures [4-9]. The one common factor in all of these more advanced methods is that they take the members subjected to fire in their operating condition and hence model frame effects. The models described in the literature have various degrees of sophistication and are modelled with and without thermal expansion. The basic modelling factors included to varying extents in these models are listed below, (only FIPCA considers all factors):

- Time dependent thermal loading;
- Scenario based fire loads;
- Temperature dependent thermal conductivity;
- Temperature dependent specific heat capacity;
- Temperature dependent heat transfer properties;
- Passive fire protection;
- Active fire protection;
- 3D thermal analysis;
- Whole structure (or significant substructure) thermal representation;
- 3D structural analysis;
- Whole structure (or significant substructure) structural representation;
- Thermal expansion;
- Temperature varying elastic modulus;
- Temperature varying yield strength;
- Temperature varying ultimate strength;
- Member initial imperfection;
- Member plasticity;
- Loss of structural members;
- Time dependency; and
- Structural cooling.

4.2.1 Basic 3D frame analysis

The basic 3D-frame analysis method considers the heating of members within the frame and their state as the fire develops. The member temperature loading for this type of analysis can be taken from single member thermal analysis or full-frame thermal analyses. If the member exceeds its allowable temperature, then the member is removed. This method is only strictly valid up to the point of failure if the thermal expansion is included. In removing the member, it is necessary to verify that the expansion did not lead to a plastic deformation of other parts of the structure. Any local plastic deformation should be modelled in the removed member analysis to maintain a reasonable validity. The process should also heat all members affected. The basic analysis method may or may not include the loss of mechanical properties as the fire develops. Evidently this method should be used with great care and the results should be carefully validated.

4.2.2 Temperature dependent 3D frame analysis

The extension of the basic methodology is to model the temperature dependent properties of the material during the analysis and to allow plastic hinges to form. The analysis is then run until a member is found to fail and causes the analysis to abort. At this point, the failed member is removed from the model and the analysis restarted from scratch. This analysis type allows the progress of a structural collapse to be monitored and if the time dependence of the thermal loading is linked to the structural analysis, then the time to collapse can be derived. The flaw in this analysis scheme is the methodology for member removal.

The load carrying capacity of a member is controlled by its elastic modulus, yield strength, ultimate strength and its straightness. A cross beam framing into a large main beam is going to push the main beam out of line. In most structures this deformation will be allied with the minor axis bending giving a destabilising effect on the main beam. If the crossbeam is removed and the only deformation of the main beam due to the expansion of the crossbeam is elastic, then the removal is valid. However, if the main beam has sustained plastic deformation due to the expansion of the removed member and the reduction in its yield strength, then removing the member and starting from scratch is non-conservative. Here the method to use is the so-called birth and death method. Element birth and death works by effectively removing failed members from the structure matrix. The usual method is to multiply the element stiffness matrix by a small number such as 10^{-6}. This effectively removes the member from the load-path, however, as it takes place during the analysis and the analysis continues from the removal time, and plastic deformation is retained and, therefore, so is any residual destabilising deformation.

3D analysis of frames has an additional benefit, that of being able to study the cooling frame. In the authors' experience many frames fail in the cooling phase as members contract. The cooling frame does not unload along the loading path, many members will have suffered plastic deformation and other members will have failed. A fire lasting 30 minutes may be such that the structure does not fail before the fire extinguishes, but it does fail in the cooling period. If this failure occurs before the required survival time, then the resistance to the fire scenario is not proven.

4.2.3 Enveloping fire scenarios

The tendency is to try to envelope fire scenarios as they are developed to reduce the quantity of analysis to be performed, the basic thinking being that a 6m diameter flame will be more damaging than a 4m diameter flame of the same temperature and flux. Experience has shown that this is not necessarily the case as the structure can often survive the larger fire more readily than the smaller fire. This is due to the effect of the flame on the structure as a whole, which is based on the resistance of the frame. It is identified above that the inclusion of the thermal expansion is not optional, it is required. A smaller flame heats a smaller region of the structure. Therefore, the heated members will be pushing against a more competent cooler structure as they expand. The expanding members will induce higher loads in the cooler members, as the loading is displacement controlled. On the other hand, if the fire also influences the supporting structure, then its elastic modulus is reduced by the fire, the induced displacement controlled forces are therefore reduced and the structure may survive. The analysis of structures subjected to fire is one area where the idea of bounding cases can be counter-intuitive. It also leads to the conclusion that designing fire protection to resist the worst case fire may not provide a scheme that will satisfy lesser events.

5 Sample analysis

The following example looks at the structural analysis of a simple portal frame. The frame has been modelled as a 6m high 14.5m span flat portal. The frame has a dead load of 15kN/m run and a live load of 40kN/m run. The beam is a 610×305UB179, the columns are 305×305UC283. The material properties are defined as follows:

Table 1. Material properties used for sample analysis

Temperature (°C)	Elastic Modulus (N/mm^2)	Yield Strength (N/mm^2)	Ultimate Strength (N/mm^2)
20	210,000	275.0	385.0
400	173,560	267.0	373.8
600	36,350	130.4	182.5

The purpose of this analysis is to demonstrate the structure response. Therefore, uniform temperatures have been used. Ten analyses have been performed using the general-purpose industry standard FEA suite ANSYS v5.5.3. The analysis has been configured to allow plasticity, temperature dependent material properties (using the Eurocode 3 methodology for the material stress-strain relationship), large deformation and stress stiffening. The beam elements used are quadratic finite strain Timoshenko beams. Load cases 1 and 2 are the normal operating case for the ultimate limit state and the serviceability limit state. Load cases 3 to 5 and 7 to 9 represent a flame that engulfs the rafter only. Load cases 6 and 10 represent a flame that engulfs the whole structure.

The analyses demonstrate the aspects of the behaviour of structures subjected to fire discussed within this paper. Load cases 1 and 2 provide the design benchmark, which are to be compared with load cases 3 to 6. The results from load case 3, the temperature loading only compare well with load case 2, the serviceability limit state. This is to be expected, as changing the elastic modulus does not result in different stresses. However, the load case 3 model assumes no thermal expansion of the rafter. If expansion is included it is seen that the rafter mid-span stress reduces but the column stresses increase. It is therefore the case that adding the expansion has moved the design problems away from the element of the structure subjected to the temperature load. The stresses induced in the columns are significant. Inclusion of the initial imperfection, load case 5, does not significantly alter the results. Allowing the columns to be heated, representing a large fire gives a significantly changed result. The rafter stresses increase slightly. However, the column stresses reduce. This is a function of the rafter response being load controlled and the column response being displacement controlled. The softening of the columns has lead to reduced induced stress.

Table 2. Load cases analysed

Load Case	Load Factors	Temperature (°C)	Coefficient of Linear Expansion (Strain/°C)	Imperfection (%Span)	Heating of Columns
1	Yes	20	0	0	No
2	No	20	0	0	No
3	No	400	0	0	No
4	No	400	$12x10^{-6}$	0	No
5	No	400	$12x10^{-6}$	0.08	No
6	No	400	$12x10^{-6}$	0.08	Yes
7	No	600	0	0	No
8	No	600	$12x10^{-6}$	0	No
9	No	600	$12x10^{-6}$	0.08	No
10	No	600	$12x10^{-6}$	0.08	Yes

Table 3. Results from analysis

Load Case	Rafter mid-span comp. stress (N/mm^2)	Top of column comp. stress (N/mm^2)	Base of column comp. stress (N/mm^2)	Rafter mid-span vertical displacement (mm)	Exceed	
					Yield	Ulti-mate
1	266.0	210.0	100.0	61.0	No	No
2	184.0	150.0	66.0	44.9	No	No
3	179.0	141.0	65.0	51.3	No	No
4	100.0	220.0	210.0	27.7	No	No
5	100.0	220.0	210.0	27.6	No	No
6	120.0	180.0	170.0	6.2	No	No
7	170.0	216.0	35.0	320.0	Yes	Yes
8	180.0	233.0	180.0	174.6	Yes	Yes
9	181.0	235.0	181.0	184.3	Yes	Yes
10	159.0	159.0	124.0	243.7	Yes	No

When the load cases 3 to 6 are repeated with a steel temperature of 600°C, the responses as expected are the same. In load cases 7 to 9, the steel stresses exceed the ultimate at 600°C. Load case 9 actually failed to complete reaching 97% of the full load, indicating global collapse of the structure. The interesting result is when the whole structure is subjected to fire, the column stresses fall below ultimate indicating that the frame would not collapse when subjected to the larger fire.

6 Conclusions

This paper has briefly reviewed the methodologies currently employed for the design and optimisation of structural fire protection. It is seen that even the simplest method, protecting the structure without looking at optimisation, can result in under-design if the thermal expansion can cause distress elsewhere in the supporting system.

All of the current methodologies have their shortcomings, even the full scale FIPCA technique. Generally, however it is evident that the more holistic approaches capture the detailed responses and can provide well optimised solutions. The major concern with analysis using FIPCA is that the fire scenarios have to be carefully reviewed and any enveloping has to be carried out with care.

The simple single beam analyses have the advantage that there are no significant data transfer issues in that the member temperature is calculated and used for a structural assessment. The holistic methods on the other hand require that the thermal data be transferred from the thermal assessment or modelling to the structural modelling. When a large structure is analysed in the time domain, this can represent the transfer of significant quantities of data. The safe and efficient application of FIPCA therefore necessitates automated data transfer.

Optimisation of fire protection is a viable process if it is carried out with care and with a clear understanding of the underlying restrictions. Although the analysis process can be involved, and validation of the models and results is essential, the cost savings can be significant. In the experience of the authors, which extends over more than ten years of practice in fire response analysis, it is possible to get a savings-to-analysis cost ratio of greater than 10:1. If the maintenance savings are included, this figure increases significantly. It is also often useful to revisit the safety case or Quantified Risk Assessment, once the fire response analysis is complete since it is often possible to revise the overall probabilities of loss of structure.

7 References

1. USFOS Users Manual, SINTEF Division of Structural Engineering, Trondheim, Norway.
2. BSI, BS 1501 Steels for fired and unfired pressure vessels-plates, British Standards Institution, London.
3. Commission for European Communities, Eurocode No. 3 (EC3): Design of Steel Structures Part 10, Structural Fire Design, Draft April 1990

4. Rogers, C.P., Bruce, R.L. and Medonos, S. (1993) Optimisation of passive fire protection on platform topsides, *12th International Conference on Offshore Mechanics and Arctic Engineering.*
5. Rogers, C.P. and Ramsden, M. (1994) Optimisation of passive fire protection for offshore structures using progressive collapse techniques, *Offshore Structural Design - Hazards, Safety and Engineering.*
6. Rogers, C.P. and Ramsden, M. (1994) Optimisation of passive fire protection measures to develop cost effective and safe steel structures, *Fire and Safety '94.*
7. Medonos, S. and Rogers, C.P. (1998) Optimisation of fire and explosion protection on FPSO, 1998 SPE International Conference on Health, Safety and Environment in Oil and Gas Exploration and Production.
8. Medonos, S. and Rogers, C.P. (1998) Optimisation of fire protection On FPSO, *17th International Conference on Offshore Mechanics and Arctic Engineering.*
9. Ossei, R. and Yasseri, S.F. (1997) The fire protection of a topsides using computational methods, *ERA Fire and Blast Engineering Conference.*

SUBSTRUCTURING IN FIRE ENGINEERING

M. KORZEN, K.-U. ZIENER and U. KÜHNAST
Fire Engineering, Bundesanstalt für Materialforschung und –prüfung (BAM), Berlin, Germany
G. MAGONETTE, P. BUCHET and Z. DZBIKOWICZ
Structural Mechanics Unit, Joint Research Centre (JRC), Ispra, Italy

Abstract
In contrast to real fires in classical fire resistance tests building elements are considered as stand alone elements without interaction with the surrounding building. In order to run a fire test in a more realistic fashion the development of special experimental techniques is required. As a contribution to overcome the situation, the application of the substructuring method, a special hybrid method known from earthquake engineering, has been adopted to fire engineering. The paper gives a presentation of current research activities.
Keywords: Fire engineering, fire resistance, hybrid method

1 Introduction

In classical fire resistance tests, building elements are considered as stand-alone elements. In a real fire, however, each building element interacts with its adjacent elements. This behaviour is also supported by numerical calculations [1]. Additionally, there is a current international trend for a change in the design procedure from *Prescriptive Methods* to *Performance Based Design*. This will require the development of special experimental techniques in order (i) to run a fire test in a more realistic manner and (ii) to get a set of continuously acquired data representing the mechanical boundary conditions and temperature field [2] of the specimen during the test.

2 Substructuring in fire engineering

Due to the concept of hybrid substructuring [3] the entire building is decomposed into two parts (Fig. 1): One part is represented by the building element under test, i.e. the

Abnormal Loading on Structures edited by K. S. Virdi, R. S. Matthews, J. L. Clarke and F. K. Garas.
Published in 2000 by E & FN Spon, 11 New Fetter Lane, London EC4P 4EE, UK. ISBN 0 419 25960 0

column specimen, whereas the remaining building environment is simulated by an analytical or numerical model. It is characterized by (i) static loadings and (ii) generally a nonlinear thermo-inelastic behaviour of the simulated substructure. Forces and moments at the boundaries of the specimen, i.e. at the upper and lower bearing of the column, are measured and utilized for the computation of the corresponding displacements and angles, which are sent to the specimen in order to keep the entire building in mechanical equilibrium with its prescribed overall boundary conditions.

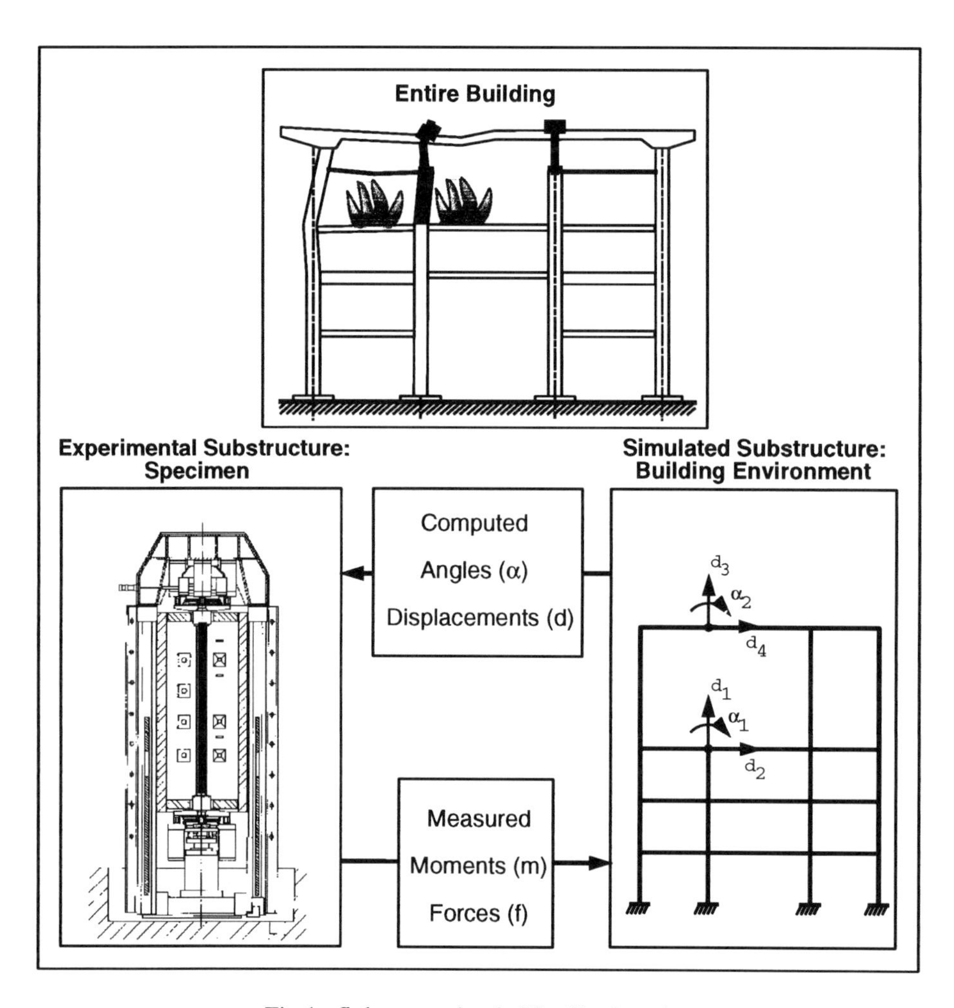

Fig.1. Substructuring in Fire Engineering

3 Project Activities

3.1 Data acquisition and control system

The justification of the establishment of an in-house data acquisition and control system [3] is twofold: It is needed (i) as the basic element of the substructuring method itself and (ii) as a working environment for the development process. The latter item requires a very flexible and modular tool in order to test different algorithmic approaches or hardware assemblies and to analyse the input and output data offline and online in a reliable and comfortable manner.

3.2 Behaviour of structural steel at high temperatures

Efforts to understand the failure mechanisms of structural elements like columns [4] or beams and slabs [5] within a building structure are related to our project in a natural way. The key ingredient of all such modelling approaches is the underlying constitutive model.

A series of tests has been run on structural steel specimen in compression and tension at temperature levels ranging from room temperature (RT) up to 600 °C. Those tests were realised in a piecewise linear fashion at changing loading rates supplemented by periods of hold time at constant strain or stress. This loading pattern is used as a tool for a more detailed analysis of the rate dependent behaviour above the yield stress, which may be of importance in the case of inelastic failure.

Other tests with the same type of specimen at constant compressive stress and linear time varying temperature from RT up to 600 °C complete the first series. The results can be utilised for the verification of constitutive models, which have been identified from data of the first test series at constant temperature.

3.3 Qualitative tests for checking basic capabilities and consistency

As an example, results of a special column test are shown in Fig. 2. During this test a thermally insulated steel column of cross section HEM 200, clamped at the top and bottom, is heated in accordance with the ISO 834 curve and mechanically driven in the axial direction under displacement control.

In more detail, the axial displacement is fixed throughout the test, i.e. the thermal elongation is restrained and the axial force grows according to the temperature rise of the specimen. Additionally, at some time instants during the test, the axial cylinder is moved with a ramp-hold-time pattern. This loading type is chosen as an example of the recently established function generator feature of the data acquisition and control system. During these characteristic segments, the rise of the axial force is composed of the restrained thermal expansion and the compression (moving up) or expansion (moving down) due to the function generator.

It is found that the magnitudes of the slopes in the graph with respect to time for compression (positive) and expansion (negative) are found to be different, except at the points 9 to 11, where the mean column temperature reaches a plateau and remains approximately constant, i.e. no further restraint forces can arise.

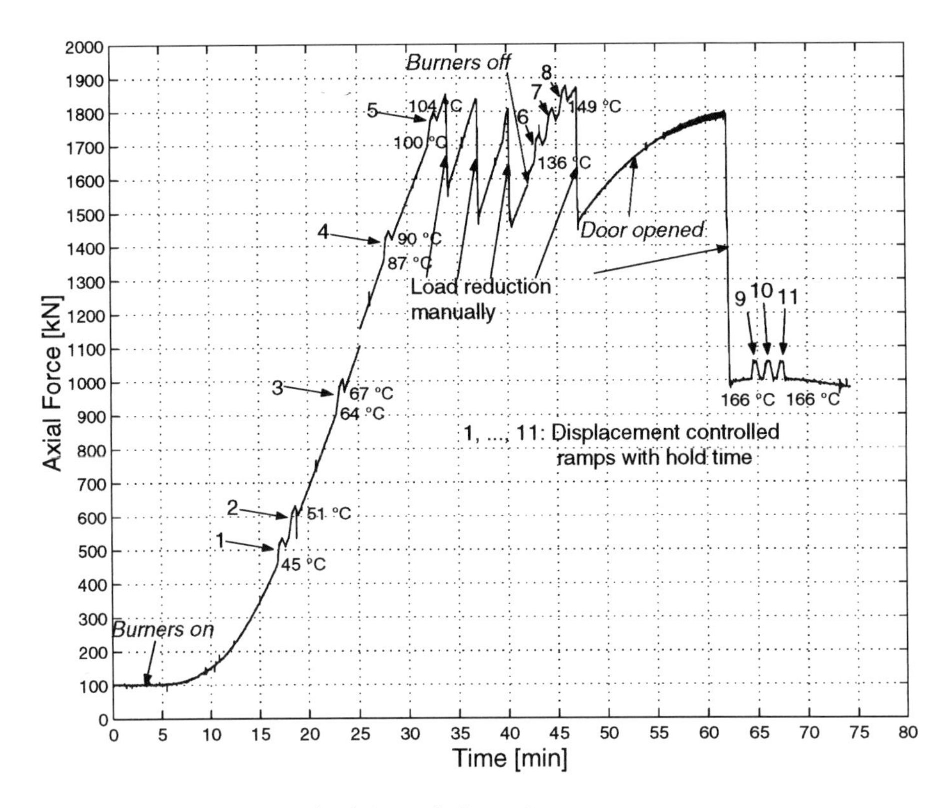

Fig. 2. Fire test with restrained thermal elongation

4 Conclusion

In recognition of recent findings, it is now considered necessary to study the interaction of an element exposed to fires with its surrounding elements in a building. In order to run a fire test in a more realistic fashion, the development of special experimental techniques has been described. The application of the substructuring method has been adopted to fire engineering. The paper gives results from recent research activities.

References

1. Bock, H.M. and Erbay, S. (1993) Die kritische Stahltemperatur von Bauteilen und Bauwerkssystemen aus Stahl. *Stahlbau*, Vol. 62, pp. 107-15.
2. Korzen, M., Schriever, R., Ziener, K.-U., Paetsch, O. and Zumbusch, G.W. (1996) Real-Time 3-D Visualization of Surface Temperature Fields Measured by Thermocouples in Fire Engineering, in *Proc. of Symposium Local Strain and Temperature Measurements in Non-Uniform Fields at Elevated Temperatures,*

Berlin, Germany, 14 - 15 March 1996 (eds. J. Ziebs et al.), Woodhead Publishing Limited, Cambridge, pp. 253-62.

3. Korzen, M., Magonette, G. and Buchet, Ph. (1999) Mechanical loading of columns in fire tests by means of the substructuring method. *Z. Angew. Math. Mech.*, Vol. 79, pp. S617-18.
4. Neves, I. C. (1995) The critical temperature of steel columns with restrained thermal elongation. *Fire Safety Journal*, Vol. 24, pp. 211-27.
5. Rotter, J.M., Sanad, A.M. and Gillie M. (1999) Structural performance of redundant structures under local fires, in *8th International Fire Science & Engineering Conference, Edinburgh, Scotland, 29th June – 1st July 1999*, Interscience Communications Limited, London, pp. 1069-80.

DEVELOPING 'HAZOP' METHODOLOGIES TO REFINE FIRE RISK ASSESSMENTS

Hazop methodologies for fire risk

R.J.TIMPSON
BAA Plc, Gatwick, Crawley, UK

Abstract

The purpose of this paper is to describe the techniques involved in the process of Hazard and Operability studies (HAZOP). The document will then describe how this process could be used to assist in the process of Fire Engineering Design and Risk Assessment (FEDRA).

Keywords: Fire Engineering, fire hazard, HAZOP, risk assessment

1 Introduction

One of the most important tests that will be applied to any design process following a structural failure or major fire, is that of foreseeability. Usually this will be defined by case law and experience, and is therefore difficult to quantify on individual designs where innovative elements are being employed. Prescriptive regulations, codes and standards give a degree of reassurance, but they are not flexible enough to cope with the synergistic effects of interacting component applications.

In most functional buildings fire is an ever-present hazard. Occupant safety can be threatened in the following ways during a fire:

- being overcome by toxic gas and smoke
- direct burning from flames, radiant and convected heat
- evacuation injuries (crushing near exits)
- falling injuries (jumping from windows)
- building collapse and falling fittings
- exposed electrical services
- explosion from tanks, gas supplies and cylinders

Abnormal Loading on Structures edited by K. S. Virdi, R. S. Matthews, J. L. Clarke and F. K. Garas.
Published in 2000 by E & FN Spon, 11 New Fetter Lane, London EC4P 4EE, UK. ISBN 0 419 25960 0

Within the design process consideration must be given to inherent fire hazards of a design, and the potential hazards that final fittings and furniture will create. Fire and smoke loading, along with burning droplet characteristics for elements of structure can be assessed using accepted tests and burning apparatus. The proposed pan European Construction Products Directive [5] will standardise these test procedures and facilitate progress in component selection.

Irrespective of this increased knowledge of structural component fire characteristics, there is still a requirement to understand the fire hazards and subsequent risks within a design in their totality. Emphasis on 'traditional' fire hazards may well lead to significant oversight and misplaced risk management.

2 Hazard and operability studies (HAZOP)

2.1 Origins of HAZOP

Within the evolution of risk management and proactive design assessment, a need arose to challenge a design for potential weaknesses and failure points. This was especially true in the process industries such as petrochemical and pharmaceuticals. Large scale disasters such as Flixborough (UK), Seveso (Italy), and Bhopal (India) had focused attention on the need to increase foreseeability, and make designs inherently safe. HAZOP was initially developed by Imperial Chemical Industries (ICI) Ltd for improving the safety of their chemical plants [1]. The procedure proved to be so successful that it gained wide acceptance within industry as a useful tool for qualitative hazard analysis.

2.2 Applications of HAZOP studies

The HAZOP technique can be applied at any time in the cycle of a design, build, commissioning, maintenance and eventual demolition. For maximum benefit it should be applied at the conceptual stage in any planning process. The outputs from the HAZOP should guide the designer away from hazardous configurations, which may not be obvious from even detailed first inspection. This assessment will also have the added benefit of reducing wasted time on design tasks which will become irrelevant following the HAZOP. From a legal perspective, the HAZOP will also help to demonstrate 'due diligence' on behalf of the designer and construction team with regard to safety of any future occupants.

2.3 Variants of HAZOP

HAZOP has been developed to serve needs in different industries [2]. The two main variants of the HAZOP process are:

Experience - where a team of suitably qualified assessors analyse a design using historical data and the experience of those assembled to determine failure modes. This is effective when a design has been developed over a prolonged period, and those involved have in-depth knowledge of system and component performance. The need for a systematic approach to the process is vital.

Guide word - where a team of suitably qualified assessors analyse a design using set phrases to provoke failure modes. Guide words can include terms like 'more', 'less', 'reverse'. These guide words when applied to elements of the design, prompt the HAZOP team to discuss what would happen. For example, they may ask what would happen if 'more' pressure is applied to a length of supply pipe. They may conclude that the pipe would rupture, the glands and seals would fail and so on. (see Table 1). This paper will focus on the use of the guide word HAZOP to assess a design.

Table 1. HAZOP guidewords

Guideword	Process Deviation
No, Not or None	The complete negotiation of the design intention
More Of	Quantitative increase of any relevant physical property
Less Of	Quantitative decrease of any relevant physical property
As Well as	Quantitative increase of any relevant physical property
Part Of	Quantitative decrease of any relevant physical property
Reverse	The logical opposite of the intention
Other Than	Complete Substitution

3 Implementation of HAZOP

3.1 HAZOP team selection

In order to achieve a satisfactory result from HAZOP, the selection and composition of team members, chairmanship, and clarity of purpose of the team is essential. The team should represent all perspectives on the design, construction, and working practicalities with regard to fire safety. Other associated safety issues will be discovered as part of this process, and these should be recorded for subsequent design action.

The HAZOP team will need to be briefed and prepared for the exercise. This briefing should include basic details of the process, aims and objectives, and consideration given to any supporting data that may be useful. For large buildings HAZOP can be extremely time consuming. Once established the HAZOP team should not be changed, and time constraints should be taken into consideration in the management process.

3.2 Preparation

In order to allow the HAZOP to be completed in a practical amount of time, the following should be prepared for the Fire HAZOP:

1. all participants should have received a briefing note, describing the task ahead (the note should include a simple case study of a completed Fire HAZOP)
2. a large 'A0' site plan overlaid with a control grid should be available in the meeting room
3. interior, plans, diagrams and manufacturers specifications
4. the designated secretary should have a scoring sheet filing system ready

5. the HAZOP team should complete a 'walkround' of the assessment site just prior to commencement
6. 'rules' regarding classification of control grid area when only a small element of a significant risk is present (from neighbouring control grid) will need to be agreed

3.3 Control Grid

In order to make the HAZOP systematic and thorough, a control grid will need to be placed over the site plan. The HAZOP team chairman will need to decide this in advance. For purposes of planning a 3m×3m grid should be adopted as a starting dimension and adjusted according to risks present (see Fig. 1). Once all of the building geometry has been identified using the control grid, the process of HAZOP can commence.

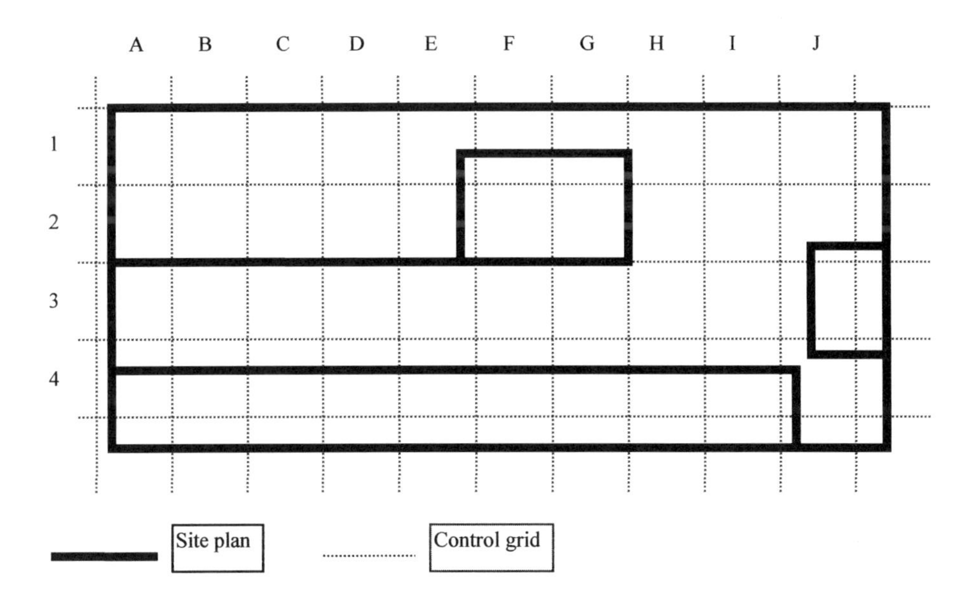

Fig. 1. Site plan overlaid with control grid

3.4 Record keeping for HAZOP

It is of the utmost importance that accurate sequential records are kept during the process, and one member of the team should be designated as secretary for data produced. The finished assessment may well sit within the safety file as produced for the Construction Design Management Regulations, this will therefore require the involvement of the planning supervisor, and any filing system being employed.

4 Applicability of HAZOP to fire safety

4.1 Background to use of HAZOP in fire safety

Like all other areas of safety management, fire safety is moving towards performance based, risk assessment solutions, and away from prescriptive codes. At the start of any risk assessment process there is a need to identify all of the significant hazards. When considering a fire risk assessment it is necessary to systematically check a design or existing premise, to identify these hazards.

Currently, most fire risk assessment schemes rely on checklists and the assessors' knowledge of the design or building. The quality of the assessment is based on the individual undertaking the task, and overall it adds little value to the business it seeks to serve.

4.2 Use of Guideword approach to fire safety HAZOP

By careful selection of relevant guidewords and scoring systems it will be possible to systematically check a design or existing building with regard to fire safety using the HAZOP methodology. While the HAZOP team members may be experts in their own areas, it is the pooled knowledge and collective inspiration which lead to a successful study. Like all good hazard identification techniques, the output is not just the written record, but also the changed views of the participants, brought about by their involvement. This will have the added benefit of increasing fire safety knowledge and understanding.

4.3 Information generated

A Fire Safety HAZOP will produce the following assessment information [3]:

- Locality of significant fire hazards in comparison with other local areas
- Quantitative comparative measurement of overall Fire Risk
- Starting points for fire scenario generation, for corresponding checks on alarm and mitigation systems
- When combined with a colour coding system, it will produce a Fire Risk 'contour map' of a design, which will assist staff to appreciate hazards present
- It will be a clear 'milestone' to demonstrate due diligence

4.4 Guide words

The process begins by identifying the node in question (A1, A2, etc.). The secretary should record findings as they are assessed. The guide words and corresponding weighting will have been agreed before the meeting. A typical range of guide words and weightings for fire safety issues is detailed in Table 2

The process begins by using the guide words in turn and applying them to the nodes, starting with ignition hazards. Ignition is split into the main headings of; hot process, lights, heating, electricity, smoking and arson. Each of these is weighted according to the hazards present (note: as a rule, temporary work or fittings should be ignored, however they can be included if the HAZOP team feels it is appropriate). A weighting guide is included, and typically will have a numeric range appropriate to the magnitude of the hazard.

Table 2. Fire HAZOP guidewords

Main	Sub cat	Weighting	Score
Ignition	hot process	0 - 10	
	lights	0 - 5	
	heating	0 - 5	
	electricity	0 - 8	
	smoking	0 - 10	
	arson	0 - 10	
Fuel	quantity	0 - 6	
	type	0 - 6	
	toxicity	0 - 8	
	highly flammable	0 - 10	
Visual access		0 - 8	
Physical access		0 - 8	
		0 - 8	
Fire Propagation	layout	0 - 6	
	ceiling height	0 - 8	
	enclosure	0 - 6	
	stopping	0 - 6	
	voids	0 - 6	
People	staff	0 - 10	
	public	0 - 10	
TOTAL SCORE			

Example

Node 3F - contains a deep fat frying unit.

Guide word ignition - hot process range 0 - 10

Consideration - history: cooking operations are a common source of fire
fire: upon ignition, large, rapid growth fire
Therefore, the score will be 10.

Guide word ignition - lights range 0 - 5

Consideration - history: fluorescent lighting used in 3F, is assessed as low hazard
fire: will probably smoulder and produce strong burning smell before actual ignition occurs. Protected by circuit breakers and ECB
Therefore ,the score will be 2.

The process then continues through the various guidewords and a final score is arrived at. A colour coding scheme can then provide a graphical representation of the fire risk. To assist in the completion of this task, large areas of the building that have similar layout and guideword scoring can be regarded as a single grid (see Fig. 2).

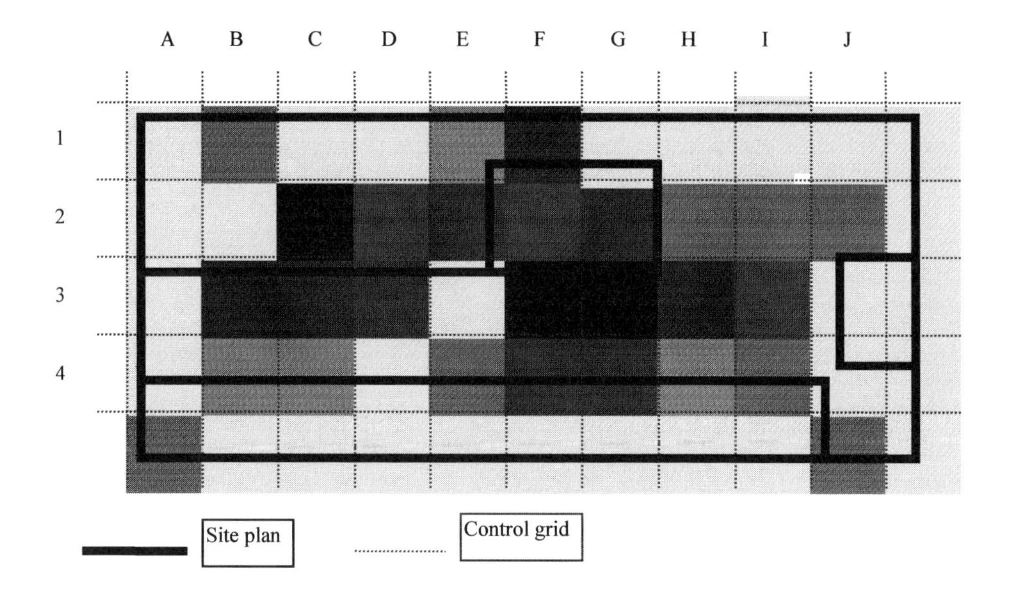

Fig 2 - Fire risk areas as highlighted by colour schemes

5 Where Fire HAZOP sits within the overall fire assessment hierarchy

5.1 Range of assessment options

The completion of a fire HAZOP is only one option for the prediction of abnormal loads on building structures during a fire. Further evaluations will have to take place when the building changes and redevelops. Currently the options for assessment fall into four categories. These are shown in Fig 3.

5.2 'Workplace' fire risk assessments

This is the starting point for the risk assessment process. The main objective of level 1 assessments is to raise awareness and encourage 'ownership' of fire safety issues. This level of assessment is sufficient for 'regular' buildings, which are code compliant. Within larger more complex buildings this assessment will usually always require progress to levels 2, 3 and 4. This technique is usually employed in existing buildings, and from December 1999, such assessments will become compulsory as part of the revised Workplace Regulations [4]. Such assessments can provide valuable insights for the design architect or engineer on the practical fire hazards faced by building occupants following construction.

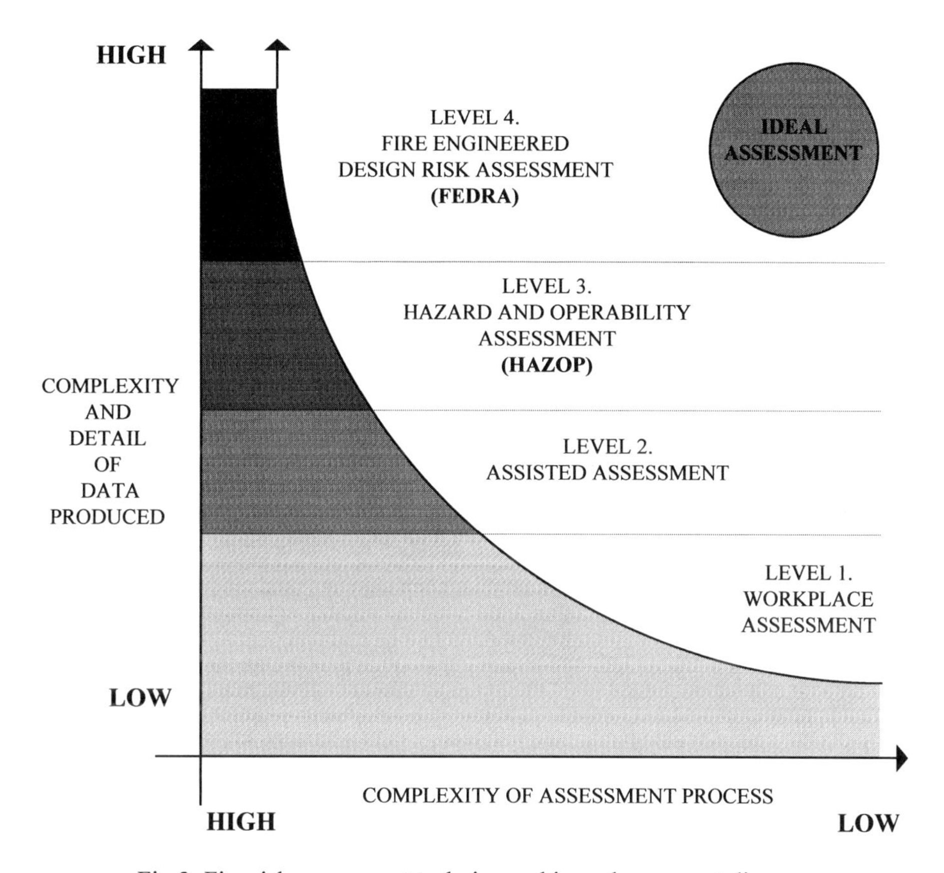

Fig 3. Fire risk assessment techniques, hierarchy concept diagram:

5.3 'Assisted' fire risk assessments

This is the second stage in the assessment process. The main objective of level 2 assessment is to begin the quantification of fire risk, and resolve technical issues raised in the level 1 assessment. This assessment is usually conducted with assistance of a 'competent' person (competence being defined as some one with the knowledge, skills, qualifications or other).

5.4 Hazard and operability (HAZOP) fire risk assessments

This is the third stage in the assessment process. The main objective of level 3 HAZOP assessment is to refine the quantification of fire risk to a level where long term decisions can be made with regard to fire strategy. To capture the fire risk profile of a building in total.

5.5 Fire engineered design risk assessment (FEDRA)

This is the fourth stage in the assessment process, and will only be commenced following a level 3 HAZOP. The main objective of level 4 advanced assessment is to provide in-depth and detailed predictions on dynamic fire related phenomena. This could include smoke movement, occupant evacuation performance, fire growth and structural integrity. This work is usually undertaken by specialist fire engineering consultants.

5.6 Output of assessment - Fire Engineered Design Risk Assessment (FEDRA)

The advanced assessment is an analysis of dynamic fire phenomena. There are three main types of advanced assessment:

1. Deterministic - using accepted mathematical techniques engineers model the evolution of fire phenomena. This is based on assumptions of rates of fire growth and smoke generation along with human escape behaviour.
2. Probabilistic - within this method, engineers try to predict fire events by means of probability analysis. This method is based on statistical data, which produces values of likelihood for fire events.
3. Stochastic - this is a combination of the two methods already detailed. A probability of fire is calculated for a given scenario, this may then form the basis for a deterministic model. When a boundary is reached, a probability study can then be used to predict further growth and so on.

The outputs from these assessments are most likely to be used in conceptual stage of design. However they can be used to validate tactical changes within the building fire strategy for purposes of approval.

6 Conclusion

One of the main causes of abnormal loading on a structure is that imposed by fire, and products of combustion. The reduction in the risk of fire at the design stage of construction will yield many long term benefits.

6.1 Fire HAZOP

The Fire HAZOP process can aid the management of fire risk. It is best utilised when a flexible team approach is adopted, and time is taken to adapt the process to the ambient conditions and equipment encountered. The next stage after the HAZOP is to take the information, and then develop fire scenarios around the derived high-risk areas. Fire sizes and growth rates can be estimated and then control systems can be analysed for adequacy and relevance.

6.2 Due diligence

Fire HAZOP will assist in the task of demonstrating due diligence with regard to the designer or architect. As society becomes more litigious, people will seek to identify shortcomings in any design process that has not given adequate consideration to inherent hazards. On a more positive note, fire HAZOP can be useful in justifying the cost of more comprehensive fire safety measures in a building for business continuity and building preservation.

6.3 Changing legislation

Buildings that employ fire engineering principles as part of their fire safety tactics are increasing in number and complexity. There is a growing movement towards 'performance' based solutions that take account of all elements of the fire strategy employed. Legislation and guidance is changing to reflect this move. The Construction Products Directive [5] will harmonise fire testing of component elements and classify them in terms of contribution to fire and smoke loading.

6.4 Corporate governance

The recently published 'Turnbull' report [6] has laid down guidelines for the corporate governance of risk within companies. All risks should be identified and appropriate measures taken to eliminate and mitigate them. These details will form a crucial element within annual reports.

6.5 Rethinking construction

The report of the construction task force [7] has stated that the industry will need to make radical changes to the processes through which it delivers its projects. Risk and waste *must* be reduced at the design stage. Architects, engineers, specifiers must embrace all new methods for testing a design so that they deliver intrinsically safe buildings. Fire HAZOP is one method that could be used to achieve this objective.

7 Acknowledgements

I would like to acknowledge the help and assistance given to me by Newcastle Airport. The presentation accompanying this paper includes video footage of a fire within Newcastle Airport Terminal, which took place in July 1999.

8 References

1. Kletz, T. (1991) *Plant design for safety,* Hemisphere Publishing, London.
2. Bird, F.E. and Germain, G.L. (1996) *Practical Loss Control Leadership,* Det Norske Veritas, Georgia, USA.
3. Jenkins, P., Mann, L., Timpson, R. (1995) *Degree design study - Sittingbourne Football Club main stand, Ignition Hazard Study,* Fire Service College Library, Moreton in Marsh.
4. Home Office (1999) *Fire safety - An employer's guide,* HSE Books.
5. Colwell, S., Colwell, R., and Smith, D.(1999) Fire Testing into the next Millennium. *Fire Safety Engineering,* Vol 6, number 5. Pp 13-16.
6. The Institute of Chartered Accountants, (1999) *Internal Control - Guidance for directors on the Combined Code.* Accountancy Books, London.
7. DETR, (1998) *Rethinking Construction.* DETR, Publication Sale Centre, Rotherham.

BAYESIAN ASSESSMENT OF STRUCTURAL RISK UNDER FIRE DESIGN SITUATION

Structural risk in fire

H. GULVANESSIAN
Building Research Establishment, Garston, Watford, UK
M. HOLICKÝ
Klokner Institute, Czech Technical University in Prague, Czech Republic

Abstract

Probabilistic concepts of fire safety analysis are based on assumed proportion of fully developed fires of all started fires, i.e. on probability of fire flashover provided that fire started. Bayesian network is used to analyse risk of structures under fire design situation. Typical nodes of the network include fire start, detection, tampering, sprinklers, smoke detection, fire brigade, fire flashover, and structural failure. Performance of a fire protective system is characterised by appropriate causal links and input conditional probabilities related to states of nodes. The decision nodes and utility nodes are implemented to enable structural risk assessment under fire design situation. It appears that Bayesian networks provide an effective tool to analyse fire protective systems and to make assessment of structural risk.

Keywords: Bayesian network, decision theory, fire design, risk analysis, structural safety.

1 Introduction

Safety in case of fire is one of the essential requirements imposed on construction works by Council Directive 89/106/EEC [1]. Experience and available data [2] [3] [4] [5] [6] [7] [8] indicate that depending on particular conditions and applied fire protection system, the probability of fire flashover (outbreak) may be expected within a broad range. Recent studies [9] and [10] attempt to show that reliability methods applied commonly for the persistent design situation may be also applied for the accidental (fire) design situation. Taking the above mentioned documents into account basic probabilistic concepts of fire safety analysis are developed here using general probabilistic rules [11] and international code documents [2] and [3].

Abnormal Loading on Structures edited by K. S. Virdi, R. S. Matthews, J. L. Clarke and F. K. Garas.
Published in 2000 by E & FN Spon, 11 New Fetter Lane, London EC4P 4EE, UK. ISBN 0 419 25960 0

The study described here is an extension of previous papers [12] [13] [14] using Bayesian networks [15] as an effective tool to find a more accurate estimate for the probability of fire flashover and to propose an optimum fire protection system in a rational way. An acceptable probability of failure due to fire is derived from the total probability of failure due to persistent and accidental (fire) design situations [4] following general probabilistic rules (see for example [11]). Illustrative examples related to office areas are provided. The use of Bayesian networks is investigated and recommended as an approach for additional investigation.

A typical network containing active fire protecting measures (sprinklers and fire brigade) is considered in the presented study. The network consists of fire start, detection, tampering, sprinklers, smoke detection, fire brigade, fire flashover and structural failure. All nodes are inter-connected by directional links corresponding to causal dependencies of relevant nodes.

2 Probabilistic concepts

2.1 General

Probabilistic concepts of structural reliability are powerful tools when determining safety factors and other reliability elements (e.g. combination factors) and risk of structural failure that may result in loss of life or other social and economic consequences. Acceptable probabilities for loss of life may be dependent upon public perception and economic and political consideration as indicated in several international documents [2] [3] [4].

ISO 2394, 1997 [2] states that "an acceptable overall individual lethal accidental rate $\approx 10^{-4}$ per year". Considering public perception with regard to structural failure ISO 2394, 1997 [2] advises that a reasonable reference probability of structural failure per year may be taken as

$$p_{\mathrm{ref},1} \approx 10^{-6} \tag{1}$$

This value also accords with Eurocode 1 [4]. Corresponding reference probability for 50 years is about $p_{\mathrm{ref},50} = p_{\mathrm{ref}} \approx 5\times10^{-5}$ (in the following text the subscript 50 indicating the time period is omitted). For 50 years period the Eurocode 1 [1] assumes the basic value of reliability index as 3.8, which corresponds to the target probability p_{t}

$$p_{\mathrm{t}} \approx 7.2 \times 10^{-5} \tag{2}$$

If the probability of fire flashover is p_{fi} then the target probability $p_{\mathrm{t,fi}}$ is approximately (assuming that failure probability under persistent design situation is less than required target probability p_{t}) given as

$$p_{\mathrm{t,fi}} \cong p_{\mathrm{t}}/\,p_{\mathrm{fi}} \approx 7.2 \times 10^{-5}/\,p_{\mathrm{fi}} \tag{3}$$

If, however, the failure probability under persistent design situation is equal to the target probability p_{t}, then, theoretically, also $p_{\mathrm{t,fi}}$ should be equal to p_{t} (this follows from the theorem for the total probability - see for example [10]).

2.2 The probability of fire flashover

An action caused by fire will occur only when a fire becomes fully developed and could then be capable of causing structural collapse. In accordance with available documents [5] [6] [7] [8], the probability of occurrence of a fire design situation per year, $p_{fi,1}$, (additional subscript 1 is used to indicate one year period) may be expressed as follows

$$p_{fi,1} = p_{fi,1,s} \times p_{fi,d} \tag{4}$$

where $p_{fi,1,s}$ denotes probability that a fire starts during one year, $p_{fi,d}$ denotes the conditional probability that fire will fully develop given the fire actually starts.

Available documents [6] [7] [8] indicate values of $p_{fi,1,s}/A$, where A [m^2] denotes the floor area, within a broad range from 0.05×10^{-6} up to 100×10^{-6} depending on type of occupancy. Even for a specified type of area, available data on $p_{fi,1,s}/A$ may considerably differ. Table 1 shows probabilities that fire starts $p_{fi,1,s}/A$ as indicated in [6] [7] [8] for office areas.

Table 1. Probability that fire starts per year, $p_{fi,1,s}/A$, for offices as indicated in International Fire Engineering Design for Steel Structures 1993 [6], DIN 18230-1 [7] and BSI DD240 [8]

Reference	Germany	USA	UK
$p_{fi,1,s}/A$	0.5×10^{-6}	1 to 5×10^{-6}	12×10^{-6}

Thus, the following interval for $p_{fi,1,s}/A$ should be reasonably considered in case of office areas

$$0.5 \times 10^{-6} < p_{fi,1,s}/A < 12 \times 10^{-6} \tag{5}$$

The conditional probability $p_{fi,d}$ of fire flashover given a fire has started is indicated in the International Fire Engineering Design for Steel Structures 1993 [6], from where the following Table 2 is taken.

Table 2. Conditional probability $p_{fi,d}$ that a fire becomes fully developed given that it has started, in Germany[6]

Extinguishing system	Probability $p_{fi,d}$
Manual by personnel	0.05 –1
Fire brigade – two units	0.10
Fire brigade – three units	0.05
Fire brigade – four units	0.01
Water spraying system	0.02
CO_2 system	0.02

Similar values may be deduced from [8], from where the following Table 3 is taken. Conditional probabilities given in Table 3 are related to various degree of damage and not directly to fire flashover and, therefore, should not be directly interpreted as conditional probabilities $p_{fi,d}$. It follows from Table 3 that about 10% of fires in buildings cause severe structural damage and about one third of this number leads to destruction of the structure, which is in good agreement with the data given in Table 2. Therefore, using the data in Table 3, the value $p_{fi,d} = 0.1$ may be expected.

Table 3. Conditional probability of damage due to fire given that it started in UK [8]

Extent of structural damage	Conditional probability
No damage of building fabric	0.700
Slight damage to building fabric	0.120
Surface damage to building fabric	0.075
Severe damage to building fabric	0.075
Destruction of building fabric	0.030

Conditional probabilities given in Table 3 are related to various degrees of damage and not directly to fire flashover and, therefore, should not be directly interpreted as conditional probabilities $p_{fi,d}$. It follows from Table 3 that about 10% of fires in buildings cause severe structural damage and about one third of this number leads to destruction of the structure, which is in good agreement with the data given in Table 2. Therefore, using the data in Table 3, the value $p_{fi,d} = 0.1$ may be expected.

Depending on the extinguishing system (see Tables 2 and 3), it appears that the following range can be generally expected for $p_{fi,d}$.

$$0.01 < p_{fi,d} < 1 \tag{6}$$

It follows from Eqns. 4-6 that for offices, the probability of the fire flashover (fire design situation) per year $p_{fi,1}/A$ related to floor area may be generally expected within an interval

$$0.005 \times 10^{-6} < p_{fi,1}/A < 12 \times 10^{-6} \tag{7}$$

This extremely broad range having an upper bound of 1.2×10^{-5} (the lower bound is practically zero) should be expected when analysing reliability related to an accidental (fire) design situation of an office area.

2.3 An example of office area

As an example let us consider an office area compartment $A = 25$ m^2. For the design working life of 50 years (subscript 50 will be omitted) it follows from (5) that the probability of fire start is

$$0.625 \times 10^{-3} < p_{fi,s} < 15 \times 10^{-3} \tag{8}$$

The corresponding probability of fire flashover $p_{fi,50} = p_{fi}$ (subscript 50 will be omitted) follows from (7)

$$0.00625 \times 10^{-3} < p_{fi,s} < 15 \times 10^{-3} \tag{9}$$

In the example considered here ($A = 25$ m^2, for period 50 years) the value

$$p_{fi} = 1.0 \times 10^{-2} \tag{10}$$

may be expected as a reasonable upper bound for the probability of a fully developed fire in an office area of 25 m^2 with a design working life of 50 years.

Note that in accordance with DIN data [7] (see Table 1), the probability $p_{fi,50}$ for the design working life of 50 years and the compartment area A = 25 m^2 is given as

$$p_{fi} = 6.25 \times 10^{-4} \tag{11}$$

in accordance with BSI data [8] for office areas (see Table 1) and assuming approximately $p_{fi,1,d} = 0.1$ (deduced from Table 3 above) leads to a value

$$p_{fi} = 1.5 \times 10^{-3} \tag{12}$$

Obviously, there are considerable differences between the above values (Eqns. 10-12). In particular cases the appropriate probability $p_{fi,50}$ should be determined on the basis of a detailed analysis of the real conditions.

A possible approach for analysing fire protective system using Bayesian (causal) network indicated in Fig. 1 is described in the following.

3 Fire protection system

3.1 Bayesian network

A typical Bayesian network representing a fire protection system of an office building is indicated in Fig. 1. The network consists of ten chance nodes (oval shape), labelled by numbers 1 to 10, two auxiliary chance nodes 3^+, 4^+, one decision node 11 (square shape), and four utility nodes 12, 13, 14 and 15 (diamond shape).

Bayesian network supplemented by decision and utility nodes is often referred to as an influence diagram. The decision and utility nodes are implemented to enable risk assessment in terms of relevant costs.

The network can also be used to make decision on optimum fire safety system by minimising the total cost due to installation of sprinkler system (node 12), sprinklers and fire brigade operation (node 13), fire flashover (node 14) and structural failure (node 15). Description of all nodes (chance, decision, and utility nodes) is summarised in the following sections.

At present all the nodes considered in Fig. 1 are described by alternative random variables (having two states only). The input data consist of conditional probabilities of node states related to appropriate states of parent nodes (if there are any). These conditional probabilities can be assessed using a combination of theoretical insight [12] [13] [14], empirical studies, and various more or less subjective estimates. However, main advantage of the network is the clear definition of all input data, which may be therefore adjusted on the basis of new information, and a very good insight into the network analysis [15].

All nodes (chance, decision and utility nodes) used in the network are briefly described below. However, not all input data (conditional probabilities) necessary to analyse the Bayesian network shown in Fig.1 were possible to include in the following text.

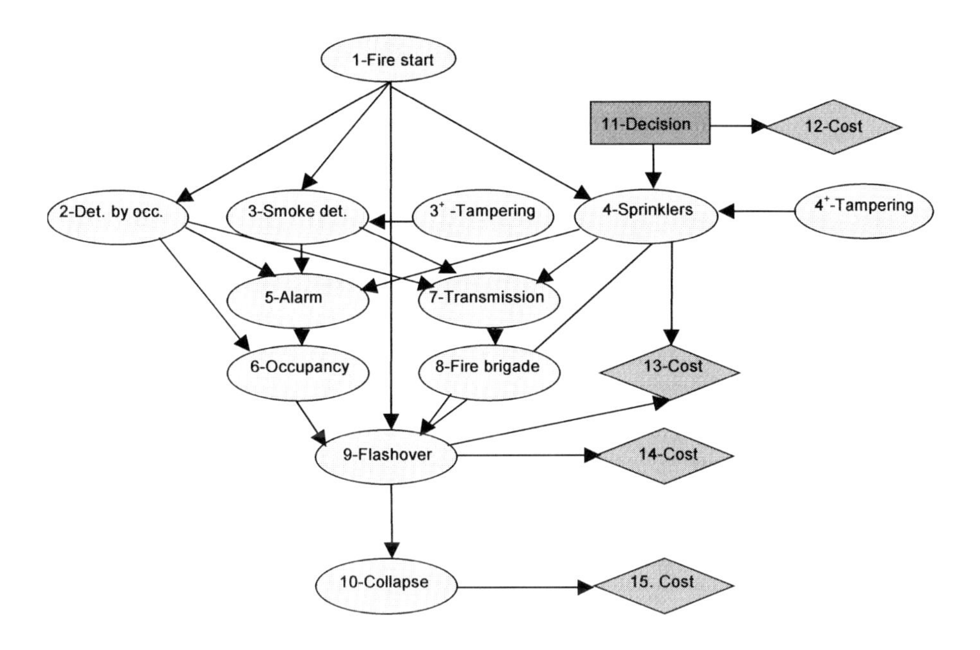

Fig. 1. Bayesian network used for fire risk analysis and assessment.

3.2 Chance nodes

The network in Fig.1 contains a total of 12 chance nodes (oval shape) numbered 1 to 10 and 3^+, 4^+ (two auxiliary chance nodes).

1 – Fire start: Initiation of a fire (probability of fire starts is dependent on the character of fire compartment and assumed design life. In the example considered below, the probability 0.01 (see Eqn. 9) is assumed for an office area 25 m^2, and the period of 50 years. This node has no parent nodes.

2 – Detection by occupancy: Discovery of a smoke by occupancy or neighbourhood within a suitable time period (conditional probability given fire starts may be high). Parent node 1.

3 – Smoke detection: Automatic smoke detection system (conditional probability given fire starts and tampering may be very high). Parent nodes 1, 3^+.

3^+ – Tampering: Interference of random factors with automatic smoke detection system described by node 3 (probability of unfavourable effect may be very low depending on maintenance of the system). No parent nodes.

4 – Sprinklers: Activation of the automatic sprinkler system if installed (conditional probability of activation given fire starts and tampering may be very high). Parent nodes 1, 4^+.

4^+ – Tampering: Interference of random factors with automatic sprinkler system described by node 4 (probability of unfavourable effect may be very low depending on maintenance of the system). No parent nodes.

5 – Alarm: Acoustic fire alarm system (conditional probability given positive state of nodes 2, 3 and 4 may be very high depending on maintenance of the system). Parent nodes 2, 3, 4.

6 – Occupancy: Activity or occupancy of the building to diminish the fire (conditional probability given positive states of nodes 2 and 5 may be low or moderate). Parent nodes 2, 5.

7 – Transmission: Functioning of manual or automatic alarm transmission to fire brigades (conditional probability given positive state of nodes 2, 3 and 4 may be very high depending on maintenance of the system). Parent nodes 2, 3, 4.

8 – Fire brigade: Operation of a professional fire brigade (conditional probability given positive state of the node 7 may be high depending on local conditions). Parent node 7.

9 – Flashover: Development of the fire (conditional probability given positive state of nodes 4, 6 and 8 may be very low depending on maintenance of the system). Parent nodes 4, 6, 8.

10 – Collapse: Structural failure (conditional probability given fire flash over may be relatively high, for example 0,01 [11]. Parent node 9.

3.3 Decision node

The network in Fig.1 contains only one decision node (of square shape) describing active measure (decision on installation of sprinklers) in design stage. This node makes it possible to compare fire protection system with and without sprinklers.

11 – Decision: Resolution in design stage concerning installation of the sprinkler system. If the state is "yes", sprinklers (node 4) are installed, if the state is "no", sprinklers are not installed. No parent node.

3.4 Utility nodes

The network in Fig.1 contains four utility nodes numbered 12 to 15 (nodes of diamond shape) describing costs related to the states of decision and relevant chance nodes.

12 – Cost: Node describing cost C_{12} of sprinkler system installation, which depends on the state of the parent node 11 (it is zero when no sprinklers are installed).

13 – Cost: Node describing damage cost C_{13} caused by sprinklers (parent node 4) and fire brigade (parent node 8) if the fire (parent node 9) will not flashover.

14 – Cost: Node describing damage cost C_{14} assuming that fire flashover (parent node 9) occurred

15 – Cost: Node describing damage cost C_{15} due to structural failure related to the states of the parent node 10.

4 Assessment of structural risk

4.1 The total cost

The total expected cost C_{tot} may be written as a sum

$$C_{tot} = C_{12} + C_{13} + p_{fi}C_{14} + p_f C_{15} \tag{13}$$

In this equation C_{12} is the installation cost depending on the state of node 11, C_{13} is the damage costs depending on the states of nodes 4, 8 and 9 (as described above). Further p_{fi} C_{14} is expected cost due to fire flashover, where p_{fi} is probability of fire flashover and C_{14} is the damage costs given the fire flashover (node 9 is in the state "yes"). Finally, $p_f C_{15}$ is expected cost due to structural failure (collapse), where p_f is the resulting probability of structural failure (which must be determined using the network) and C_{15} is the damage costs given the structure fails (node 10 is in the positive state "yes").

4.2 Effect of sprinklers

The decision node 11 and four utility nodes 12, 13, 14 and 15 supplement the Bayesian network in Fig. 1. This influence diagram enables to study efficiency of sprinkler installation considering expected total cost given by Eqn. 1. However, assessment of input costs described by utility nodes (as a rule in relative currency units) is needed.

Fig. 2 shows an example (analysed in detail in the previous study [12]) of the total expected cost C_{tot} versus the cost of structural failure C_{15} assuming the decision node 11 is in positive state of "yes" with sprinklers as well as in the negative state "no" without sprinklers. Both costs C_{tot} and C_{15} are expressed in a relative currency unit.

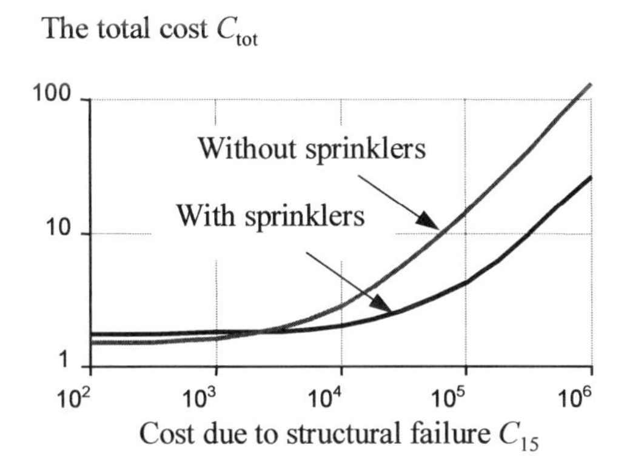

Fig. 2. The total expected cost C_{tot}.

It follows from Fig. 2 that for the cost of structural failure $C_{15} < 10^4$ of currency units, the total expected cost C_{tot} is about 2 units only, and almost independent of the cost C_{15} and the decision concerning installation of sprinklers. For the cost of structural failure $C_{15} > 10^4$ of currency units, the total expected cost C_{tot} is rapidly increasing, and considerably greater for the system without sprinklers than for the system with sprinklers.

Thus, the cost C_{15} due of structural failure seems to be an important quantity affecting economical efficiency of the fire protection system. Obviously other technical (evacuation time) as well as political and social aspects should also be considered.

4 Conclusions

Submitted study indicates that available data on fire starts and fire flashover have considerable scatter and more detailed and comprehensive studies are needed in order to introduce probabilistic concepts in structural design under fire design situation. It appears that Bayesian network may be used to analyse efficiency of fire protective system, and to find most effective arrangements in actual conditions.

The network supplemented with decision and utility nodes (influence diagram) can be also used for risk analysis and assessment. More information on input data (conditional probabilities and costs) is however needed.

The following most important conclusions may be drawn from the presented study in particular.

1. There are considerable, and worrying, differences in authoritative international documents when comparing available information on fire starts and fire flashover in office areas.
2. Bayesian networks seem to provide very logical, well-defined and effective tools to analyse probability of fire flashover under various fire protective conditions.
3. Considering available international data, the probability that fire starts in office area of 25 m^2 within the period of 50 years seems to be less than 0.015.
4. Analysis of relevant Bayesian networks based on available input data indicates that the conditional probability of fire flashover, given fire starts, is greater than 0.015 (0.015 with sprinklers, 0.12 without sprinklers).
5. Acceptable target probability of structural failure at the fire design situation adequate to large fire compartment areas may be rather low (for example less than 0.001).
6. To decrease probability of fire flashover and increase appropriate target probability of structural failure under fire design situation it may be necessary to rearrange protection system. For this purpose, Bayesian network seems to be a very effective tool.
7. Bayesian network supplemented by decision and utility nodes (influence diagram) enables to minimise the expected total costs due to fire flashover.
8. Further comprehensive studies to analyse effects of various fire protection measures on the probabilities of fire occurrence and fire flashover are needed.

5 Acknowledgement
This research has been partly conducted at the Klokner Institute of the Czech Technical University in Prague, Czech Republic, as a part of the research project CEZ: J04/98/210000029 "Risk Engineering and Reliability of Technical Systems".

6 References

1. European Communities. (1988) *The Construction Products Directive*. EC, Brussels. Directive 89/106/EEC.

2. International Standard Organisation. (1997) *General Principles on Reliability for Structures*. ISO, Zurich. ISO 2394.
3. International Standard Organisation. (1995) *Fire resistance - General requirements*. ISO, Zurich. ISO 834.
4. European Committee for Standardisation. (1994) *Basis of Design and Action on Structures – Part1: Basis of Design*. CEN, Brussels. ENV1991-1.
5. International Council for Building Research, Studies and Documentation. (1993) *Action on structures - Fire*. CIB, Rotterdam. CIB Report, Publication 166.
6. International Iron and Steel Institute. (1993) *International Fire Engineering Design for Steel Structures. State of the Art*. IISI, Brussels. 1993.
7. Deutsche Industrie Norm. (1996) *Bualicher Brandschutz im Industriebau, Teil 1: Rechnerischerforderliche Feuerwiderstandsdauer*. DIN, Berlin. DIN 18230-1.
8. British Standard Institution. (1997) *Fire Safety Engineering in Building. Part 1: Guide to the application of fire safety engineering principles*. BSI, London. DD 240.
9. ECCS (1996) Working Document No. 94. *Background document to Eurocode 1 (ENV1991-2-2)*. ECCS, Luxembourg.
10. ECSC Semestrial Reports No. 1 to 8 (1999) *Natural fire safety concept for buildings*. CEC Agreement 7210-SA/522. ECSC, Luxembourg.
11. Ang A.H-S. and Tang W.H. (1975) *Probabilistic Concepts in Engineering Planning and Design,* John Wiley, London.
12. Gulvanessian H., Holický M., Cajot L.-G. and Shleich J.-B. (1999) *Probabilistic analysis of fire safety using Bayesian causal network*. In: ESREL99, Munich.
13. Holický M. and Vorlíček M. (1999) *Risk assessment of fire safety using Bayesian causal networks*. In: First international conference on advanced engineering design, Prague.
14. Gulvanessian H., Holický M., Cajot L.-G. and Shleich J.- B. (1999) *Reliability Analysis of a Steel Beam under Fire Design Situation*. In: EUROSTEEL 99, Prague.
15. Hugin Expert A/S. (1998) *Software Program HUGIN EXPLORER, version 5.2*. Aalborg.

INFLUENCE OF STRAIN GRADIENT ON THE DUCTILITY OF A RC COLUMN

Ductility of a Reinforced Concrete Column

F. AKINTAYO
Laboratory of Mechanics and Materials, Polytechnic School,
Aristotle University of Thessaloniki, Thessaloniki, Greece
P.G. PAPADOPOULOS
Division of Structural Engineering,
Aristotle University of Thessaloniki, Thessaloniki, Greece
E.C. AIFANTIS
Laboratory of Mechanics and Materials, Polytechnic School,
Aristotle University of Thessaloniki, Thessaloniki, Greece
Center of Mechanics of Materials,
Michigan Technological University, Houghton, USA

Abstract

Part of a reinforced concrete column between two consecutive sets of transverse reinforcement bars, is simulated by a plane truss model with bars obeying nonlinear uniaxial stress-strain laws of concrete or steel. The equilibrium and stiffness conditions are written with respect to the deformed structure, in order to account for structural instability phenomena. First, this part of the column is subject to a gradual monotonic compressive axial loading, and then a gradual monotonic compressive deformation, which produces strain gradient across the column section. In both cases, the loading continues until the appearance of a global structural instability of concrete. In the second case, a more ductile behavior of the RC column is observed, that is, the global instability is delayed and finally occurs for a larger value of peak compressive strain of concrete. This is found in satisfactory approximation with recently published experimental results. However, in the case of cyclic loading, the RC column does not exhibit a ductile behavior, which is also in agreement with recent experimental evidence.
Keywords: Buckling, cracking, ductility, strain gradients, structural instability, truss model.

1 Introduction

In a reinforced concrete (RC) building with the usual case of vertical loads only, the columns are mainly subject to compressive axial loads and do not exhibit significant shear and flexural behavior. However, under an unusual loading, e.g. a strong earthquake, significant shear and flexural behavior appear in the columns of RC buildings, which produce large strain gradients across the column sections.

Abnormal Loading on Structures edited by K. S. Virdi, R. S. Matthews, J. L. Clarke and F. K. Garas.
Published in 2000 by E & FN Spon, 11 New Fetter Lane, London EC4P 4EE, UK. ISBN 0 419 25960 0

Recently, the influence of a significant strain gradient on the ductility of RC columns has been experimentally investigated. It was observed by Scott et al [1], Sheikh and Yeh [2], and Sakai and Sheikh [3], that in the case of monotonic loading, the strain gradient increases the ductility of a reinforced concrete column, that is, the failure occurs for a larger value of peak compressive strain of concrete. However, in the case of cyclic loading, it has been observed by Watson et al [4], and Watson and Park [5], that the strain gradient across such a column section is not favourable.

In the present work, this influence of strain gradient on the ductility of a RC column is studied computationally using the plane truss model of Akintayo et al [6], and Papadopoulos and Xenidis [7], which simulates a part of the RC column between two consecutive sets of transverse reinforcement. The bars of the truss model obey nonlinear uniaxial stress-strain laws σ-ε of concrete or steel. The equilibrium and stiffness conditions are written with respect to the deformed structure, in order to account for structural instability phenomena, local as well as global ones. As such, the influence of strain gradient across a RC column section on the ductility of the column is interpreted in terms of structural instability of columns.

2 Proposed truss model

2.1 Description of the model

A part of a reinforced concrete column is considered (Fig. 1a), with rectangular section b×d, without concrete cover. Distributed around the perimeter of the column section are longitudinal reinforcing bars, while transverse reinforcing bars run parallel to each principal direction of the column section. The column is subject to a concentric vertical load N, a shear force V and a uniaxial bending moment M, with respect to the principal axes of the column section.

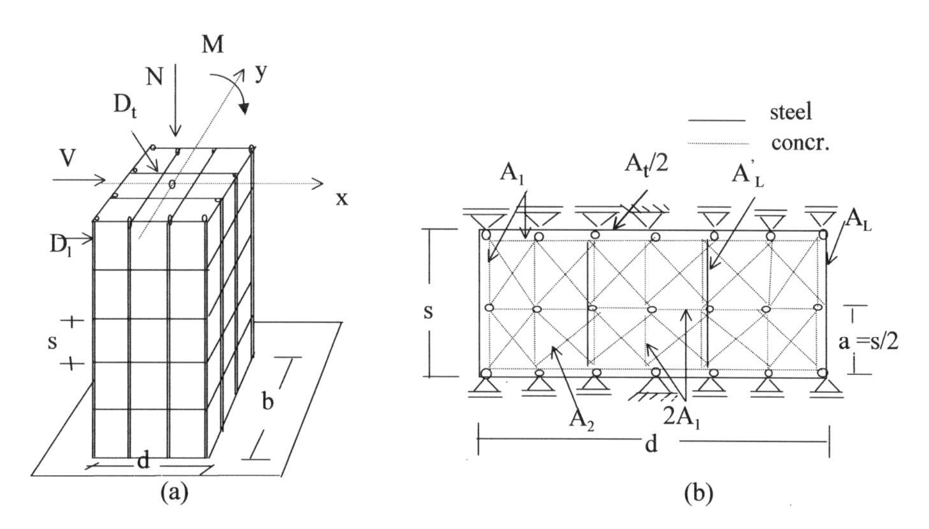

Fig.1 (a) Reinforced concrete column under simulation. (b) The proposed truss model (broken lines represent concrete bars and thick lines steel).

A part of the column shown in Fig.1a, between two successive sets of transverse reinforcement, is simulated by a plane truss with concrete and steel bars. The whole truss model consists of equal elementary square trusses, as shown in Fig. 1b.

In Fig. 1b, the vertical steel bars of the model (longitudinal reinforcement of the column) are not connected with the concrete bars of the model at the intermediate nodes, located at mid-height between the two successive sets of transverse reinforcement. This is mainly due to spalling of external concrete by arching action. The connection of vertical steel reinforcing bars with the concrete bars of the model is reasonably realised only at the levels of transverse reinforcement.

2.2 Determination of the bar sections of the model

The sections of the vertical and horizontal steel bars of the truss model of Fig. 1b are determined, in a simple and obvious way as the sum of sections of the corresponding real longitudinal and transverse reinforcing bars of the RC column under simulation, as explained in Fig. 1.

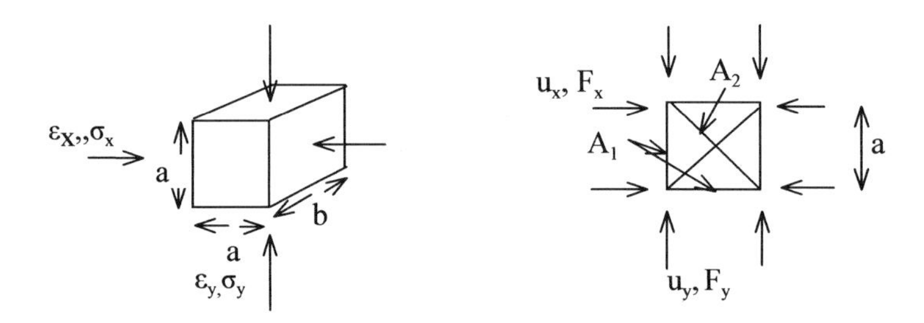

Fig. 2. (a) Square concrete prism under simulation (b) Elementary square truss model of concrete only.

The problem of determining the concrete bar sections of the model remains, which reduces to the simulation of a square prism of concrete under biaxial stress state (Fig. 2a) by an elementary square truss (Fig. 2b).

By combining the stress-strain relations of the concrete square prism of Fig. 2a, in the initial linear elastic isotropic state, with the force-displacement relations of the elementary square truss of Fig. 2b, the concrete bar sections, of the elementary truss model of Fig. 2b, are obtained with respect to lateral section ab (where a = s/2 and s the spacing of transverse reinforcement) of concrete prism of Fig. 2a :

$$A_1 = \frac{5}{12}ab \quad ; \quad A_2 = \frac{5}{12\sqrt{2}}ab. \tag{1}$$

2.3 Stress-strain laws of the bars of the truss model

The bars of the proposed truss model obey non-linear uniaxial stress-strain laws of concrete and steel. The σ-ε law of the concrete bars (Fig. 3a), includes tensile cracking,

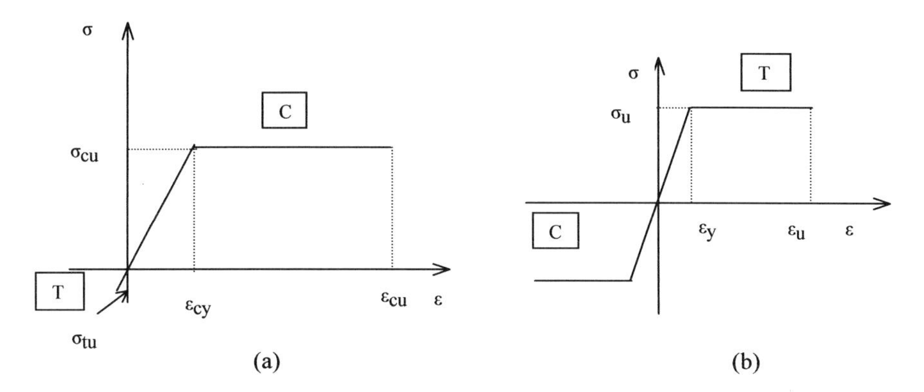

Fig. 3. Nonlinear uniaxial stress-strain curves of the bars of truss model (where C is compression and T is tension.): (a) concrete (b) steel.

compressive yielding and fracture, hardening, and represent the behavior of confined. Concrete, whereas the σ-ε laws of steel bars (Fig. 3b) include yielding and fracture in tension and compression.

2.4 Computer program for incremental loading of truss model

As mentioned in Section 1.1, the RC column under simulation is subject to axial loads N, shear forces V and bending moments M, which can follow various monotonic or cyclic loading histories. It is assumed that this loading is applied gradually by strain control. In order to describe this gradual loading of the RC column, the truss model is subject to corresponding incremental prescribed displacements of the supports.

For the nonlinear static analysis of the truss model of these prescribed displacements of supports, a simple and short computer program has been developed which consists of six subroutines and a total of about 300 Fortran instructions.

Within each step of the incremental loading algorithm, the coordinates of the nodes are updated. So the equilibrium and stiffness conditions are written with respect to the deformed truss, in order to allow for structural instability phenomena to be described. As the problem of confinement of a RC column by transverse reinforcement is mainly a problem of structural instability, local instability due to spalling of concrete by arching action, as well as global instability due to buckling of internal vertical concrete struts, will be described in the next section.

3 Application

The proposed truss model of a RC column and the computer program for its incremental loading are applied on a typical RC column of which a part is shown in Fig. 4. The section of the column is square with side 40 cm and no concrete cover is considered for the sake of simplicity. Twelve longitudinal reinforcing bars of diameter

D_l = 20 mm are distributed around the perimeter of the section. There are four transverse reinforcing bars of diameter D_t = 10 mm parallel to each principal direction.

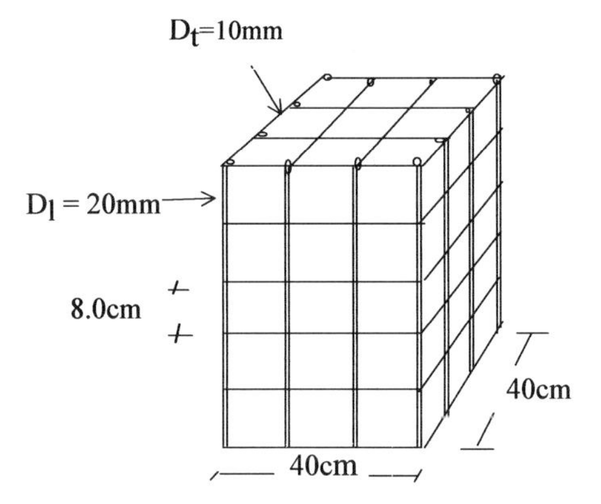

Fig. 4. Input data of the application: The geometry of the RC column

The spacing between adjacent sets of transverse reinforcement is s = 8.0 cm. The compressive strength of unconfined concrete is σ_{co} = 30 Mpa. From the given data (of confinement) of the RC column under consideration and by use of the confinement model of Mander et al [8], the stress-strain curve of confined concrete is determined and subsequently approximated by a simplified elastoplastic σ-ε curve as shown in Fig. 5a. As ultimate compressive strain, the strain corresponding to a 20% strength drop is assumed. For both longitudinal and transverse reinforcement, the simplified elastoplastic stress-strain law of Fig. 5b is considered.

As mentioned previously, a part of the column between two consecutive sets of transverse reinforcement, is simulated by the plane truss of Fig. 6. The sections of concrete and steel bars, determined as described in Section 2.2, are noted in Fig 6. Also shown is that, the upper and lower nodes of the truss are provided with the proper supports which allow the lateral expansion of the column.

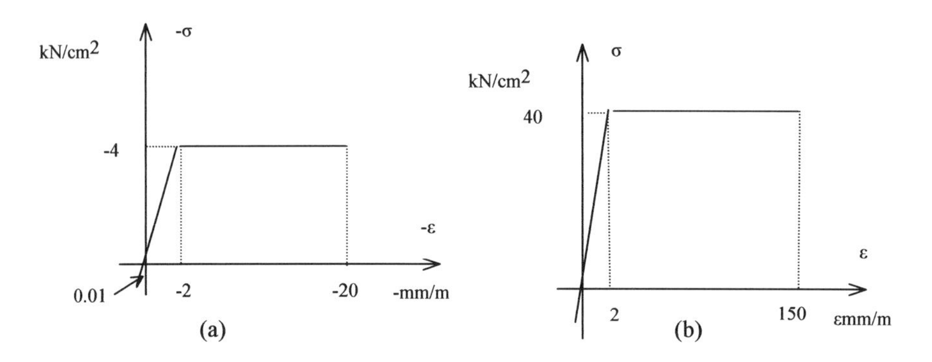

Fig.5. Stress-strain curves (a) concrete (b) steel

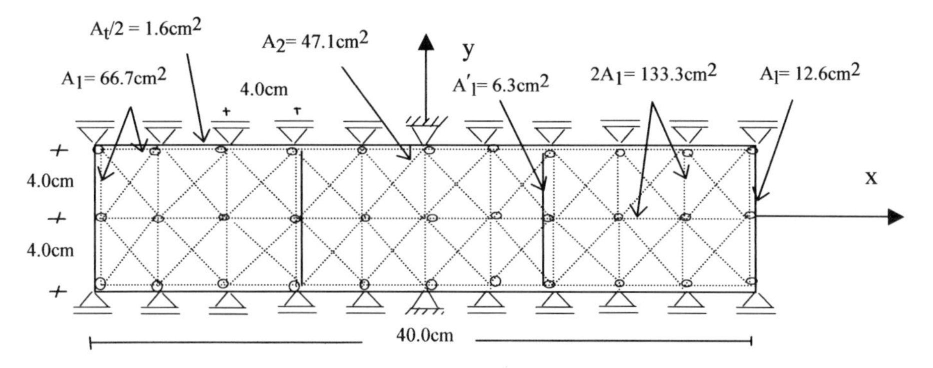

Fig. 6. Truss model for application (broken lines represent concrete bars and thick lines steel).

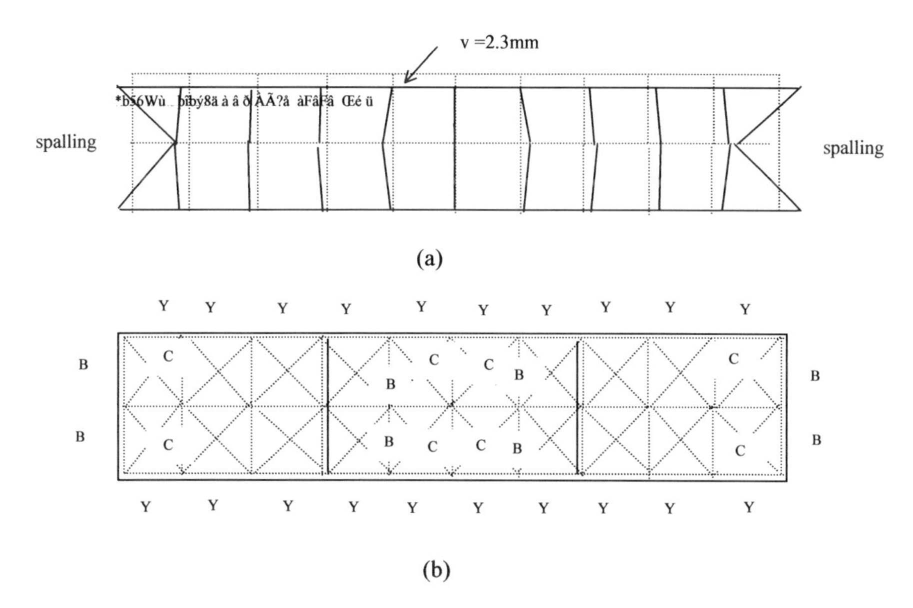

Fig. 7. Results of the application: Concentric axial loading failure mode. (a) Deformed configuration (b) Diagram showing steel bars in tensile yield noted by Y, concrete bars are in tensile cracking noted C (all the horizontal concrete bars are in tensile cracking), and concrete bars tending to buckle noted by B.

By using the computer program mentioned in Section 2.4, the truss model is first subjected to a gradual monotonic compressive axial loading and then to a gradual monotonic compressive loading produced by clockwise rotation of the upper section of column around its left edge.

In the first loading case, the common prescribed displacement increment downwards of the upper support-nodes of the truss is Δu = 0.012 mm and, in the second loading case, the rotation increment of the upper horizontal line of truss is $\Delta\varphi$ =

0.012 mrad. In the first loading case at the left and right sides of the truss, and in the second loading case at the right side only, for a rather small vertical compressive strain $\varepsilon \approx 0.004$, a local instability mode is observed, which corresponds to the spalling of the outer part of the confined concrete core, the so called arching action.

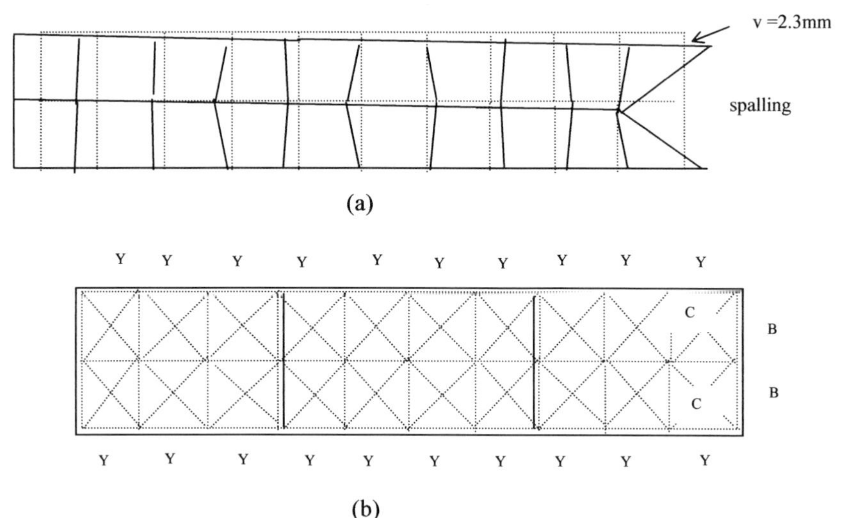

Fig. 8. Results of the application: Loading with strain gradient. State before failure. (a) Deformed configuration (b) Diagram showing steel bars in tensile yield (Y), concrete bars tending to buckle (B) and all the horizontal concrete bars are in tensile cracking.

As shown in Fig. 7, at each edge of the column, the intermediate horizontal concrete bar and two external diagonal concrete bars are cracked in tension thus leaving free two external concrete struts to buckle.

In the first loading, for total displacement downwards of the upper supports v = 2.3 mm, that is for a uniform axial compressive strain of the column $\varepsilon \approx 0.0276$, a global structural instability mode of the RC column is revealed. Because of the heavy compressive axial load and the subsequent lateral expansion of the column, the transverse reinforcement is in tensile plastic yield, and an extensive cracking of horizontal and diagonal concrete bars occurs. So, internal vertical concrete struts are formed. For the formation of these struts, in an axially compressed concrete column, there exists recent experimental evidence by Jansen & Shah [9].

These internal vertical concrete struts carry heavy axial compressive loads and as they have a loose lateral support, and because of concrete cracking, they tend to buckle by pushing the whole part of the column beyond them, thus producing a global structural instability of the RC column.

The state of the truss model when the above global instability mode occurs is represented in Fig 7. The deformed configuration of the structure is drawn in Fig 7a, through the deformed elementary squares of the model in comparison with the undeformed ones. In Fig 7b, the transverse steel bars, which are in tensile yield and the concrete bars, which are cracked in tension, are noted. In addition, Fig. 7b also shows the internal vertical concrete struts that tend to buckle. So, in this figure, the global instability mode of the RC column is clearly described.

In the second loading case, where a significant strain gradient across the column section is produced, it is observed that the column does not fail for a total displacement downwards of the right upper node of v = 2.3 mm, that is, for a peak compressive concrete strain $\varepsilon = 0.0275$, a value for which the first loading case failed. As explained in Fig. 8, in the most compressed region of the column, the strain gradient prevents the diagonal concrete bars from exhibiting an early tensile cracking. These, in turn prevent the internal vertical concrete struts from exhibiting a premature buckling. Thus, the global structural instability of the RC column is delayed.

For a quite larg peak compressive concrete strain $\varepsilon = 0.0513$, corresponding to a total displacement v = 4.1mm downwards of the extreme right upper node of the truss, the second loading with strain gradient, finally fails. From the deformed configuration of the structure shown in Fig. 9a, and from Fig. 9b (diagram of steel yield, concrete crack and buckling), it is observed the lateral expansion is larger because of the quite heavier loading so more diagonal concrete bars crack thus, initiating the buckling of some vertical concrete struts and the subsequent global instability of the RC column.

The comparison between Figs. 7 and 9 shows that in the case of a monotonic loading, the strain gradient over the RC section is favourable because it increases the ductility of the RC column, that is the failure occurs for larger peak concrete compressive strain.

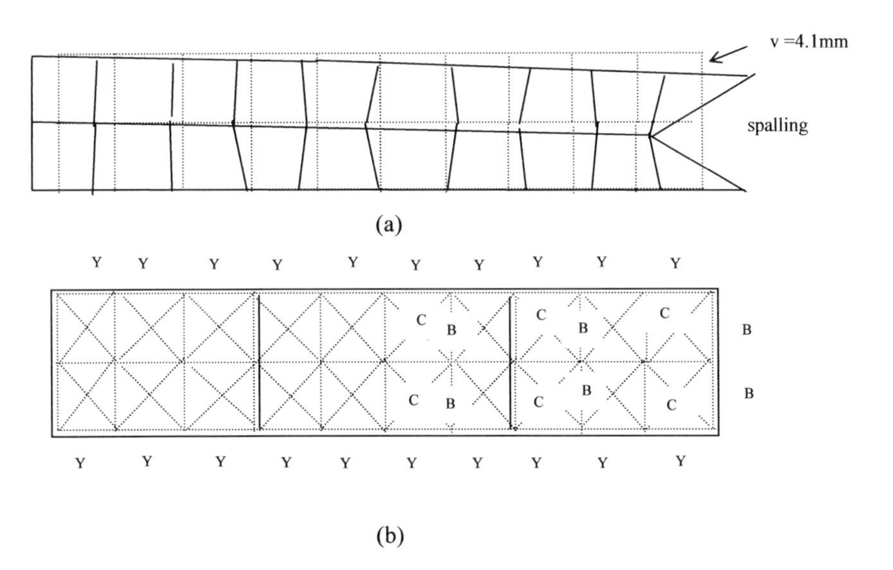

Fig. 9. Results of the application: Loading with strain gradient. State before failure. (a) Deformed configuration (b) Diagram showing steel bars in tensile yield Y, concrete bars in tensile cracking (C) concrete bars tending to buckle B and the rest of the concrete bars are in tensile cracking.

The above trends of the proposed truss model are found in satisfactory approximation with recent published experimental tests on RC column specimens with similar geometric, mechanical and loading characteristics by Scott et al [1]. However, in the case of cyclic loading, numerical examples by the truss model show that the strain gradient is no more favourable: After the first half loading cycle, because of the permanent plastic strains, the strain gradients do not any more prevent the early cracking of diagonal concrete bars and these, in turn do not prevent the premature buckling of internal vertical concrete struts and the subsequent global instability of the RC column. These trends are also in agreement with recently published experimental results of Watson & Park [5].

4. Conclusions

A part of a concrete column, between two consecutive sets of transverse reinforcement, was simulated by a plane truss with bars obeying nonlinear uniaxial stress-strain laws of concrete or steel. The equilibrium and stiffness conditions are written with respect to the deformed structure, so as to take into account structural instability phenomena.

A local instability mode of the RC column represented by the proposed truss model is the spalling of an outer part of the confined concrete core by the so-called arching action.

A global instability mode of the RC column is revealed by the proposed truss model, as a result of a heavy axial compression and the subsequent lateral expansion of the RC column, the transverse reinforcing bars are in tensile plastic yield and an extensive cracking of concrete occurs. Because of this cracking, vertical concrete struts are formed which carry heavy axial loads. These concrete struts tend to buckle, producing a global structural instability of the RC column. For the formation of these vertical concrete struts in an axially compressed concrete column, there exists recent experimental evidence by Jansen and Shah [9].

Numerical applications by the proposed truss model show that when a significant strain gradient appears across a reinforced concrete column section in case of monotonic loading, the diagonal concrete bars of the model in the most compressed side of the column are prevented from exhibiting early cracking. These, in turn prevent the vertical concrete struts to exhibit premature buckling. So, the global structural instability of the column is delayed and finally occurs for a larger value of the peak compressive strain of the reinforced section.

This observed ductile behavior of a reinforced concrete column, due to a significant strain gradient, in case of monotonic loading is found to be in satisfactory approximation with recently published experimental results.

On the other hand, in the case of cyclic loading, numerical applications by the truss model show that, after the first half loading cycle, the strain gradient no longer prevents the early cracking of the diagonal concrete bars in the most compressed region of the column. This is because of the permanent plastic strains present. So, these cracked diagonal bars in turn do not prevent the premature buckling of the vertical concrete struts and the subsequent global instability of the column. This observed nonductile behaviour of a reinforced concrete column in case of cyclic loading, even when a significant strain gradient exists, also agrees with recently published experimental results.

5 Acknowledgement

The financial support of A.U.T–Project code: 7845 is gratefully acknowledged.

6 References

1. Scott, B.D., Park, R. and Priestley, M.J.N (1982) *Stress-Strain Behaviour of Concrete Confined by Overlapping Hoops at Low and High Strain Rates*, ACI Journal, (Jan.-Feb.), pp.13-27.
2. Sheikh, S.A. and Yeh, C.C. (1986) *Flexural Behaviour of Confined Concrete Columns*, ACI Journal, (May-June), pp.386-404.
3. Sakai, K. and Sheikh, S.A (1989) *What Do We Know About Confinement in Reinforced Concrete Columns? (A Critical Review of Previous Work and Code Provisions)* ACI Structural Journal, (March-April), pp.192-207.
4. Watson, S., Zahn, F.A. and Park, R. (1994) *Confining Reinforcement for Concrete Columns*, Journal of Structural Engineering ASCE, June, pp.1798-1824.
5. Watson, S., and Park, R. (1994) *Simulation of Seismic Load tests on Reinforced Concrete Columns*, Journal of Structural Engineering ASCE, June, pp.1825-1849.
6. Akintayo F., Papadopoulos, P.G. and Aifantis, E.C. (1998) *Simulation of Uniaxial Compression of a Concrete Column by a Truss Model with Instability*, Proc. of 5th National Congress on Mechanics, Ioannina, Greece, Vol. 2, pp.899-906.
7. Papadopoulos, P.G. and Xenidis, H.C. (1998) *Amount of Confinement Preventing Global Instability of a Concrete Column*, Proc. 11th European Conference of Earthquake Engineering, Paris, Vol. 3.
8. Mander, J.B, Priestley, M.J.N and Park, R. (1988) *Theoretical Stress-Strain Model for Confined Concrete*, Journal of Structural Engineering ASCE, August, pp.1804-1826.
9. Jansen, D.C. and Shah, S.P. (1997) *Effect of Length on Compressive Strain-Softening of Concrete*, Journal of Engineering Mechanics ASCE, Vol. 123, January pp. 25-35.

HIGH VOLUMETRIC PRESSURES AND STRAIN RATE EFFECTS IN CONCRETE

Constitutive laws for concrete

J. EIBL, B.C. SCHMIDT-HURTIENNE and U.J. HÄUSSLER-COMBE
Institut für Massivbau und Baustofftechnologie, Universität Karlsruhe (TU), Germany

Abstract
Practical needs demand constitutive laws for concrete under extremely rapid loading. A report is given on research projects done at the University of Karlsruhe concerning extremely high pressures and the mostly unavoidable adiabatic heating at fast loading as it occurs at impact, blasting e.g. Also a new physical approach to model the strain rate effect is presented including a constitutive model and its first experimental verification.
Keywords: Blast loading, constitutive laws, dynamic loads, extreme volumetric compression, shockwaves, strain rate.

1 Introduction

There is a growing interest in understanding the effect due to blast loading, and of impact due to missiles, vehicles, ships, aeroplanes or avalanches upon concrete structures like bridges, offshore structures, tanks, chemical factories, power plants etc. In cases of so-called hard impact when striker and struck body both contribute to the contact force as well as in cases of applied contact charges volumetric pressures up to 20,000 MPa may be exerted at strain rates ranging up to $\dot{\varepsilon} > 100\,[1/s]$ and even more.

2 Research on high volumetric pressures

2.1 Static pressure

As the maximum endurable deviatoric stresses are orders of magnitude smaller than the volumetric ones, the latter were investigated first and are treated in the following.

Abnormal Loading on Structures edited by K. S. Virdi, R. S. Matthews, J. L. Clarke and F. K. Garas.
Published in 2000 by E & FN Spon, 11 New Fetter Lane, London EC4P 4EE, UK. ISBN 0 419 25960 0

We started applying static volumetric pressures up to 1,000 MPa to sealed concrete cylinders surrounded by a special mercury liquid within a thick steel tube. The liquid was stressed by the piston of a 15 MN hydraulic machine (see Fig. 1). The so gained relations are given in Fig. 2.

To reach even higher pressures in a short time we then started using shockwaves which were generated by contact charges.

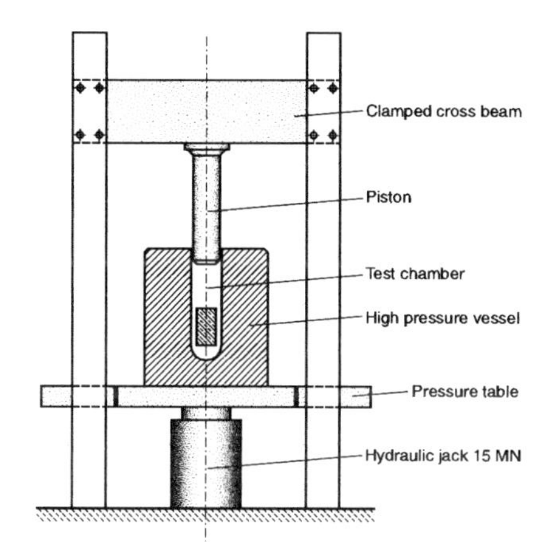

Fig. 1. Test setup

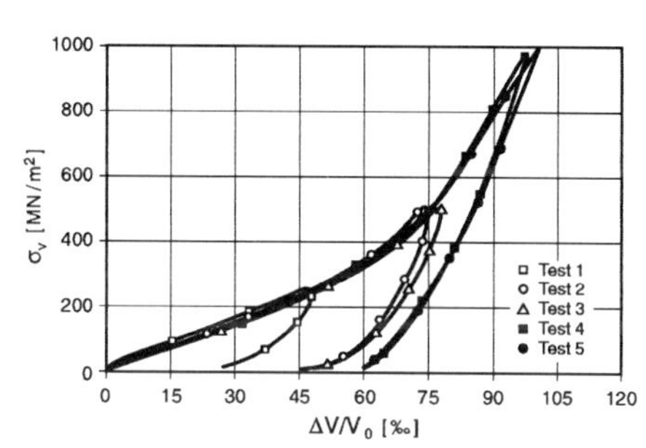

Fig. 2. Volumetric pressure versus vol. strain [1]

2.2 Dynamic pressure

The occurrence of such shockwaves may be explained as follows. The wave velocity resulting from the dynamic equilibrium condition – wave equation – is also valid for nonlinear material behaviour:

$$c = \sqrt{\frac{H}{\varrho}} \tag{1}$$

where H is the inclination of the volumetric stress-strain relation $H = d\sigma_V/d\varepsilon_V$ (Fig. 3).

When a wave of rising volumetric stress is initiated, the different sections of this constitutive relation (see Fig. 3) travel with different velocities according to (1), the part with the highest condensation overpassing all others, so that finally a discontinuity travels with supersonic speed through the material.

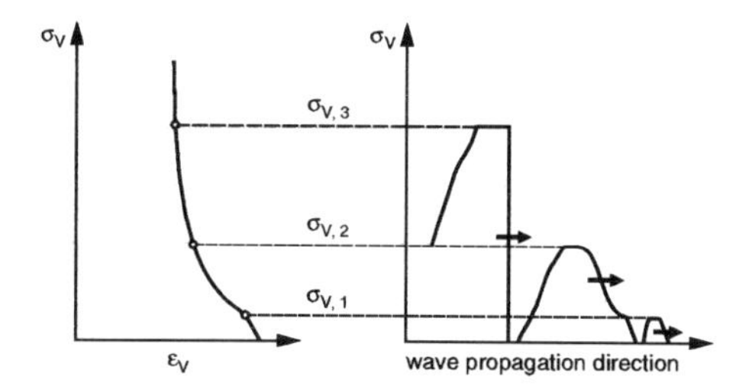

Fig. 3. One-dimensional shockwave

At discontinuity front such as this, the laws of conservation for mass, momentum and energy, originally formulated for an ideal gas, must also hold (for details see Ockert [2]). They must be supplemented by the so called Hugoniot experiments which give the adiabatic relations $p = f(V/V_0)$ at different levels of internal energy (see Figs. 4 and 9). In order to find finally the isotherm behaviour of concrete one has to eliminate this temperature effect which is different at any stage of loading.

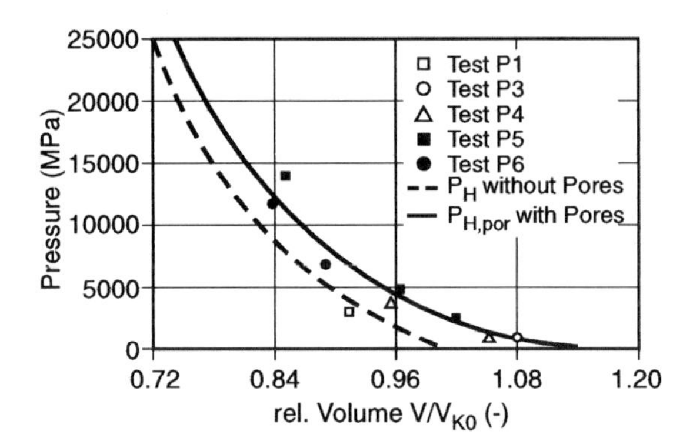

Fig. 4. Calculated Hugoniot curves versus experimental results (First series)

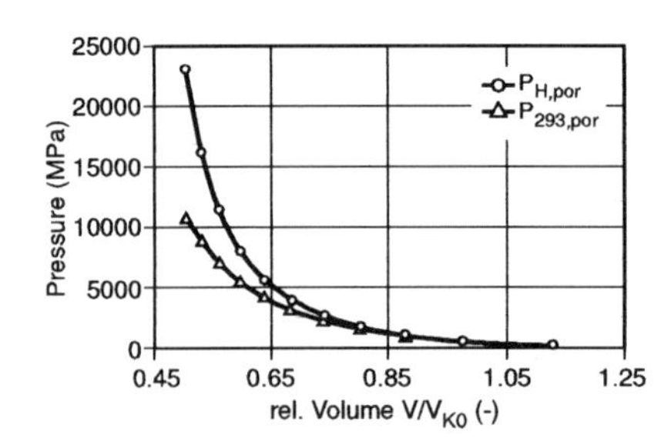

Fig. 5. 293^0 K-isotherm and porous Hugoniot curve (calculated).

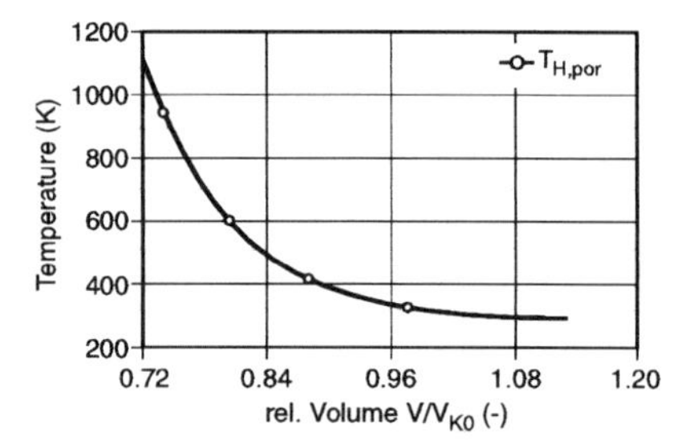

Fig. 6. Temperature for porous Hugoniot curve (calculated).

2.3 Experiments – Hugoniot-relations

To find such Hugoniot curves, concrete plates up to 150 × 150 × 50 cm with a concrete strength f_c of about 35 N/mm^2 (see Fig. 7) have been loaded by explosives at their midpoints perpendicular to the upper surfaces. A pyramidal shape of the contact charge (TNT) was chosen to generate plane stress waves.

Directly below the center of the charge Allen-Bradley pressure gauges of 470 Ohm-coal-mass-resistors were installed one above the other. Their resistance change was registered by fast transient recorders allowing thus to measure pressure-time curves. The shockwave velocity was determined by means of the individual arrival times of the pressure signals at the different gauges and their relevant distances. So different points of the Hugoniot curve could be found at every test (Fig. 8).

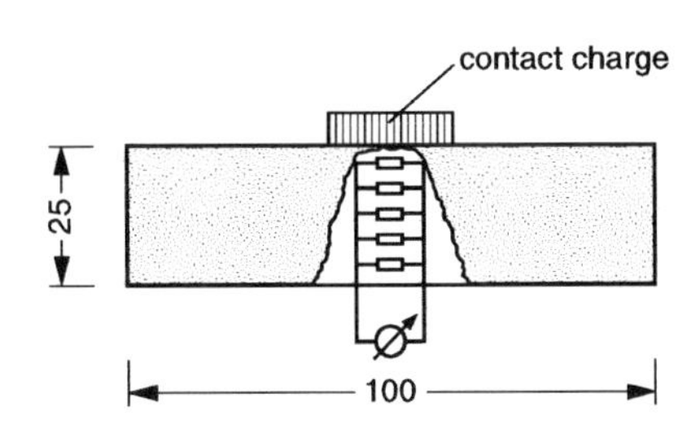

Fig. 7. Experimental setup

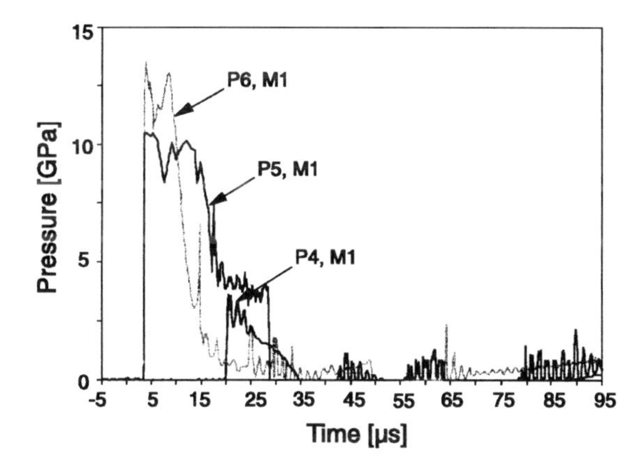

Fig. 8. Typical pressure time history in test

In all cases a very fast pressure decay with a maximum of about 24 GPa near the upper surface could be monitored. Maximum shockwave velocities of more than 5,000 m/s were registered which is much more than the subsonic wave velocity of concrete. In the meantime further experiments were performed which gave similar results. All tests are collected in Fig. 9 and are in good agreement with those of Gregson [3] and Grady [4]. They had already measured pressures up to 52 GPa, however, at very small test samples of 0.6 cm to 1.5 cm thickness.

Starting from the measured Hugoniot relation (Figs. 4 and 9) with and without pores the internal energy, the pressure isotherm for 293^0 K, the temperature and the bulk modulus versus V/V_0 (see Figs. 5 and 6) could be recalculated for control purposes. The temperature curve shows that the concrete specimen in our test experienced a heating of about 250^0 C caused by the adiabatic compression.

Further experiments to measure the calculated adiabatic heat are in progress at the institute as well as to study the influence of deviatoric stresses, although the latter is of secondary importance.

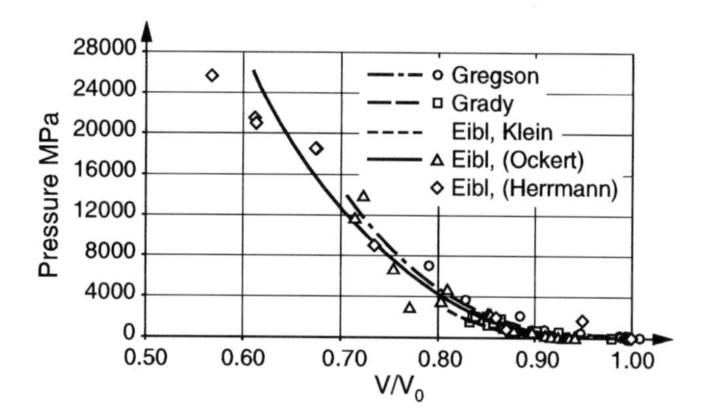

Fig. 9. Hugoniot curve for concrete - first and second series in comparison with other authors

3 Strain rate effects

3.1 Physical background

In the past, strain rate dependency of materials like concrete used to be taken into account by defining a stress-strain relation $\sigma = f(\varepsilon)\big|_{\dot{\varepsilon}=const.}$. Therefore, experimental efforts were focused on finding the relations $\sigma = f(\varepsilon)$ at different constant strain rates $\dot{\varepsilon}$. Dynamic strength was defined as the maximum stress $f_{dyn}(\dot{\varepsilon}) = \sigma_{\max,\, \dot{\varepsilon}=const.}$.

However, it can easily be shown [6] that such an approach is insufficient for many design problems. To demonstrate this, a simple beam is considered to be loaded dynamically as shown in Fig. 10. Getting to the point of maximum deflection where $d\sigma/dt = 0$ and therefore $\dot{\varepsilon} \cong 0$, the beam is expected to resist only by its static strength according to the assumptions mentioned above. However in an experiment the beam shows a much higher dynamic bearing capacity. Obviously the material "knows" that there was a high strain rate just before reaching the maximum strain. The material has a "memory". A momentary zero strain rate does not always correspond to static strength.

Therefore, in a constitutive law for fast dynamic effects the full interdependence of strain and stress history $\varepsilon(t) = f[\sigma(t)]$ has to be considered. A new understanding of strain rate effects and dynamic strength is necessary.

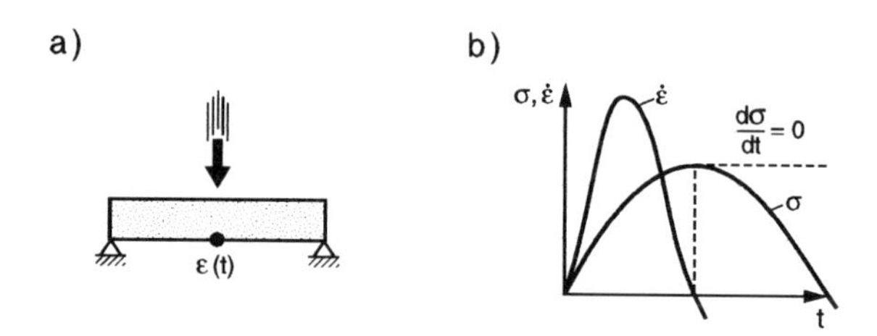

Fig. 10. Dynamically loaded concrete beam

Experiments by Bischoff et al. [7] and Zheng et al. [8] at our institute demonstrate that specimens dynamically and statically loaded to the same compressive stress level of 26 MPa show a different inner damage accumulation. For example, in Fig. 11, the beginning of new acoustic emission (AE) in the concrete specimen, which had been statically preloaded, was observed in static reloading at about 26 MPa. However in a dynamically preloaded specimen the onset of AE occured much earlier at a lower stress level of about 21 MPa. This indicates that dynamic loading up to a certain stress level leads to less internal damage than static loading. This is in accordance with the observation that axial and lateral strains of dynamically loaded samples were smaller than those of specimens loaded statically to the same level.

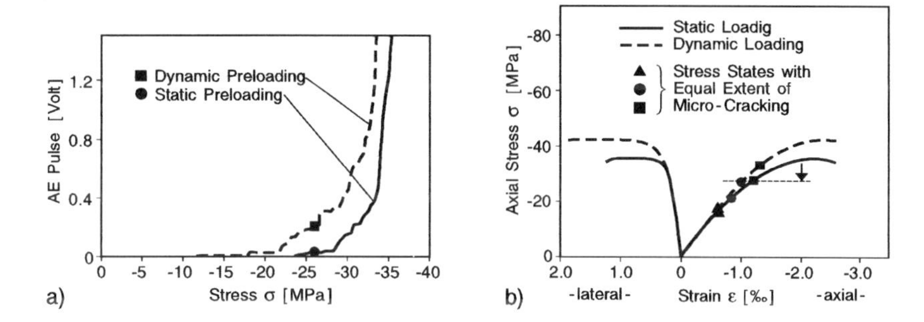

Fig. 11. Acoustic emission and strains versus stress [7]

There are strong indications that at lower strain rates $|\dot{\varepsilon}| < 10^1$ $[1/s]$ effects [10], a growing tendency to cross-aggregate cracking or moisture content may control the observed momentary strength increase (see [9], [11]-[14]). However at high strain rates generally above $|\dot{\varepsilon}| \geq 10^1$ $[1/s]$, damage formation is more and more controlled by inertia effects. Every internal damage evolution leading to crack propagation needs a finite time.

3.2 A constitutive law – the mechanical model

To derive a constitutive relation considering the fore mentioned experimental research we started from a static damage model on microscale (Fig. 12), as was developed by Curbach [15] at the institute. It shows springs for the reversible part of the internal energy and friction elements which are to simulate the irreversible part. This model may be described as follows

$$\sigma(\varepsilon) = [1 - D_1(\varepsilon)] \cdot E \cdot \varepsilon + my_{Fr} \cdot D_1(\varepsilon) \cdot [1 - D_{2,Fr}(\varepsilon)] \qquad (2)$$

with the scalar damage function $D_1(\varepsilon)$ characterizing the heterogeneous strength of concrete and derived from it a different function $D_{2,Fr}(\varepsilon)$ to model the internal friction of the springs failed according to D_1. For both functions D_i Weibull distributions were chosen, as local extreme strength values are responsible for a concrete failure. The chosen functions are:

$$D_i = 1 - \exp\left[-\frac{1}{2}\left(\frac{\varepsilon/\varepsilon_0 - 1}{a_i}\right)^2\right] \qquad \varepsilon \geq \varepsilon_0 \,;\; i = 1,2;\; 0 \leq D_i \leq 1 \tag{3}$$

where a_1, a_2, ε_0 and μ_{Fr} are material parameters.

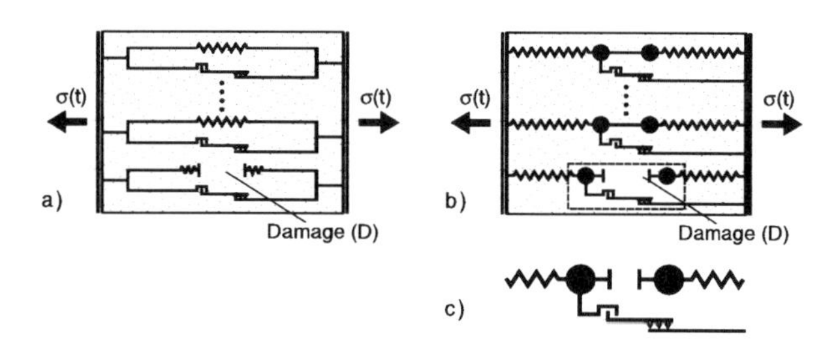

Fig. 12. Static and dynamic model including failed bi-masses

This model is now expanded for dynamic analysis purposes to include bi-mass elements (Fig. 12b) which in case of damage are to break and produce a time delay until the already spontaneous separation of the two masses becomes slowly active on macroscale. The whole system is self-equilibrated so that these masses do not contribute to the continuum approximation on macroscale.

This effect of failure delay may then be incorporated as follows

$$\sigma(t) = E \cdot \varepsilon(t) \cdot \left([1 - D(\varepsilon)] + \int_{\tau=0}^{\tau=t} \frac{\partial D}{\partial \varepsilon} \cdot \frac{\partial \varepsilon}{\partial \tau} \cdot g(t-\tau)\, d\tau \right) + \Gamma_{Fr} \tag{4}$$

where the friction term in Eqn. 2 is abbreviated as Γ_{Fr} and D_1 as D.

The integral term in Eqn. 4 expresses that every new damage increment ΔD needs a certain time $0 \leq g(t-\tau) \leq 0$ until its full equivalent internal resistance is lost. From details of Fig. 12c, function $g(t-\tau)$ may be derived in a rather consistent manner (for details see [10]) as follows:

$$g(t-\tau) = 1 - \int_{\theta=0}^{\theta=\xi} \omega \cdot e^{-\bar{c}(\xi-\theta)} \sin\omega\,(\xi-\theta)\, d\theta \quad . \tag{5}$$

where $\xi = (t-\tau)$ and ω and $\bar{c}$ are material parameter.

However, this model can only lead to an understanding of these phenomena in principle only, as the internal damage expansion is much more complicated. The loss of internal material resistance caused by a time dependant spreading of microcracks is a constitutive property and can only be determined experimentally within such a general framework. Therefore, it is also justified to use a simpler formulation with a similar characteristic if it can be verified sufficiently by tests.

Hopkinson bar experiments by Zheng [18] at our institute have shown that a simple approach with a decay function as

$$g(t-\tau) = 1-\frac{t-\tau}{t_0} \qquad for \quad t-\tau < t_0$$
$$= 0 \qquad for \quad t-\tau > t_0 \qquad (6)$$

gives sufficient results to simulate concrete behaviour.

In the meantime this rate approach using a dynamically retarded damage activation is also incorporated in a general three-dimensional constitutive law which will be published soon.

3.3 Experiments

In order to verify these constitutive laws (see also [17]) altogether 63 Split-Hopkinson-Bar (SHB) experiments were carried out for 12 different load histories in tension as well as in compression, many of them with an Extended Split Hopkinson Bar (ESHB) device (see Fig. 13 and [18] and [19]). The latter was developed as the comparatively longer concrete test specimens do not return satisfying results with the classical test setup.

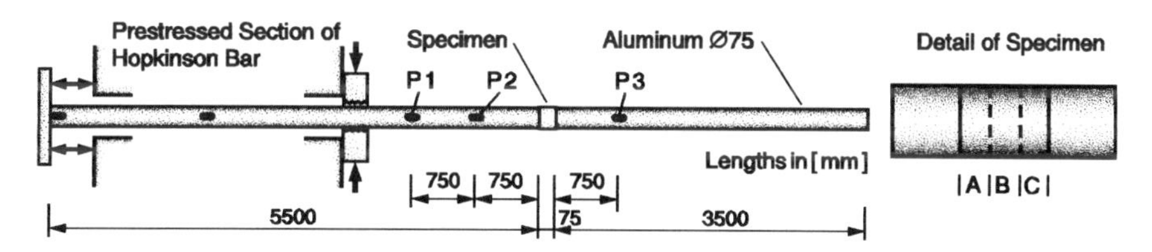

Fig. 13. Extended split Hopkinson bar; test setup

During these ESHB-tests, deformations and accelerations were measured along the specimen's length in three sections A, B and C, from which the internal stresses in A, B and C were evaluated by a new method described in [18].

Figs. 14a and 14b present a comparison of tests with numerical simulations using the one-dimensional formulation of Eqns. 4 and 6, and the values $a_1 = 18{,}9$; $a_2 = 189$; $E = 35{,}000$ N/mm^2; $\mu_{Fr} = 10$ N/mm^2; $\varepsilon_0 = 0.1$ ‰ and $t_0 = 75$ µs.

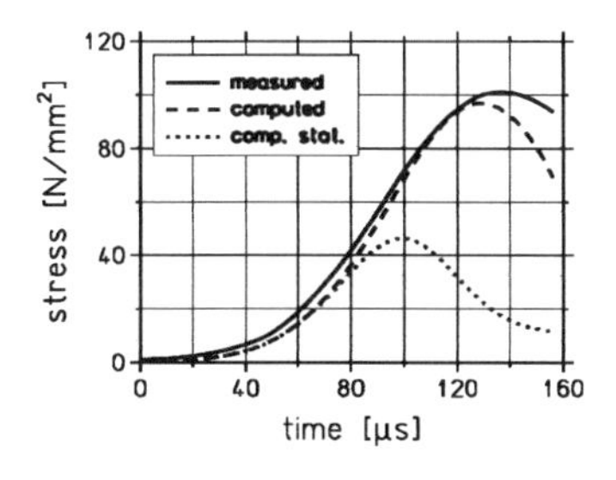

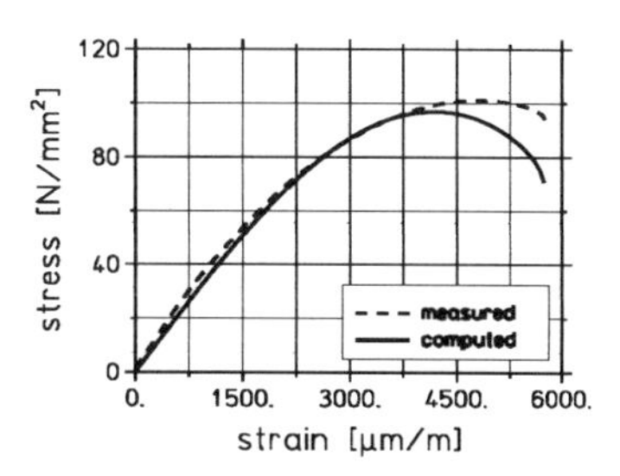

Fig. 14. One-dimensional calculation of compression test; a) stress versus time; b) stress versus strain

The computed stress history in Fig. 14a also shows the static response of the constitutive law in addition to the dynamic one to underline the dynamic strength increase due to compressive strain rates of about $\dot{\varepsilon} = 100\ 1/s$.

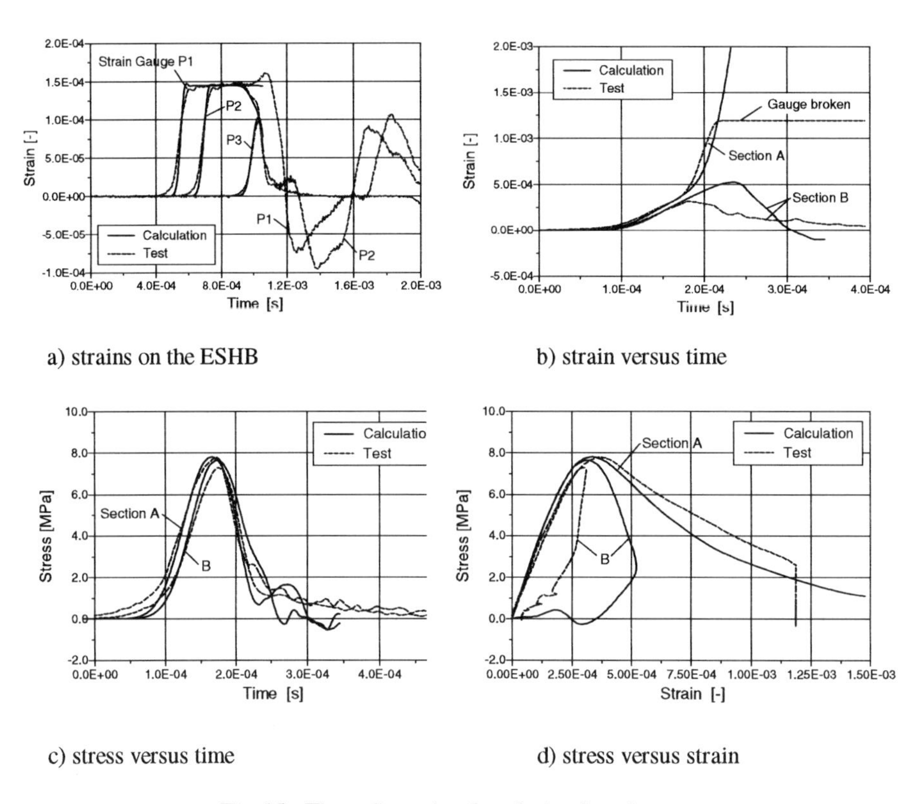

a) strains on the ESHB

b) strain versus time

c) stress versus time

d) stress versus strain

Fig. 15. Three-dimensional analysis of tension test

Figure 15a gives a comparison between an ESHB tensile test and a FE-simulation based on the three-dimensional constitutive formulation with strain data at the strain gauge positions P1, P2 and P3 (see Fig. 13). Figs. 15b, 15c and 15d give the strain and stress versus time curves and the stress strain curves of section A, where the incoming strain arrives first, and of section B in the middle of the specimen, which does not soften when the splitting occurs in section A. In all cases a rather good agreement between numerical simulation and tests was found.

4 Conclusion

A static test setup and shockwave experiments were used to determine isothermal volumetric stress strain relations by means of so called Hugoniot curves. A new concept is presented to explain the physical basis of strain rate effects, which at extremly high rates reduce the internal damage development due to inertia phenomena. A one-dimensional stress strain relation for concrete in compression as well as in tension including high strain rate effects has been developed, which in the meantime is extended to a three-dimensional version. The proposed constitutive law was successfully tested and verified for different load histories on one-dimensional split Hopkinson bar experiments with different load histories [16],[18].

5 References

1. Eibl, J. et al.: *Verhalten von Betonkonstruktionen bei harten Stößen. Teil 1, Betonverhalten bei hoher hydrodynamischer Beanspruchung.* Forschungsbericht 94/4 des Instituts für Massivbau und Baustofftechnologie der Universität Karlsruhe, 1994
2. Ockert, J.: *Ein Stoffgesetz für die Schockwellenausbreitung im Beton.* Dissertation und Heft 30 der Schriftenreihe des Instituts für Massivbau und Baustofftechnologie der Universität Karlsruhe, 1997
3. Gregson, V.R. Jr.: *A Shock Wave Study of Fondu-Fyre WA-1 and Concrete.* General Motors Materials and Structures Laboratory, Report MSL-70-30,1971
4. Grady D.E.: *Impact Compression Properties of Concrete.* Sixth Int. Symposium on Interaction of Nonnuclear Munitions with Structures, Panama City Beach, Florida, 1993, 172-175
5. Bischoff, P. H.: *Compressive Response of Concrete to Hard Impact.* Dept. of Civil. Eng., Imperial College of Science and Technology, Ph.D. Thesis, London, 1988
6. Eibl, J.: *Ein neuer Ansatz für ein Stoffgesetz zur Berücksichtigung großer Dehngeschwindigkeiten bei zugbeanspruchtem Beton.* Waubke-Festschrift, BMI 9/96, Schriftenreihe des Instituts für Baustofflehre und Materialprüfung der Universität Innsbruck, 1996
7. Bischoff, P. H., Bachmann, H., Eibl, J.: *Microcrack Development during High Strain Loading of Concrete in Compression.* Proc. of the European Conference on Structural Dynamics, Eurodyn'90, Bochum, A.A. Balkema Rotterdam Brookfield, 1991
8. Zheng, S., Häußler-Combe, U., Eibl, J.: *Dynamic Compressive Strength of Concrete under Varying Load Histories - A New Experimental and Theoretical Approach.* 1999, (in press)
9. Weerheijm, J.: *Concrete under Impact Tensile Loading and Lateral Compression*: Dissertation, Prins Maurits Laboratorium TNO Rijswijk, 1992
10. Eibl J., Schmidt-Hurtienne, B.: *A Strain Rate Sensitive Constitutive Law for Concrete.* 1999, (in press)
11. Körmeling, H.A., Zielinski, A.J., Reinhardt, H.W.: *Experiments on Concrete under Single and Repeated Uniaxial Impact Tensile Loading.* Report 5-80-3, Delft University, 1980
12. Zieliński, A. J.: *Fracture of Concrete and Mortar under Uniaxial Impact Tensile Loading.* Delft University Press, 1982
13. Gödde, P.: *Rechnerische Untersuchung zur Betonfestigkeit unter hoher Belastungsgeschwindigkeit.* Dissertation Universität Dortmund, 1986
14. Ross, C.A., Jerome, D.M., Tedesco, J.W., Hughes, M.L.: *Moisture and Strain Rate Effects on Concrete Strength.* ACI Materials Journal 93, 1996, 293-300

15. Eibl, J., Curbach, M.: *An Attempt to Explain Strength Increase due to High Loading Rates.* Nuclear Engineering and Design, Vol. 112, Elsevier Science Publishers B.V. North Holland, Amsterdam 1989, 45-50
16. Eibl, J., Schmidt-Hurtienne, B.: *Stress history versus strain history.* Proceedings of the EURO-C 1988 Conference on Computational Modelling of Concrete Structures. A.A. Balkema, Rotterdam 1998, 613-622
17. Eibl, J., Ockert, J.: *Ein Stoffgesetz für die Schockwellenbeanspruchung von Beton.* Bauingenieur 70 (1995) Springer-Verlag, 307-311
18. Zheng, S.: *Beton bei variierender Dehngeschwindigkeit, untersucht mit einer neuen modifizierten Split-Hopkinson-Bar-Technik.* Dissertation und Heft 26 der Schriftenreihe des Instituts für Massivbau und Baustofftechnologie, Universität Karlsruhe, 1996
19. Bachmann, H.: *Die Massenträgheit in einem Pseudo-Stoffgesetz für Beton bei schneller Zugbeanspruchung.* Dissertation und Heft 19 der Schriftenreihe des Instituts für Massivbau und Baustofftechnologie, Universität Karlsruhe, 1996

NON-LINEAR NUMERICAL MODELLING OF SLENDER BRIDGE PIERS
Design of slender bridge piers

W.B. CRANSTON and C.J. TOOTH
Department of Civil, Structural & Environmental Engineering,
The University of Paisley, Paisley, UK

Abstract
The provisions of BS5400 for the design of slender reinforced concrete piers are conservative. As an alternative, rigorous numerical methods of analysis can be used which more accurately model the behaviour and lead to a more efficient design.

The analysis used in this paper takes an assumed deflected shape of the column as a starting point. The column is divided into a number of segments and each cross section at the ends of segments is divided into a number of elements. Each element follows an appropriate stress-strain curve. A solution is found by an iterative procedure, which gives the forces, deflections, slopes, moments, axial strains and curvatures at the ends of each segment. Successive solutions are found corresponding to defined increments of deflection at a specific point along the column.

Two reinforced concrete bridge piers, typical of current practice, are examined and a comparison is made between the requirements of BS5400 and the results of analyses using a non-linear numerical computer method. This confirms that the method given in BS5400 for slender bridge piers is conservative.

Implications for the design of slender bridge piers are considered including the possibility of an 'integral' design approach for longer bridges.
Keywords: Non-linear analysis, slender bridge columns, integral bridges.

1 Introduction

It has been acknowledged for some time that the provisions of BS5400 for the design of slender reinforced concrete columns are conservative. P.A. Jackson, in his report "The buckling of slender bridge piers and the effective height provisions of BS5400:Part 4" [1], states the following:

Abnormal Loading on Structures edited by K. S. Virdi, R. S. Matthews, J. L. Clarke and F. K. Garas.
Published in 2000 by E & FN Spon, 11 New Fetter Lane, London EC4P 4EE, UK. ISBN 0 419 25960 0

"The introduction of shorter (and more realistic) effective heights for some types of cantilever pier is . . . only a small step towards making design methods more representative of the actual behaviour of slender piers. Where slenderness moments are significant, it will usually be possible to gain a substantial additional economy by calculating the actual moments on the pier when it is subjected to the actual design loads. This will automatically take account of slenderness, provided that the effects of displacements are included in the analysis."

The report [1] outlines a rigorous numerical analysis procedure for examining the behaviour of reinforced concrete piers bent about both axes. The program used in this paper incorporates such a procedure.

The procedure is used to analyse two typical realistic reinforced concrete bridge piers and comparisons are made with the requirements of BS5400: Part 4 [2].

2 Brief description of program

The method divides the column into a number of segments and each cross section, at the ends of segments, into a number of elements. A control point on the column is selected which it is estimated will experience the greatest horizontal movement. That point is assumed to have deflected by a specified small amount. A deflected shape for the column is proposed by the program, estimated to be compatible with the specified 'control deflection'. A proposed fraction of the design loading is then applied to the column and the axial forces and bending moments are calculated, as are the axial strains and curvatures due to these force actions in each cross section at segment ends. From these axial strains and curvatures, deflections are calculated which then allows comparison between the specified movement of the 'control' point and the calculated value.

On the first iteration the two values will differ, but by varying the fraction of the design loading applied (and any other restraint forces) an acceptable solution can be obtained for both the 'control' point deflection and the deflected shape of the column. This solution satisfies both equilibrium and compatibility and is therefore valid. Convergence procedures are used to achieve a solution in a few cycles. The control deflection is then increased by a small amount until a second solution is found and so on. The resulting analysis is therefore similar, as far as the results generated are concerned, to a physical test carried out on the column with gradually increasing load. The analysis is non-linear in that 'Pδ' effects are automatically taken into account, and pre-defined stress/strain relationships for steel and concrete are followed.

The analytical approach is developed from a single axis version described in [3] and validated in [4]. A full description is available in [5]. A broadly similar approach, extended to consider the effects of fire producing non-uniform temperatures across the cross-section, and subsequent changes to material properties has recently been developed and validated at the University of Liège [6].

3 Bridge piers analysed

A bridge viaduct of the type shown in Fig. 1 is assumed.

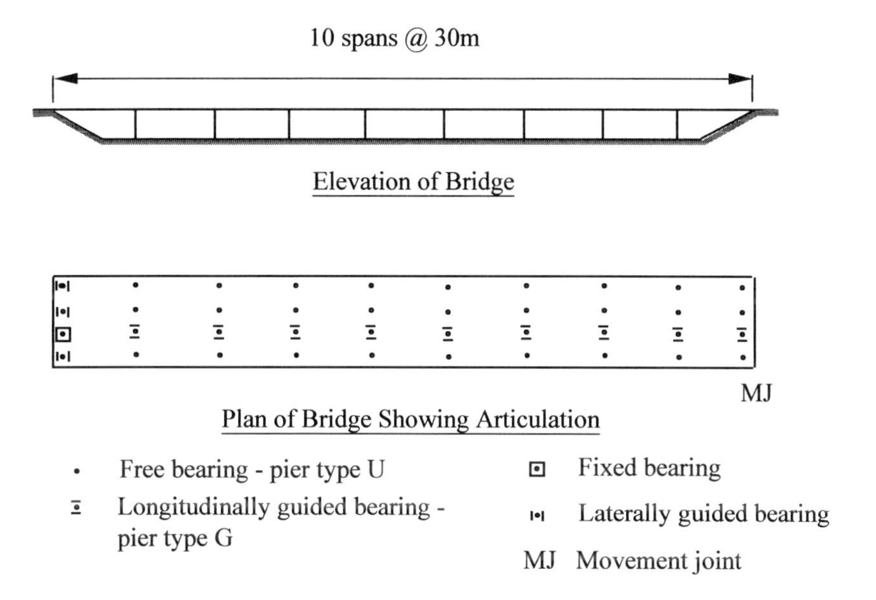

Fig. 1. General arrangement of bridge

Two types of bridge pier have been analysed; one with free bearings, type U, and one with longitudinally guided bearings, type G. The arrangement of these piers is shown in Fig. 2 below. The piers have been divided into segments and elements for the purposes of analysis. This is also shown in Fig. 2. While BS 5400 lays down the various ultimate load combinations that must be considered, only a single idealised load case is analysed for each column in this paper.

For the column with a free bearing, axial loading arising from extreme traffic loading and factored dead load is considered along with lateral loads of 10 kN applied about both axes at the column top. This produces bending moments at ultimate at the base of the cantilever column of 130 kNm, equivalent to an accidental out of plumb of around 20 mm in a diagonal direction.

For the column with the longitudinally guided bearing, the condition analysed is that of lateral wind loading on the viaduct, where it has been assumed that each column has to take a uniform share of the total wind load. This assumption is discussed below in the light of the results of the analysis.

The loads applied to the piers are shown in Fig. 3 and Fig. 4. The loads are ULS (ultimate limit state) values, both for dead and imposed loads. The factored dead load is applied from the start of the analysis procedure and the imposed load is multiplied by the 'load factor' which is determined during the analysis. This means that when the analysis load factor is 1.0, the full ULS loading is applied.

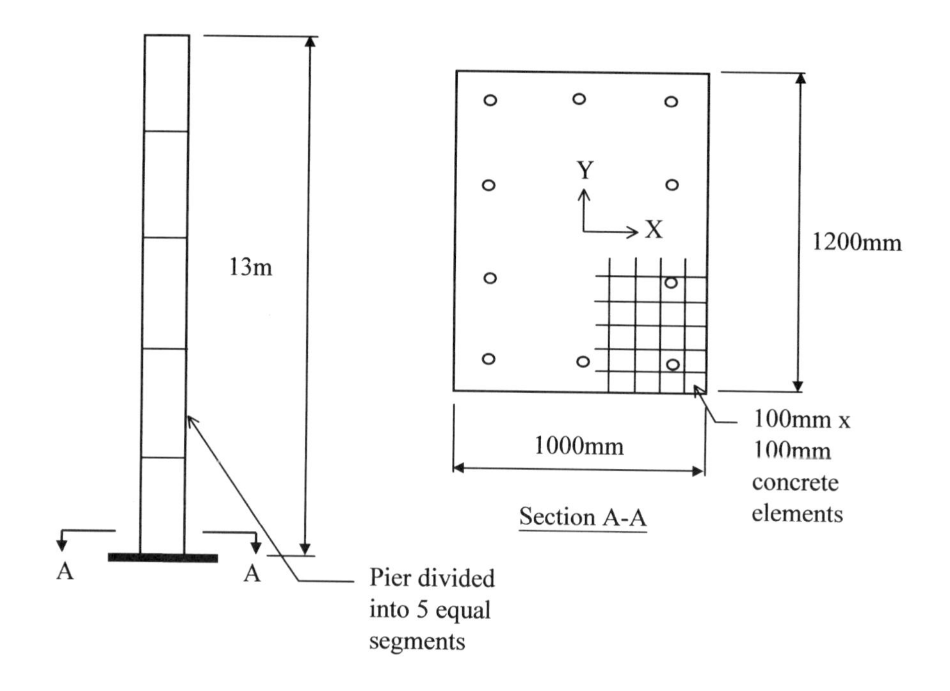

Note:- Both pier types U and G have the same dimensions. However, the main steel in type U consists of 28 No. 25mm dia. bars and for pier type G consists of 28 No. 40mm dia. bars. In both cases 28 bars have been represented in the analysis as 10 correspondingly larger bars giving the same total cross sectional area. Reinforcing steel $f_y = 460N/mm^2$ and concrete is C40. Horizontal link steel has been omitted for clarity.

Fig. 2 Arrangement of piers U and G.

4 Assessment of strength using BS5400: Part 4

Assessment of design capacity was carried out for both piers. The piers were assumed to have an effective height l_e of $2.3l_o$, (2.3 x 13m = 29.9m) i.e. 29.9m. The slenderness (l_e/b = 29.9/1 = 29.9) is 29.9 which is less than the maximum permitted value of 30. The piers are classified by BS5400 as slender. Additional moments are calculated using equations of the form shown below:

$$M_{tx} = M_{ix} + \frac{Nh_y}{1750}\left(\frac{l_e}{h_x}\right)^2\left(1 - \frac{0.0035l_e}{h_x}\right) \quad \text{(Equation 20, BS5400: Part 4)}$$

The design capacities obtained from the calculations for the load cases considered show the piers are overloaded. In the case of pier type U, by almost 100% and in the case of pier type G, by approximately 20%. Thus according to BS5400, these designs are unacceptable.

5 Results from rigorous analysis

Analysis using the program described here gives values (for each loading stage and corresponding control deflection) of ; the load factor; the forces, deflections and slopes at the ends of each segment; the moments, axial strains and curvatures at the ends of each segment. Graphs have been plotted of load factor against deflection and these can be seen in Fig. 3 and Fig. 4.

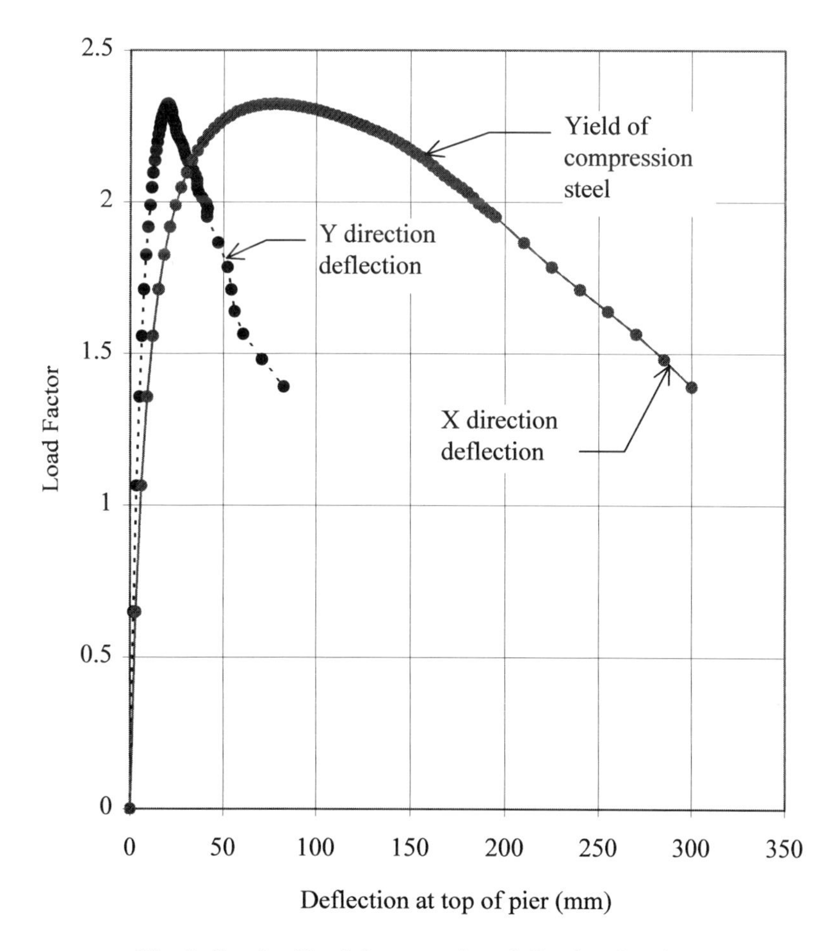

Fig. 3 Graph of load factor against deflection for pier type U.

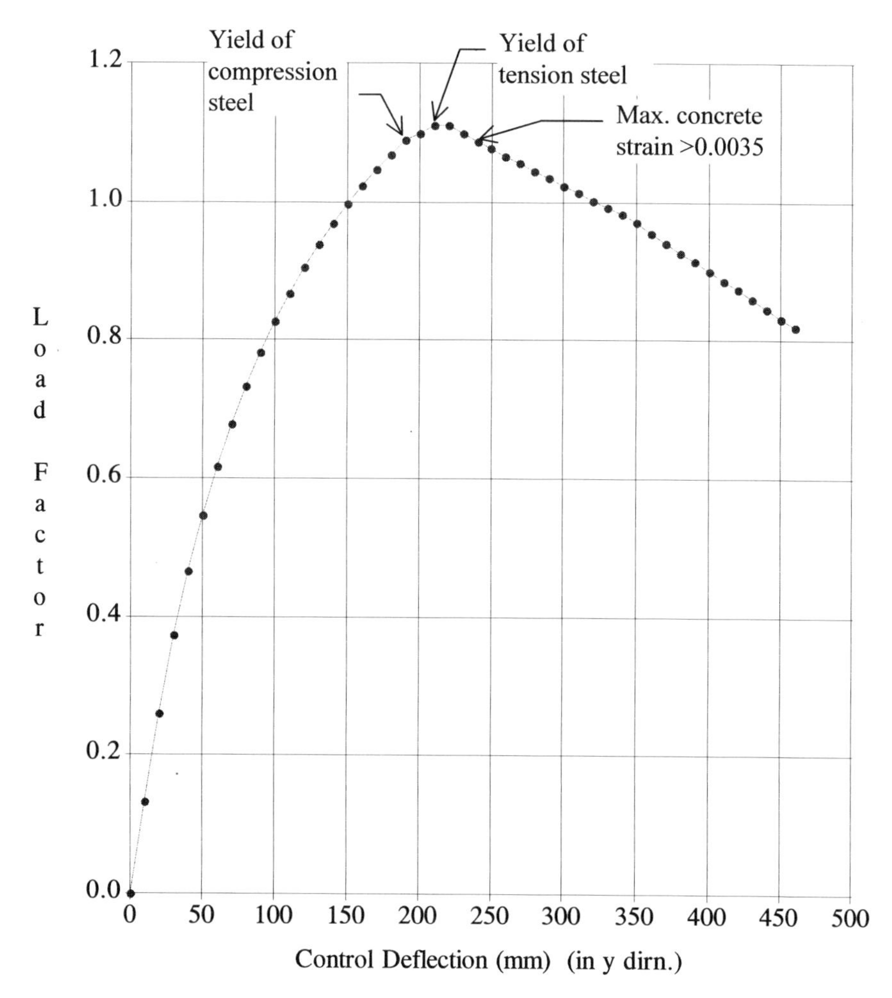

Fig. 4 Graph of load factor against control deflection for pier type G.

6 Discussion of results

6.1 Pier type U

As stated already, pier type U is subject to predominantly axial force and a relatively small amount of bending about both X and Y axes.

It will be seen that deformation dominates about the weak axis, despite the column cross-section being only slightly asymmetrical (Fig. 2). Failure is due to instability effects, with the material strength capacity not being reached until well beyond peak load - yield of the compression steel does not occur until a deflection of 150mm, approximately twice the deflection at peak load. The maximum load factor from Fig. 3

is 2.32. Given that at a load factor of 1.0 the ULS load is applied, this represents a pier capacity more than three times that predicted by the BS5400 method. Thus in practice an even more slender column would suffice.

6.2 Pier type G

Pier type G is subject to axial force together with bending about the major axis. Notable points marked in Fig. 4 are the onset of yield in the tension and compression steel and the attainment of the ultimate concrete strain. The program does not halt analysis at a concrete strain of 0.0035, but assumes a continuation of the stress-strain relationship However, practically, a column would fail at this point. The maximum load factor from Fig. 4 is 1.11. This indicates that the pier capacity is adequate whereas the BS5400 method indicated a 20% shortfall in strength.

6.3 Consideration of calculated deformations at the column top, pier type U

As axial load increases up to ultimate the column top moves in both X and Y directions relative to the bridge deck above. The analysis has effectively assumed the coefficient of friction resisting this movement to be zero. In practice the coefficient is unlikely to be less than 0.01, leading to lateral restraint forces of 100 kN or more. This would mean that the column could be restrained against lateral movement, making buckling effects negligible.

Alternatively, it is possible that the free bearing might not be installed level. An unintentional slope of 5mm/m about both axes would imply eccentricities of load of 0.005 x 13000 = 65 mm at the column base - much greater than those assumed in the analysis.

Studies of the behaviour of bearings in practice would appear to be appropriate.

6.4 Consideration of calculated deformation at the column top, pier type G

The analysis shows that the axial plus wind loading gives a lateral column deformation of around 150mm at ultimate. However this deformation can only take place if the deck also deflects 150mm laterally. But excessive traffic loading will only affect two spans of the viaduct at the most. Under wind loading and light traffic loading it can be inferred from Fig. 4 that the column top deformation will be of the order of 50 mm or even less. Thus spans adjacent to the spans with excessive traffic load (see Fig. 1) will almost certainly restrain the lateral deformation.

It seems clear (see Fig. 1) that under abnormal load to ultimate, the behaviour of the bridge as a lateral girder should be taken into account.

7 Conclusions and implications for design of slender piers

The rigorous analyses carried out above confirm that the design method given in BS5400: Part 4 is conservative, particularly for piers having high slenderness and relatively small bending moments (pier type U). If longitudinal girder action of bridge decks is considered, it is possible that type G piers, also shown to be conservatively designed, may be found to be particularly so.

From the results obtained above, it is suggested:

- that studies of actual bearing performance be initiated in order that the implications of frictional forces and out of level installation can be properly assessed, and
- that studies of longitudinal lateral girder behaviour of continuous span bridges supported by slender columns be made.

Looking at the problem more radically the analyses in this paper show that for longer bridges, such as the one shown in Fig. 1, there is clearly a possibility of omitting bearings and making the piers integral with the bridge deck. A choice would need to be made between providing fixity of the viaduct to one abutment (which would provide resistance to braking and impact forces at minimal cost) with a large movement provision at the other abutment, or providing fixity to a central row of 'strong columns' with movement provisions at both abutments. The deformation based analysis used in this paper would be ideal for the design of such structures.

8 References

1. Jackson, P. A. *The buckling of slender bridge piers and the effective height provisions of BS 5400: Part 4.* Wexham Springs, Cement and Concrete Association, 1985. Technical Report 561.
2. British Standards Institution. *Steel, concrete and composite bridges.* BS5400: Part 4: Codes of Practice for the design of concrete bridges: 1990.
3. Cranston, W. B. *A computer method for the analysis of restrained columns.* London, Cement and Concrete Association, April 1967, Technical Report 402.
4. Cranston, W. B. *Analysis and design of reinforced concrete columns.* Wexham Springs, Cement and Concrete Association, 1972. Research Report 20.
5. Cranston, W. B. *Analysis of slender biaxially loaded restrained columns.* Wexham Springs, Cement and Concrete Association, 1983. Prepared for presentation to meeting of Buckling Commission of the European Committee for Concrete, Munich, November 1982.
6. Fransen, J. M., Schleich, J. B. and Cajot, L. G. *A simple model for the fire resistance of axially loaded members according to Eurocode 3.* Journal of Constructional Steel Research, V.35, 1995, pp. 49-69.

TESTING DEVICES FOR INVESTIGATING STRUCTURES UNDER ABNORMAL LOADING

Testing devices for abnormal loading

K. BRANDES
B A M Berlin, Gemany

Abstract

Abnormal Loading very often implies accidental loading, that is, loading of short duration generating dynamic and strain rate effects. We are dealing with high intensity of loading taking the structure far beyond its limits of load bearing capacity in terms of static loading for a very short period of time. It is a matter of fact that investigations in that case very often demand full scale or large scale testing in the real time domain. This, in turn, requires large scaled testing equipment, which need to respond at high speed under controlled conditions. At BAM, testing facilities, which have been developed for high speed tests on large specimens up to its failure after large deflections (Kinetic Load Bearing Capacity) are presented. They have been installed for tests on reinforced concrete structures under impact and impulsive loading and for tests on seismically forced structures. The paper discusses the aims of the tests and the range of parameters which are measured in the tests, such as force, velocity and dimensions of specimens.

Keywords: Dynamic Loading, impact, impulsive loading, load carrying capacity, reinforced concrete structures, steel structures, testing devices.

1 Introduction

More and more, civil engineering structures have to resist impact or impulsive loading [1], or load reversals by earthquake excitation. The straining of the structural elements exceeds the limits of elastic response by far. The duration of loading is about 0.1 s or the relevant frequency of vibration is between 0.1 cps and 20 cps. Considering the material behaviour, strain rate effects and the influence of strain rate history have to be taken into account [2]. Consequently, experimental investigations must be performed in the real time domain and on large-scaled specimens because bond of reinforced

Abnormal Loading on Structures edited by K. S. Virdi, R. S. Matthews, J. L. Clarke and F. K. Garas.
Published in 2000 by E & FN Spon, 11 New Fetter Lane, London EC4P 4EE, UK. ISBN 0 419 25960 0

concrete or the instability behaviour of thin-walled steel structures can only be investigated on a scale similar to the real dimensions.

Fig.1 Test equipment for impact like tests on large scale specimens

In order to perform tests under these conditions, we have developed test equipment which enable us to carry out such tests under controlled conditions.

2 Impact and impulsive loading

Impact loading is caused by moving bodies hitting a structure, e.g., a reinforced concrete bridge pier, a harbour structure or a containment of a nuclear power plant. The momentum (or impulse), in physical terms, mass times velocity,

$$I = m \times v$$

can be caused by a high speed of the hitting body (e. g. an aircraft) or by a big mass (a ship). The duration of the force transmission process is between 30 ms and 5 s, if the striking body is classified as crashworthy [3].

Fig. 2 High speed test on a reinforced concrete beam up to rupture of the reinforcing bars and the crushing of the concrete in the compression zone

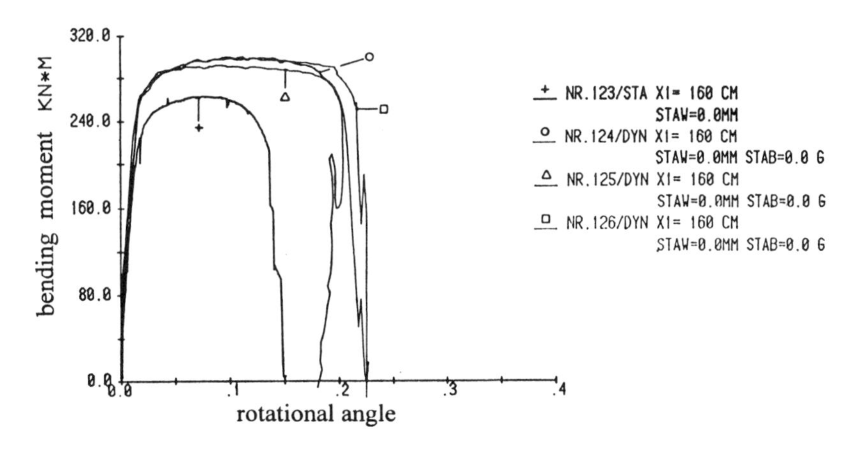

Fig. 3 Results of tests on simply supported beams of 3.20 m span, loaded at the centre, statically in test no. 123 (+) and dynamically in the three other tests no 124, 125 and 126. The rotational angle is the deflection at mid-span divided by 4, the bending moment includes the inertia of mass effects.

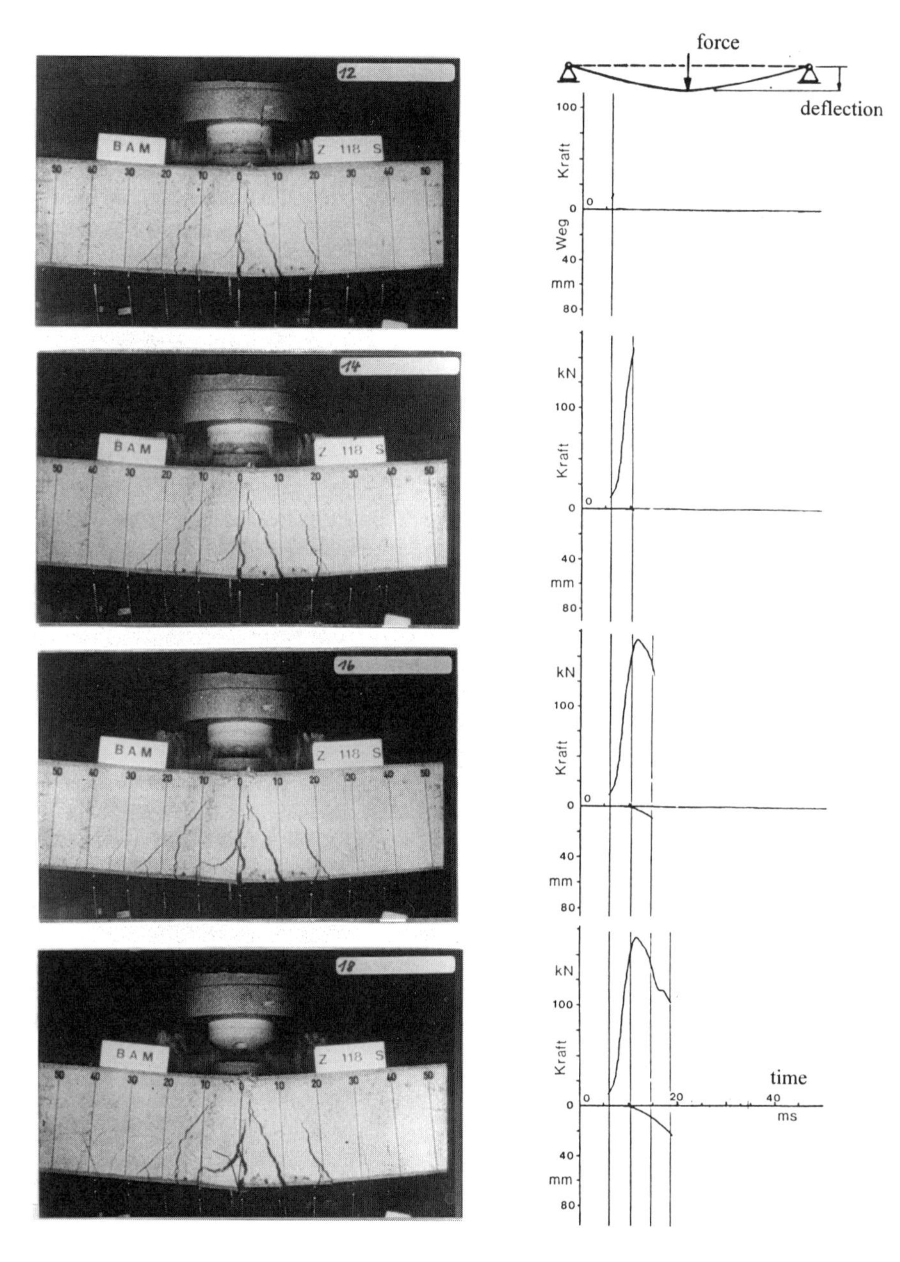

Fig. 4 High speed test on a reinforced concrete beam, high speed camera filming (500 pict./sec). The time difference between two pictures is about 4 milliseconds. On the diagram on the right hand side, the time history of the deflection at the beam's center and the bending moment there are plotted

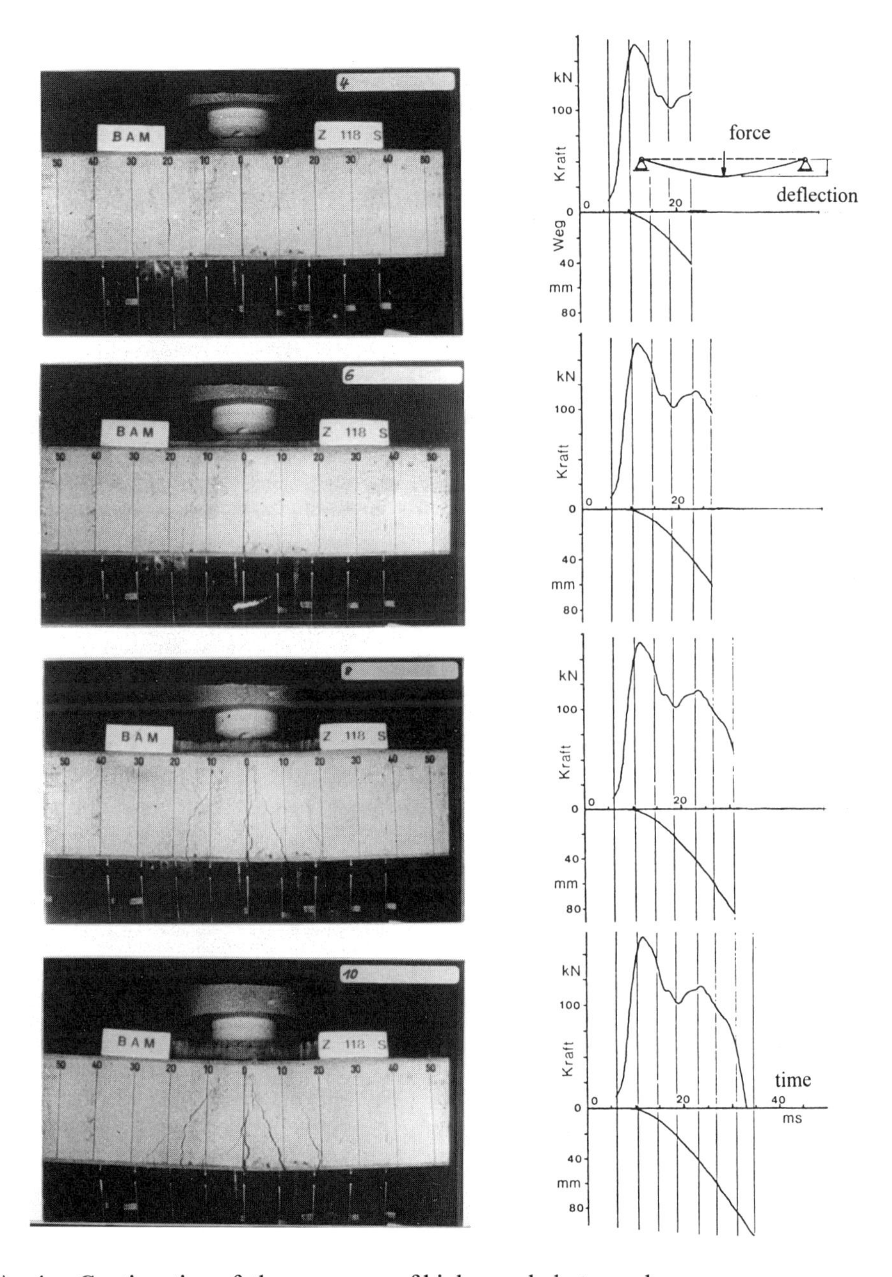

Fig. 4 Continuation of the sequence of high speed photographs.

We do not deal with compact solid bodies like cannon balls being shot on a wall, which is the subject of "Terminal Ballistics". In that case, the mechanical response is governed by across the thickness wave propagation [4]. In the past, the experimental investigation of impacted structures has been focussed on falling or flying mass tests, originating from Terminal Ballistics. However, when dealing with crashworthy striking bodies which absorb the biggest amount of energy during the impact process, another approach is much more promising.

For some decades now, specially designed electronically controlled servohydraulic equipment (Fig. 1) has been used to carry out high speed tests on large scaled specimens. The deformation velocity of the specimen at the point of impact reaches about 2 m/s to 5 m/s. This means that the failure of the specimen (for example, a reinforced concrete structural member) occurs after about 50 ms (Fig. 2). It is an advantage of that type of test that the velocity can be altered from static load conditions (v = ca. 1mm/min) to high speed load conditions.

Considering the plastic hinge, which occurs in a reinforced concrete beam, or a yield line in a slab, it is obvious that together the response to blast load is included. The most interesting result of some of the tests on large scaled reinforced concrete beams was the distinct influence of strain rate, Fig. 3.

In the design of the test equipment, besides the high speed servohydraulic actuator and the special large servo valve, the extreme rigidity of the load frame is worth mentioning. Also, for high speed testing, the eigenfrequency of the relevant vibration mode of the entire system must be larger than 100 Hz because of the properties of the closed loop control system.

To give a real impression of a high speed test, in Fig. 4 is given a sequence of photographs taken with a high speed camera (500 p/s), together with the development of the displacement and force at the specimen's centre.

Reinforced concrete structures suffer two types of failure: failure in bending and failure because of transfer shear, more found when slabs are loaded locally. The strength of concrete is highly influenced by the rate of straining as can be recognised from tests on slabs. In Fig. 5 is shown a slab of 16 cm thickness clamped at the edges in a very stiff edge beam, after an impact test.

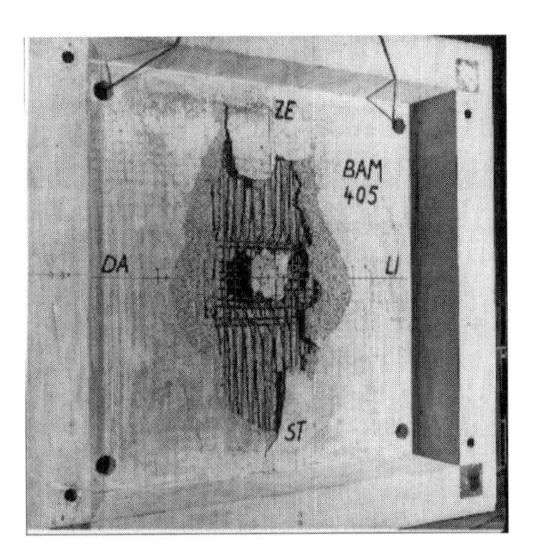

Fig. 5 Reinforced concrete slab after an impact test, loaded centrally. Slab thickness: 16 cm Slab dimensions inside clampingedge beams: 230 cm x 230 cm

Fig. 6 A cut reinforced concrete slab tested in one of the "Meppen Tests". Slab thickness was 60 cm. The punching shear cone is clearly exposed.

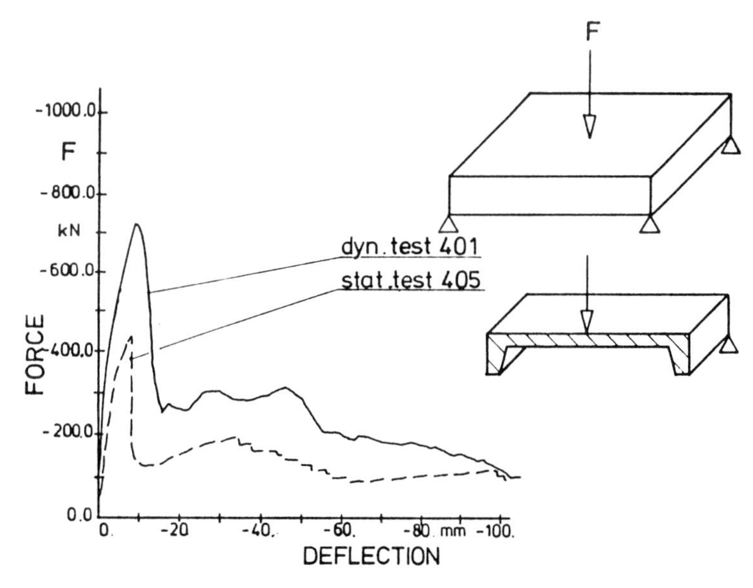

Fig. 7 Influence of rate of loading on the punching shear resistance of reinforced concrete slabs. The tests were made on specimens like those given in Fig. 5.

Punching shear failure occurred in a test of the same type that has been carried out on beams. There was no shear reinforcement in the slab. From a cut specimen of the well known "Meppen Tests" [5], the mechanism of punching shear can be recognised, as shown in Fig. 6.

The influence of the strain rate is shown in Fig. 7. The slabs had no shear reinforcement. The resistance of the slabs to high-speed loading reaches nearly double the value for static loading. About the same effects have been verified for concrete strength in Delft [6].

3 Resistance to seismic loading

Severe seismic excitations load structures far beyond the limits of elastic response. Considering reports from earthquake catastrophes, steel structures exhibit very good aseismic behaviour if the connections are strong enough. Connections are designed to be stronger than the adjacent structural members' material strength. However, as we had found for reinforcement bars, the yield stress of steel is strain rate sensitive. When dealing with the design of earthquake resistant steel constructions, the "overstrength" of the steel material can cause an early failure of certain types of connections like endplate connections, which are often used in Europe for regions of low seismic risk.

In tests on specimens as shown in Fig. 8, we applied load reversals under static conditions (about half an hour for one cycle) and dynamic conditions (about 1 Hz). The influence of the strain rate is considerable and reaches about 10% in terms of forces of hysteretic loops [7]. Also these tests have been performed in a servohydraulic test equipment which has the capacity for large deformation of the specimens in high frequency tests. The result of the mentioned test series is to recommend strengthening of the standardised end plate connections for seismic regions.

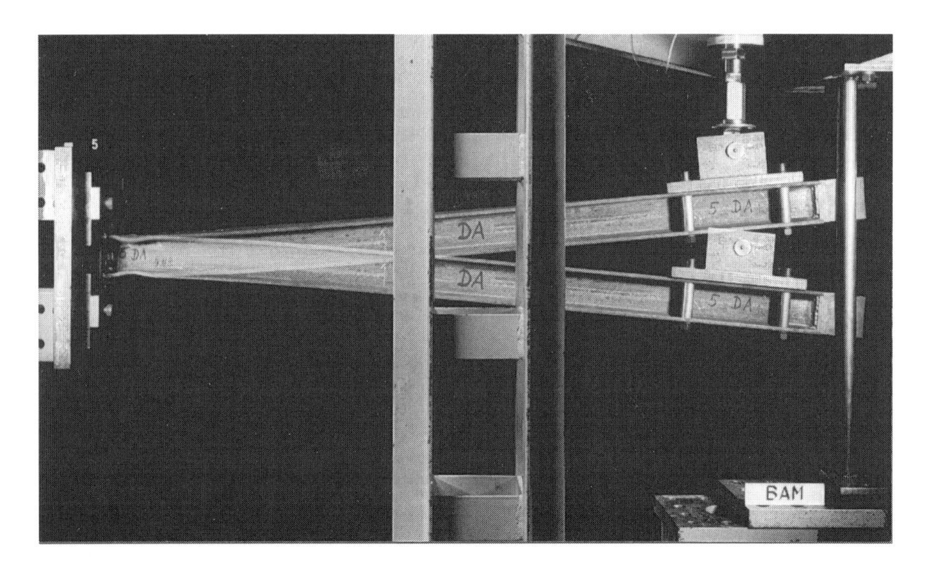

Fig. 8 Trick shot of a test on a steel beam with end-plate connection. The beam has been deformed cyclically +/- 9.5 cm having a length of 135 cm. The buckling of the flanges near the end plate is clearly visible.

4 Buckling of Damaged Shell Structures

Imperfections of geometric shape influence the buckling resistance of shells such as those used as piles in offshore structures. Steel piles suffer from impact of ships when approaching the structure. Large indentations caused by such an accident reduce the resistance of the piles to vertical loading. The indentations grow larger and larger during loading and unloading. Together with Rostock University, we are currently investigating the shake down behaviour on models scaled down by 1:3, as shown in Fig. 9.

Fig. 9 Cylindrical tube of 5m length after a shake down test. The tube with internal longitudinal stringers had a large geometric imperfection.

5 Summary

The investigation of abnormal loading of civil engineering structures in large scale experiments requires especially designed test equipment which has been developed at BAM over a period of about two decades. Some examples show that tests can help in solving the severe problems occurring when abnormal loading must be taken into account.

6 References

1. Plauk, G. (ed) (1982) Concrete Structures under Impact and Impulsive Loading. RILEM et al Interass. Symposium, Berlin. Introductory Report, Proceedings
2. Herter, J., Limberger, E. and Brandes, K. (1985) Experimental and numerical investigation of reinforced concrete structural members subjected to impact load. In: Armer, G. S. T. et al (eds): Design of concrete structures - The use of model analysis. London.
3. Johnson, W. and Mamalis, A. G. (1978) Crashworthiness of vehicles. London.
4. Kennedy, R. P. (1976) A review of procedures for the analysis and design of concrete structures to resist missile impact effects. Nuclear Engineering and Design 37, 183-203
5. Sage, E. and Pfeiffer, A. (1981) Response of reinforced concrete targets to impacting soft missiles. Transactions 5th SMiRT Conference, Paris, Vol. J
6. Reinhardt, H. W. (1980) Zugfestigkeit von Beton unter stoßartiger Beanspruchung . In: Eibl, J.(ed): Stoßartige Belastung von Stahlbetonbauteilen. Universität Dortmund.
7. Brandes, K. (1990) Strain rate sensitive cyclic behaviour of steel bolted joints. In: Krätzig et al (eds): Structural Dynamics, Rotterdam.

CONTACT FORCE RESPONSE OF CONFINED CONCRETE CYLINDERS SUBJECTED TO HARD IMPACT

Concrete cylinders under impact

S.J. PRICHARD and S.H. PERRY
Department of Civil, Structural and Environmental Engineering,
Trinity College Dublin, Ireland.

Abstract

The response of structures and materials to impact loading is an important area of research for civil and military engineers because, although severe impacts occur infrequently and in random locations, they can, occasionally, lead to catastrophic structural failure. When impacted, concrete exhibits an increase in its uni-axial compressive strength compared to its equivalent static strength. Also, if concrete is confined it exhibits an increase in its load carrying capacity. Experimental work has been undertaken on concrete cylinders confined by tubes of mild steel, aluminium and PVC plastic, using a low velocity hard impact test facility in Trinity College Dublin. The presence of confinement has resulted in considerable increase in the compressive strength of the impacted concrete cylinders. This paper describes how Hertzian contact theory is used to model the contact force response of concrete cylinders and shows a comparison between the model and experimental results.

Keywords: Concrete, contact force, hard impact, Hertz's Law, tri-axial confinement.

1 Introduction

Engineers have worked through most of the twentieth century to solve the outstanding problems in both structural mechanics and the response of materials to a variety of loading conditions. This has resulted in a vast amount of knowledge about the response to static loading, but many questions on the response of structures and materials to high strain rate loading remain unanswered. This has occurred because the work only started mid-century and also because the response of structures and materials to this form of loading is much more complicated, being both difficult to measure and to predict. However, although this form of loading occurs randomly and

Abnormal Loading on Structures edited by K. S. Virdi, R. S. Matthews, J. L. Clarke and F. K. Garas.
Published in 2000 by E & FN Spon, 11 New Fetter Lane, London EC4P 4EE, UK. ISBN 0 419 25960 0

infrequently, it can lead to catastrophic structural failure, disproportional to the severity of the original load.

"High strain rate loading" occurs when a load is suddenly applied to a material, and the stresses and strains, which it induces, do not have time to equilibrate within the period of loading. It can occur as a result of either impact or blast loads. Under these unusual forms of loading, the properties which concrete and other materials exhibit can be significantly different to those under static or quasi-static loading [1]. For example, there is a considerable increase in the uni-axial compressive strength of concrete if impacted when compared to its equivalent static strength. This increase can be found quantitatively by measuring the contact force during the impact.

Research is underway to enhance the response of materials, such as concrete, to impact loading. In an attempt to achieve this, the material can be confined in two directions to prevent lateral dilation under a load in the third, perpendicular, direction. This has been shown statically to increase the ultimate compressive strength of the confined concrete. Often the confinement is provided by steel links surrounding longitudinal reinforcement, which prevent the concrete from dilating when loaded. However, experimental work by the present authors using a low velocity hard impact test facility has shown that the presence of a passive confining cylinder of steel, aluminium, or even plastic, results in a considerable increase in the contact forces measured during hard impact.

A simple model is being proposed which uses Hertz's Contact Law to model the contact force response of confined concrete cylinders to hard impact. This paper gives an outline of the model and compares it to the experimental results.

2 Impact testing of confined concrete

2.1 Impact test rig

A series of seventy impact tests have been completed on both confined and unconfined concrete using a low velocity hard impact test rig. The facility was described in detail in a previous paper [2] when it was used for preliminary tests, so only a brief description will be given here.

The test rig consists of a guide tube, through which an impactor falls under gravity (see Fig. 1). The maximum drop height is 3.125 m, which allows contact velocities of up to 7.8 m/s. The impactor is a solid, cylindrical mass, 150 mm in diameter, tapered at the lower end to 100 mm and weighing 82.5 kg. It incorporates a load cell, which measures the contact forces generated as a result of the impact, attached between the tapered end of the impactor and the slightly rounded, hardened end striker (Fig. 2). The load cell consists of a strain gauged hollow cylinder, which was calibrated in a Losenhausen static testing machine. A 500 mm cube of concrete, firmly bolted to the 0.9 m thick concrete strong floor, acts as a solid base for the test specimen.

Electric resistance strain gauges, placed laterally and axially on the circumference of either the confining cylinder or the unconfined concrete, allow the deformational response of the specimen to be recorded. A high speed data acquisition system is used to record both the strains and the load cell response during the test, at a rate of 30 kHz from each of the channels. The results were stored and post-processed using a commercial graphical programming package on a desktop computer fitted with a high speed analogue to digital (A-D) conversion card.

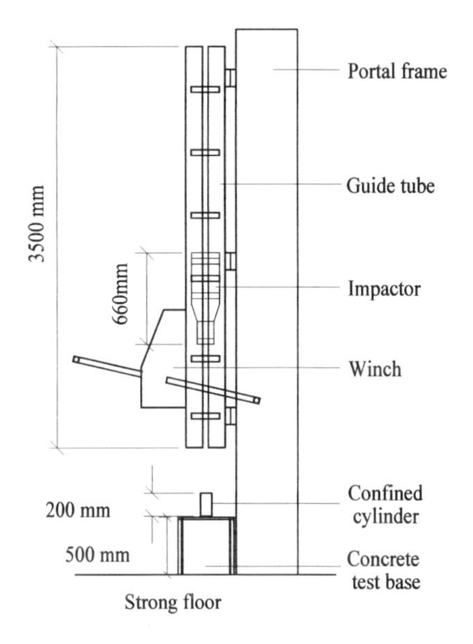

Fig. 1. Impact test rig

Fig. 2. Impactor load cell

2.2 Confined concrete cylinders

Concrete filled composite columns are a simple method of providing uniform bi-axial confinement to concrete. Confinement is commonly provided using long steel rods or bars with surrounding links, or more occasionally, by steel spirals, where the bars resist both elongation and buckling when the concrete begins to dilate laterally under a load. However, concrete filled columns have an added advantage in that they act as formwork for the concrete whilst it is being poured and also exhibit greater ductility, flexural stiffness, and toughness than unconfined, or even conventionally reinforced concrete [3]. They have the ability to carry high loads, even when of small dimensions, because the concrete is put into tri-axial confinement due to the presence of the confining section. Research has shown that if concrete is confined and prevented from expanding laterally in two out of three mutually perpendicular directions, then the ultimate compressive strength of the material in the third direction increases considerably, as cracks, which would form in unconfined concrete under an axial load, are prevented from opening [4],[5].

The confinement for the experimental work was provided by structural hollow cylindrical sections of steel, aluminium or PVC plastic pipe, each 200 mm high, within which the concrete was cast [6]. The properties of the confining sleeves can be found in Table 1. Six series of confined cylinders were cast and a seventh, of unconfined cylinders, made in steel moulds 100 mm in diameter and 200 mm high, formed a reference series. This aspect ratio (approximately 2, depending on the cylinder in question) has been shown to be sufficiently large to minimise the end effects and to get a good representation of the material properties, whilst minimising the stress variation along the length due to stress wave propagation [1].

Table 1. Properties of confining materials (as given by manufacturer)

	Internal Dia. (mm)	Thickness (mm)	Density (kg/m^3)	Young's Modulus (kN/mm^2)
Steel 1	111.9	1.20	7800	209
Steel 2	108.3	3.00	7800	209
Steel 3	104.3	5.00	7800	209
Steel 4	101.7	6.30	7800	209
Aluminium	108.5	3.25	2800	70
PVC Plastic	103.1	3.60	1420	3

Table 2. Properties of concrete at 28 days (based on seven test results)

Air Dried Density (kg/m^3)	2278
Wave Velocity (m/s)	4325
Young's Modulus (kN/mm^2)	42.6
Compressive Strength (N/mm^2)	49.2

The concrete, which was poured into the hollow cylinders, had a 28 day design strength of 40 N/mm^2. It was wet cured for seven days and was then stored in the laboratory for the remainder of the time before testing. The average properties of the concrete cubes at 28 days can be found in Table 2.

2.3 Impact test results

The impact tests were conducted on confined and unconfined cylinders, at three different drop heights - 0.5 m, 1 m and 2 m. The unconfined cylinders failed in a uniform manner regardless of the velocity of impact, splitting into either three or four sections longitudinally from the centre of the top face. The failure of the aggregate and that of the aggregate-cement bond were both evident in the cracked cylinders. At the lowest velocity, the concrete cylinders confined by steel or by the aluminium, showed no visible cracks for any of the four thickness values. However, at higher contact velocities some small radial cracks became evident in the top faces of the concrete, which was confined by the thinner two thicknesses of steel and by the aluminium. Even at the lowest velocity, the concrete confined by the plastic cylinders exhibited some top face cracking and permanent deformation of the plastic, which remained visible after the test. At the 1 m drop height, the plastic cracked and the concrete crushed, whilst at the highest velocity, the plastic ruptured and resulted in a complete failure of the concrete.

The impactor load cell was used to find the contact force during the period of impact and graphs were plotted showing the contact force response against time for each of the types of confinement. All of the graphs show a distinctive peak, which is a characteristic of hard impact. Table 3 shows the average maximum contact forces for each of the different types of test, at each of the three contact velocities. As would be expected, the increase in contact force is related to the Young's Modulus of the confining material. However, as can be seen from the response of the steel confined cylinders, although the contact force does increase as the thickness increases, the increase is not linear.

Table 3. Average maximum contact forces (in kN)

Velocity	3.19 m/s	4.44 m/s	6.27 m/s
Steel 1	387	588	797
Steel 2	403	593	1062
Steel 3	418	569	998
Steel 4	424	611	1103
Aluminium	394	539	844
PVC Plastic	283	360	444
Unconfined	192	316	346

3. Modelling of the contact force response

3.1 Model 1 - a sine curve

This paper describes a model in which the experimental results were used to find empirical values to describe the impact of two masses, which were then used in an equation to model the contact force response. This model was then compared to the experimental results to see how well it fitted the original curves.

On examination of the contact force-time graph for the confined tests, it was evident that the contact force response was almost sinusoidal, and that the collision was, therefore, showing evidence of being elastic which was confirmed by the considerable rebound of the impactor after the impact. The elastic impact of two masses can be described using Hertz's Contact law, which was originally derived for static contact, if the vibrations produced by the impact can be neglected, [7],[8]. This law relates the deformation of the areas, which are in contact, to the impact force,

$$F(t) = K\delta^{1.5} = -\frac{m_1 \times m_2}{m_1 + m_2} \times \frac{d^2\delta}{dt^2} \qquad (1)$$

where, $F(t)$ is the contact force, δ is the deformation of the contact zone and K is a contact parameter, which takes account of the stiffness values at the location of impact, and relates the elastic and geometrical properties of the contacting bodies. The masses of the bodies are m_1 and m_2. The initial assumption is that a collision is never between two perfectly flat bodies and, therefore, causes indentation to occur in one of the surfaces, thereby absorbing some of the kinetic energy of the impact.

Equation 1 can be solved by integration with the initial conditions

$$\frac{d\delta}{dt} = v_0 \ at \ \delta = 0 \quad and \quad \frac{d\delta}{dt} = 0 \ at \ \delta = \delta_{max} \tag{2}$$

where v_0 is the velocity of the impactor just prior to the impact. The maximum load, P_{max}, and the contact time, t_{max} at which P_{max} occurs can be found using:

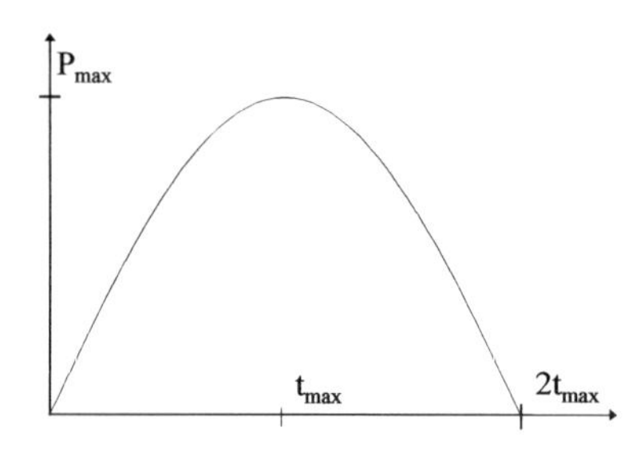

Fig. 3. Elastic impact

$$P_{max} = \left(1.25 \times \alpha \times v^2 \times K^{2/3}\right)^{\frac{3}{5}} \tag{3}$$

and $$t_{max} = 1.47 \times \left[1.25 \times \frac{\alpha}{v^{1/2} \times K}\right]^{\frac{2}{5}} \tag{4}$$

where α is the effective mass = $(m_1 \times m_2) / (m_1 + m_2)$ and v is the velocity at impact. The idealised elastic impact, modelled by a sine wave because the integration of Eqn. 1 involves elliptic integrals, is shown in Fig. 3, with P_{max} , t_{max} and $t_0 = 2t_{max}$ as indicated. The contact parameter (K_a) for approach is equal to that for restitution (K_r) . The curve can be described by

$$F(t) = F_{max} \sin\left(\frac{\pi t}{2t_{max}}\right) \tag{5}$$

This is an idealised model of impact, but has been used by many researchers to model the response of the impact zone. It is appropriate to use it for well confined impact where it can be assumed that there is only minimal cracking in the top face of the impacted concrete, and the contact force curves are nearly sinusoidal. However, as will be shown in this paper, it is probably inappropriate in the case of unconfined impact where cracking occurs in the concrete and results in a considerable change to the material properties.

Values for the maximum contact force, F_{max}, the time at which this maximum force was reached, t_{max}, and the total duration of impact, t_0, were measured for each of the impact tests. It was initially assumed that the target was comprised of the confined cylinder and the solid concrete block on which it was sitting, and the effective mass was calculated using their combined mass and the mass of the impactor. These values were substituted into Eqns. 3 and 4, to find a single optimum value of K for each of the types of confined and unconfined cylinders (see Table 4).

The values of K were used to find both F_{max} and t_{max}, which were in turn used in Eqn. 5 to find the contact force-time curve for each type of cylinder at each of the three contact velocities. Fig. 4 shows some of the curves measured during the impact tests. The curves generated using Eqn. 5 are overlaid on the contact force response curves as "model 1". These curves are sine curves and form a reasonable approximation for the measured plots. However, because a single value of K was used, the graphs exhibit a considerable overestimation in some areas, predominately in the calculation of the duration of impact.

Table 4. Values of K calculated for models 1 and 2

	K (kN/mm$^{3/2}$)	K_f (kN/mm$^{3/2}$)	$K_{t\infty}$ (kN/mm$^{3/2}$)	K_{t0} (kN/mm$^{3/2}$)
Steel 1	126.3	76.3	290.2	162.5
Steel 2	139.6	115.8	324.7	144.6
Steel 3	171.9	111.7	320.3	165.5
Steel 4	132.4	131.5	248.0	154.8
Aluminium	114.6	80.8	358.5	140.4
PVC Plastic	65.4	24.0	243.1	87.7
Unconfined	21.2	12.0	673.9	50.5

3.2 Model 2 - incorporation of an infinite mass and a quasi-sine curve

The assumptions which were made in the case of model 1, resulted in a curve which was simple but did not accurately represent the contact force-time response during the impact. In practical cases of impact, even when confined, the material response is not perfectly elastic, as some of the energy is absorbed by the concrete through microcracking, which can result in slightly different values for the contact parameters for approach (K_a) and restitution (K_r).

Secondly, the modelling of the collision between the falling impactor and the concrete cylinder can be greatly simplified if account is taken not only of the solid concrete block, but also of its being bolted to the strong floor of the laboratory. The mass of the whole "target", i.e. the floor, the reaction block and the confined cylinder is, therefore, practically infinite and much greater than the mass of the impactor (82.5 kg), and hence, the movement of the target can then be assumed to be limited to local deformation in the top of the concrete cylinder. As the mass of the impacted body is then assumed to be infinite, the ratio α reaches a limit at the mass of the striker, m_s [9]. Eqns. 3 and 4 can, therefore, be rewritten as

$$P_{\infty} = \left(1.25 \times m_s \times v^2 \times K^{2/3}\right)^{\frac{3}{5}} \tag{6}$$

$$\text{and} \quad t_\infty = 1.47 \times \left[1.25 \times \frac{m_s}{v^{1/2} \times K} \right]^{\frac{2}{5}} \tag{7}$$

where the subscript ∞ is used to indicate that the mass of the target is assumed to be infinite, a limiting case of the problem. The following equation can then be used to generate a reasonable approximation for the contact force history

$$F(t) = F_\infty \sin\left(\frac{\pi t}{2t_\infty} \right) \tag{8}$$

As before, the values for the maximum contact force, the time at which this maximum force was reached and the total duration of impact for each test, were used. The maximum contact force values were used in Eqn. 6 to find an optimum value for the contact parameter, K_f for each of the types of confinement, that is, the four types of steel and the aluminium, plastic and the unconfined cylinders. The times at which the maximum force was reached and the total duration of impact were both used in Eqn. 7 to find two other values for the contact parameter, $K_{t\infty}$ and K_{t0} respectively. These least squares, optimum values can be found in Table 4. However, 'K' is a physical property associated with the colliding bodies and should be similar, regardless of whether it is calculated using the maximum force or the maximum time. The huge differences were probably associated with the assumption of infinite mass and because of microcracking in the concrete.

Each of K_f and $K_{t\infty}$ calculated and presented in Table 4 was then used in Eqns. 6 and 7 to calculate values for F_∞ and t_∞. Eqn. 7 was also used with K_{t0} to find a value for t_0. These values, F_∞, t_∞ and t_0, were in turn used in Eqn. 8 to find the contact force-time curve. Different values of K were calculated for the approach and the restitution, as the concrete properties changed during the impact. Account was taken of this by changing from the value for $K_{t\infty}$ to K_{t0} in the calculation of t_∞, when the maximum force was reached, resulting in an overall curve formed by two different sine curves. The curves generated by Eqn. 8, are also shown in Fig. 4 as "model 2".

Figures 4 (a), (b) and (c) are of the contact force responses of one of the thicknesses of steel (3 mm) at the three different velocities of impact. Model 2 provided a curve which was considerably closer to the curves acquired during the impact than model 1, for all three velocities of impact, predominantly because the curves are not a true sine curve. The difference between the two models is not substantial and would suggest that the difference between the assumptions of effective mass and an infinite mass do not have a large influence on the results. Similarly, Figs. 4 (d) and (e), which show the models for steel (thickness 1.2 mm) and aluminium, are evidence that model 2 is reasonably accurate for different types of confinement. However, in Fig. 4 (f) the response of the concrete cylinders confined by plastic is shown, an example of where the concrete itself fractures and neither model 1 nor model 2 are able to predict the response. Model 1 overestimates both the time and the maximum contact force, and model 2, whereas it models the maximum contact force, is inaccurate with respect to both the time taken to reach the maximum force and the time for the total impact event.

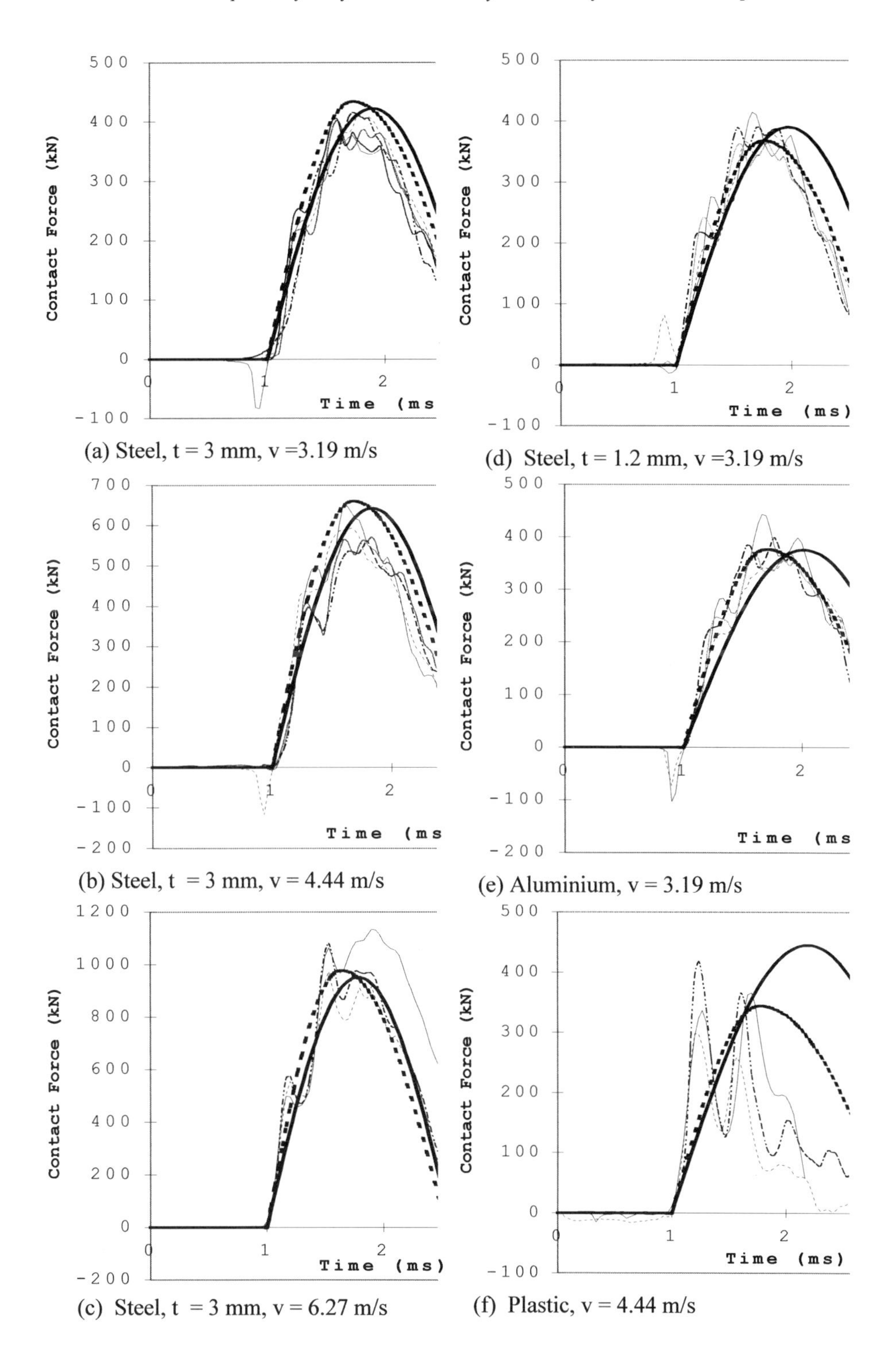

(a) Steel, t = 3 mm, v =3.19 m/s

(d) Steel, t = 1.2 mm, v =3.19 m/s

(b) Steel, t = 3 mm, v = 4.44 m/s

(e) Aluminium, v = 3.19 m/s

(c) Steel, t = 3 mm, v = 6.27 m/s

(f) Plastic, v = 4.44 m/s

Fig. 4. The measured and modelled contact force response of impacted concrete
This is hardly surprising as the assumption of elasticity is invalid as the concrete failed completely. The assumptions of infinite mass and that the curve should be described using different parameters for approach and restitution are, therefore, not unreasonable in the case of well confined impact.

4. Conclusions

This paper has shown that Hertz's Contact Law, which was originally derived for the static contact of two parabolic surfaces, can be applied to an impact situation, even when permanent deformations are produced. An initial attempt to apply the contact law produced a sinusoidal model, which gave a reasonable indication of the shape of the contact force curve. However, the preliminary model generally overestimated the maximum force, the time to reach the maximum force and the duration of impact. A second model was produced in which the curves are not perfectly symmetric, as the parameters for approach and restitution are different, because the impact of the steel impactor onto the concrete target leads to some degree of cracking and fracture of the concrete. The model is based on a realistic assumption that the mass of the target was infinite. The second model gave a very good estimation of the response when the concrete response was elastic.

It can be concluded that Hertz's Contact Law can provide an accurate model for the contact force response of concrete cylinders subjected to impact, if the confinement provided is sufficient to prevent concrete fracture and ensure that the response is nearly elastic. Where this is not the case, the law overestimates considerably the contact force response of impacted concrete.

5. References

1. Bischoff, P.H. and Perry, S.H. (1991) Compressive behaviour of concrete at high strain rates. *Materials and Structures / Matériaux et Constructions*, Vol. 24, pp. 425-450.
2. Prichard, S.J. & Perry, S.H. (1998) Hard impact testing of confined concrete cylinders. *Proceedings of Structures Under Shock and Impact V (SUSI V)*, (eds. Jones, N., Talasidis D.G., Brebbia C.A. & Manolis, G.D.), Computational Mechanics Publications, Southampton, pp. 493-502.
3. Gardner, N.J. & Jacobson, E.R. (1967) Structural behaviour of concrete filled steel tubes. *American Concrete Institute. Journal*, pp. 404-413.
4. Kotsovos, M.D. & Perry, S.H. (1986) Behaviour of concrete subjected to passive confinement. *Materials and Structures / Matériaux et Constructions*, Vol. 19, No. 112, pp. 259-264.
5. De Nicolo, B., Pani, L. & Pozzo E. (1997) The increase in peak strength and strain in confined concrete for a wide range of strengths and degrees of

confinement. *Materials and Structures / Matériaux et Constructions*, Vol. 30, pp. 87-95.
6. Prichard, S.J. & Perry, S.H. (1999) A comparison between experimental and predicted contact forces of confined concrete under impact, *Proceedings of the 9th International Symposium on the Interaction of the Effects of Munitions with Structures*, Berlin, May 1999, pp. 355-362.
7. Goldsmith W. (1960) Impact, The Theory and Physical Behaviour of Colliding Solids, E.Arnold, London.
8. Hughes G. & Speirs D.M. (1982) An Investigaton of the Beam Impact Problem, CeCA Technical Report No. 546.
9. VanMier J.G.M., Pruijssers A.F., Reinhardt H.W. & Monnier T. (1991) Load time response of colliding concrete bodies, Journal of Structural Engineering, Vol. 117, No. 2, pp. 354-374.

BEHAVIOUR OF DAMAGED REINFORCED CONCRETE FRAMES
Damaged reinforced concrete frames

E.H. DISOKY
Military Technical College, Cairo, Egypt
M.K. ZIDAN
Faculty of Engineering, Ain Shams University, Cairo, Egypt
A.N. SIDKY
Housing and Building Research Center, Giza, Egypt

Abstract
The behavior of reinforced concrete frames is examined for different cases of damage to supporting elements, giving details about the changes in vertical deflections and the distribution of internal forces (bending moments and normal forces) induced in the beams and columns of the frame members duo to each case of damage. In particular the paper considers the effect of failure of columns in such frames. Also, the effect of the existence of infill panels in different places in the frame on this behavior is studied by assuming an equivalent strut to replace the infill panel. The best locations of infill panels in each case of damage are investigated to withstand the induced load due to the damage of a supporting element.
Keywords: Damage, frame, infill, reinforced concrete.

1 Introduction

Building structures that are designed to resist the effects of ordinary loads may, however, be exposed to additional local effects arising from various accidents. In completed structures, pressure loads induced by the explosion of natural gas or detonation of bombs and impact loads caused by vehicle collisions, are examples of loads that are usually not considered in structural design. Previous studies [2,4], show that the probability of structural failure due to these abnormal loads is of the order of 10^{-4}, which may exceed failure probabilities associated with unfavourable combinations of ordinary design loads which are of the order of 10^{-5} or less.

If the structure is not designed to dissipate the energy of the accidental load or to absorb the damage, a catastrophic chain reaction of failures that propagates horizontally and vertically throughout a major portion of the structure may ensue. Although absolute safety cannot be achieved, occupants of multistorey buildings should expect to enjoy a reasonable protection against this type of failure. Thus, it is

Abnormal Loading on Structures edited by K. S. Virdi, R. S. Matthews, J. L. Clarke and F. K. Garas.
Published in 2000 by E & FN Spon, 11 New Fetter Lane, London EC4P 4EE, UK. ISBN 0 419 25960 0

useful to understand the limits and capabilities of ordinary structures to resist propagation of damage due to abnormal loads.

2 Objectives of this work

The main aim of this work is to study the behavior of reinforced concrete multi-storey frames when failure of columns in different locations of these frames occurs, Also studied is the effect of brickwork infill panels on this behavior. Another objective of this work is to make an assessment of the performance of structural components as well as the architectural features of reinforced concrete buildings to withstand these effects.

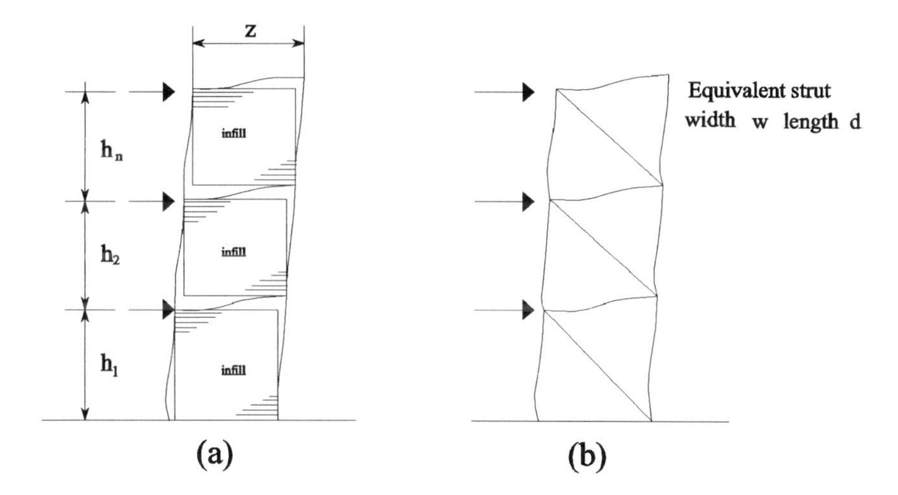

Fig. 1. (a) Lateraly loaded infilled frame , and (b) Equivalent frame.

3 Behavior of infilled frames

Fig. 1. shows the typical behavior an infilled frame subjected to horizontal loads. It is assumed for this structure that the infill and frame are not constructed integrally, nor are they deliberately bonded together, which is the common case in real life.

When the load is applied, the infill and frame are separated over a large part of the length of each side and contact remains only adjacent to the corners at the ends of the compression diagonal as shown in Fig. 1a.. In effect the infill behaves as a diagonal strut and the analogous structure shown in Fig. 1b may be postulated, with equivalent struts replacing the infill panels [6].

4 Studied models

The study is carried out mainly on a 4-bay, 6-storey reinforced concrete frame M10. All dimensions and cross sections of this frame are shown in Fig. 2.. Applied loads are vertical uniformly distributed loads with an intensity of 4 t/m (40kN/m)' on all storeys. The modules of elasticity for steel, concrete and infill are taken 2100, 210 and 60 t/cm^2, respectively. Damage cases are shown in Fig. 3. Case (a) relates to the central column in the ground storey (frame M11) and case (b) to an exterior column in the ground storey (frame M12).

To study the effect of the existence of infill panels, different locations for these panels (shown in Fig. 3) are investigated. Equivalent strut is used in place of each infill panel. The effective width of these struts is taken as equal to its length divided by 3 [5,7,8], while its breadth is the same as the thickness of the infill panel (25 cm).

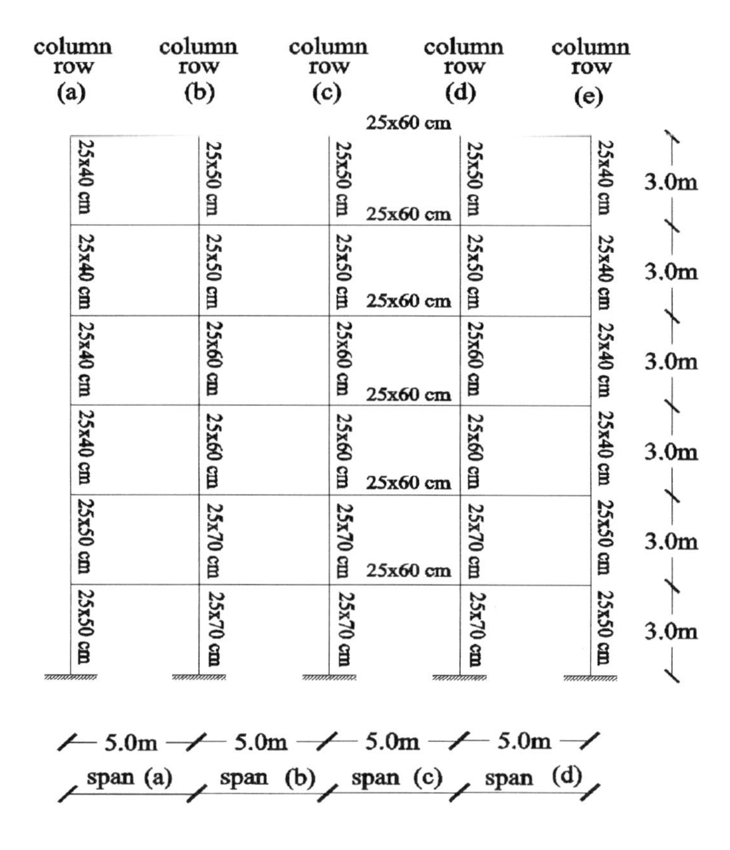

Fig. 2. 4-bay, 6-storey reinforced concrete frame M10

5 Analysis of results

5.1 Behavior of bare frames under damage of structural elements

The bare frame M10 was analyzed at first without any damage and the obtained vertical deflection, normal force and bending moment in the structural elements are used as reference lines in Figs. 4 and 5, to clarify the changes which may occur in each case of damage. The hatched areas in these figures represent the increase in the values of the concerned function. This analysis was made using a standard finite element package (SAP90) [10], using the well known frame element with 3-degrees of freedom at each node (2 translation and 1 rotation) for the frame members and a truss element with 2 degrees of freedom at each node (2 translation) for the equivalent strut elements.

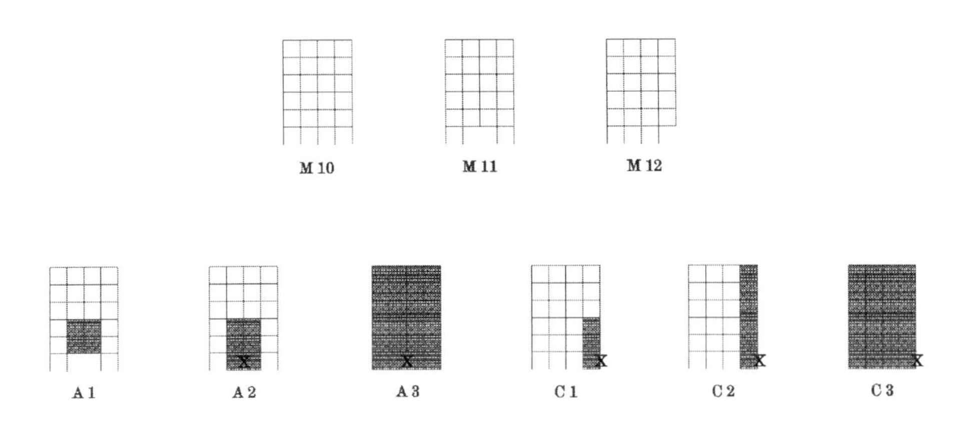

Fig. 3. Studied models

5.1.1 Damage of the central column in the ground storey (M11)

Change in vertical deflection (Fig. 4a)
Points over the damage columns (at column row c) have the maximum vertical deflection with an average value of 1.75 cm which is about 6.6 times the average of the original values at that column row. In column rows b and d, the vertical deflection increases to 1.51 and 1.47 times the original values in the $1^{st.}$, $6^{th.}$ storeys respectively. There is no increase in vertical deflection at column rows a and e)

Change in bending moment (Fig. 4b)
In sections over the damaged column, the bending moment inverses its direction relative to the original case to be positive. Bending moments increase in value to 2.9 and 1.6 times the original values in the $1^{st.}$ and $6^{th.}$ storeys, respectively. The negative.bending moments in beams over the damaged zone (spans b and c) increase in value to 4.6 and 2.9 times the original values in the $1^{st.}$ and $6^{th.}$ storeys, respectively. The negative bending moments in spans a and d increase by an average of 1.33 times the original values. There is an increase in bending moments which appears in column rows b and d to an average value of 11.4 t-m with a maximum value of 14.6 t-m at the $2^{nd.}$ storey.

Change in normal force (Fig. 4c)
In columns, the only increase in normal force relative to the original case is in rows b and d. The new values average at about 1.45 times the average of the original values, with a maximum of 1.51 and minimim of 1.4 times the original values in the $1^{st.}$ and $6^{th.}$ storeys, respectively. The rest of the columns have a decrease in normal force

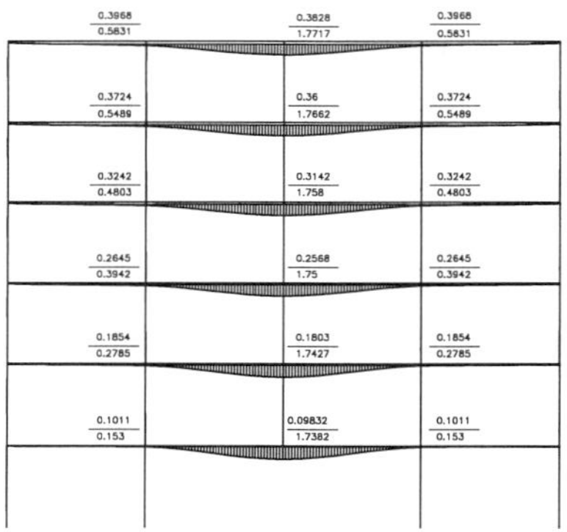

fig.4a. Vertical Deflection (cm)
(Exaggerated 25 times)

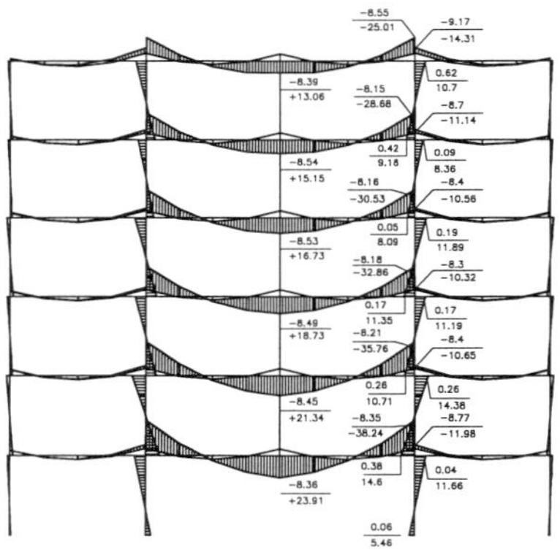

fig.4b. B.M. (m.t.)

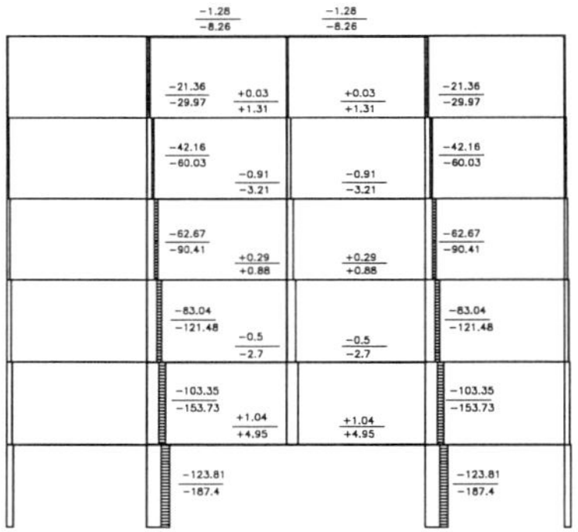

fig.4c. N.F. (t.)

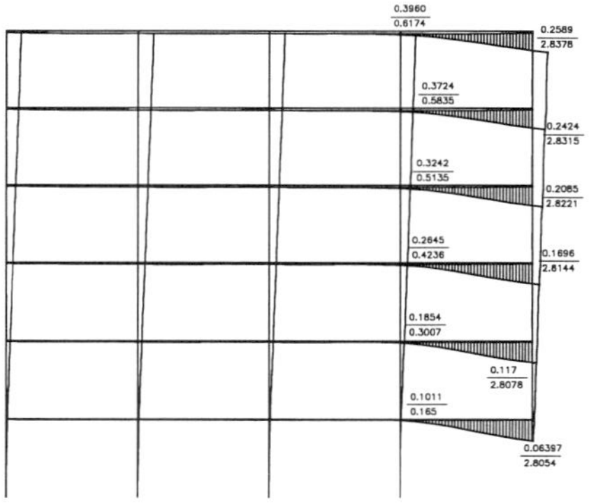

fig.5a. Vertical Deflection (cm)
(Exaggerated 25 times)

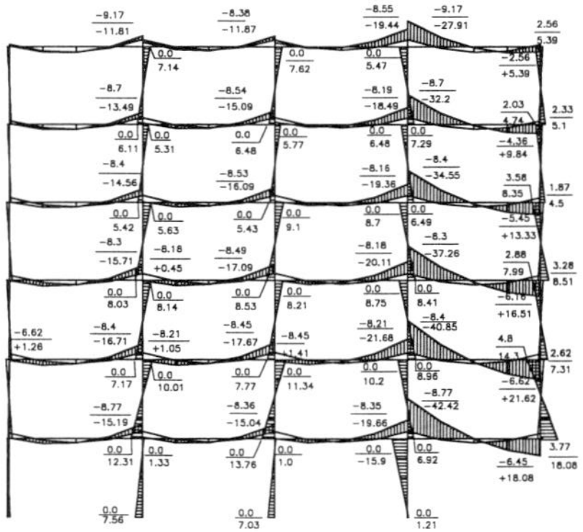

fig.5b. B.M. (m.t.)

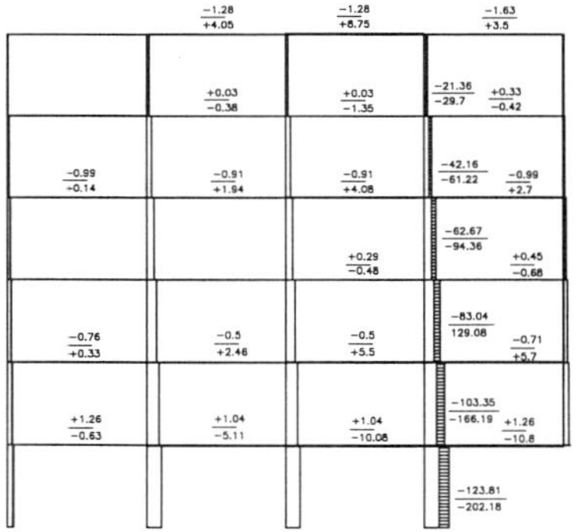

fig.5c. N.F. (t.)

*	Original value
*	After damage value
*	After damage value

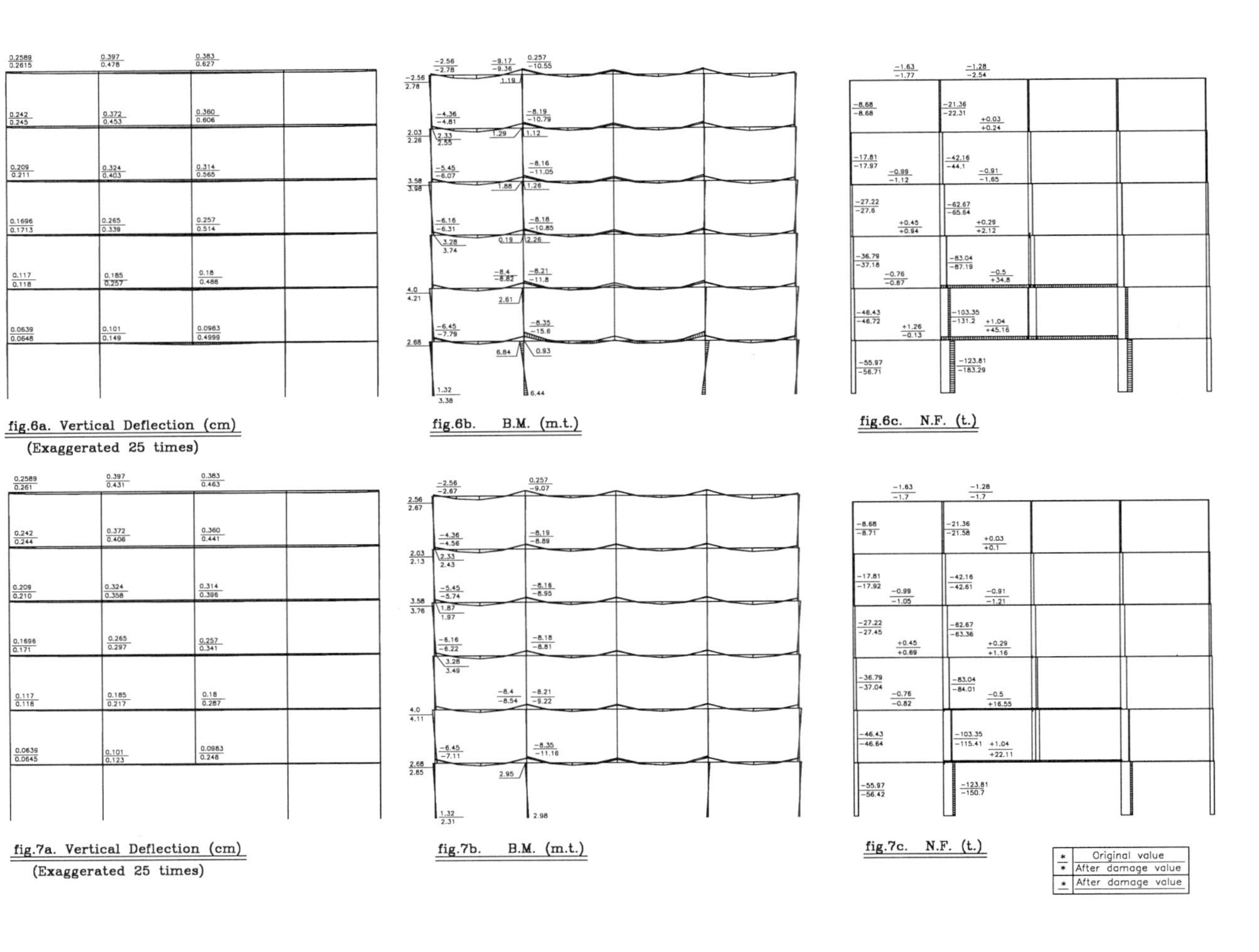

fig.6a. Vertical Deflection (cm) (Exaggerated 25 times)

fig.6b. B.M. (m.t.)

fig.6c. N.F. (t.)

fig.7a. Vertical Deflection (cm) (Exaggerated 25 times)

fig.7b. B.M. (m.t.)

fig.7c. N.F. (t.)

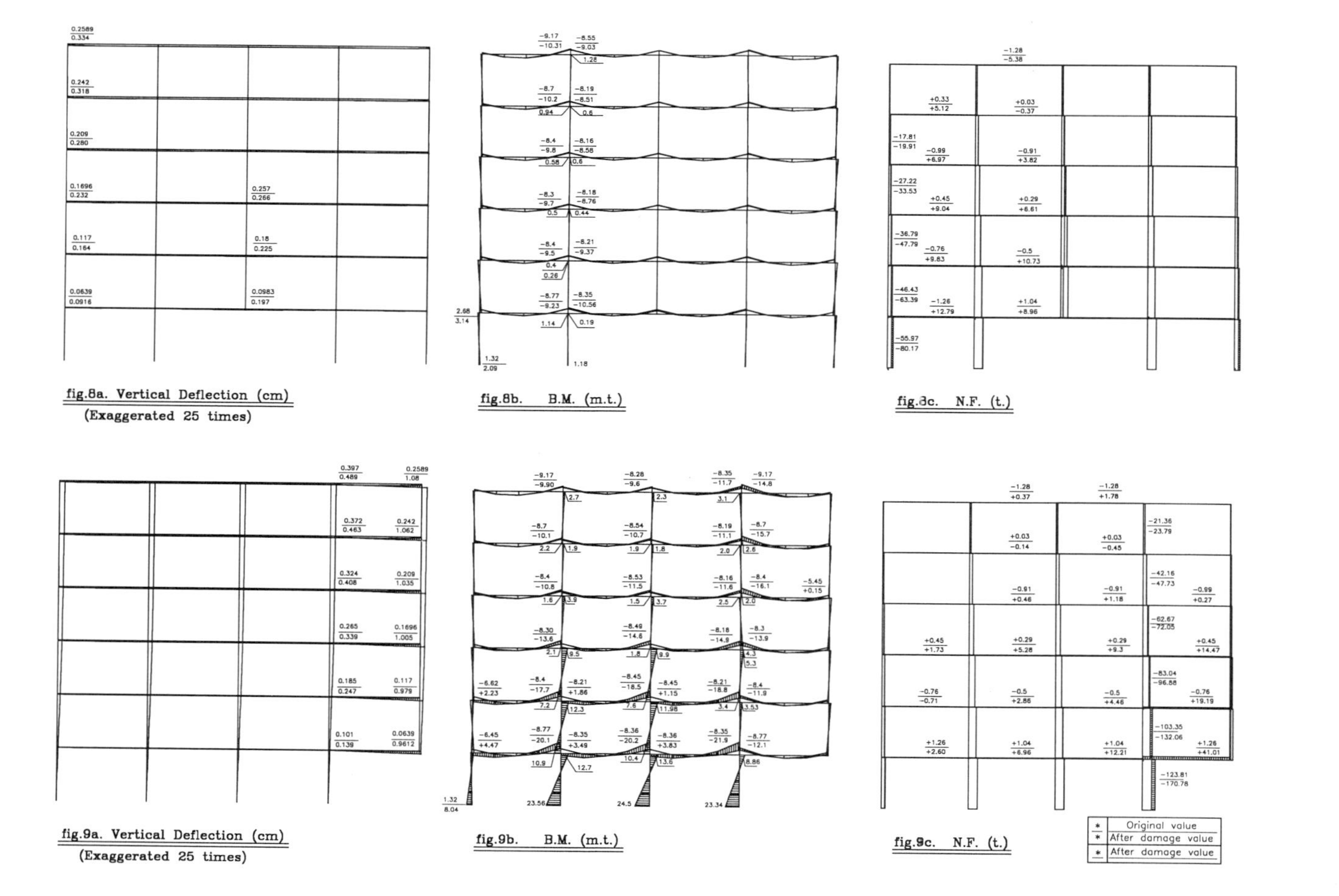

fig.8a. Vertical Deflection (cm)
(Exaggerated 25 times)

fig.8b. B.M. (m.t.)

fig.8c. N.F. (t.)

fig.9a. Vertical Deflection (cm)
(Exaggerated 25 times)

fig.9b. B.M. (m.t.)

fig.9c. N.F. (t.)

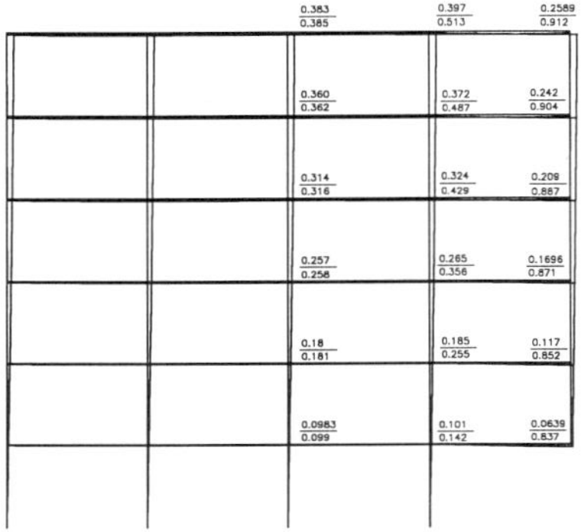

fig10a. Vertical Deflection (cm)
(Exaggerated 25 times)

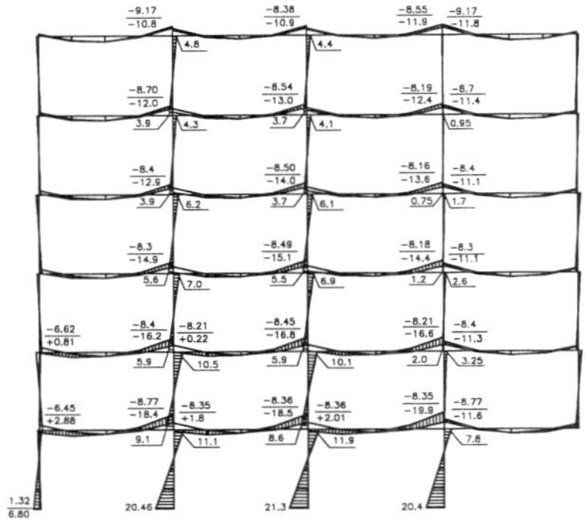

fig.10b. B.M. (m.t.)

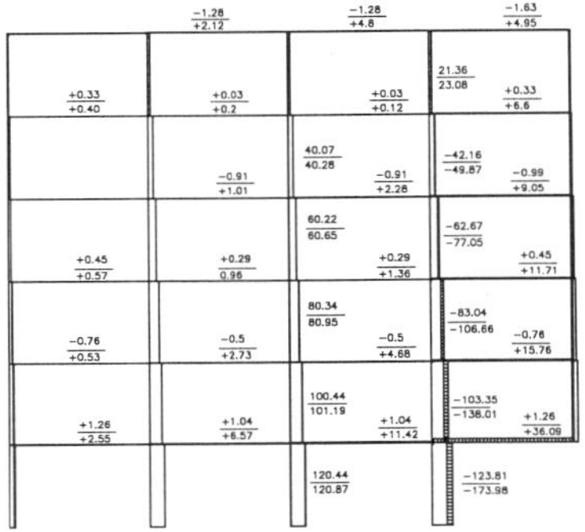

fig.10c. N.F. (t.)

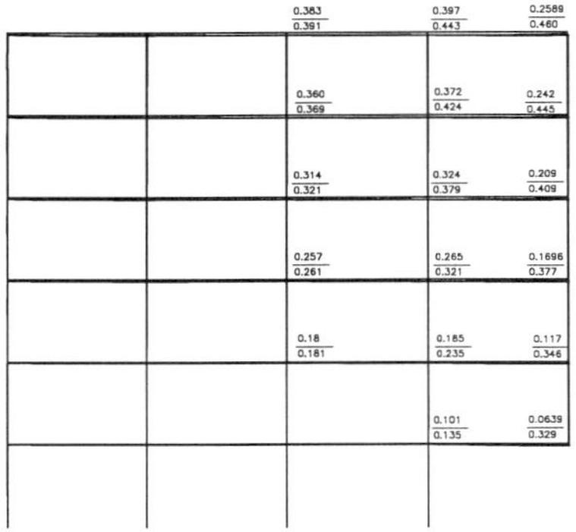

fig11a. Vertical Deflection (cm)
(Exaggerated 25 times)

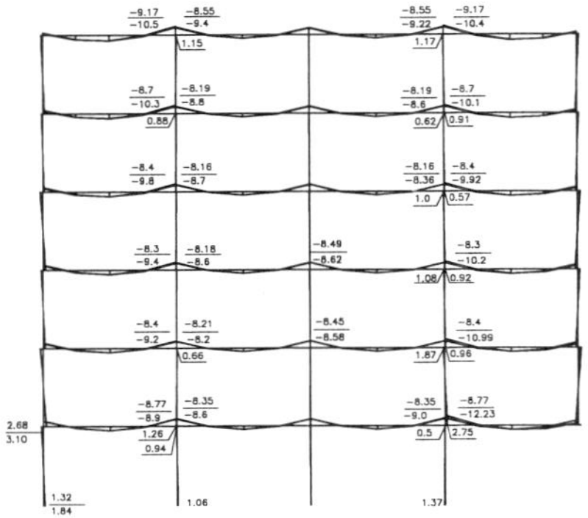

fig.11b. B.M. (m.t.)

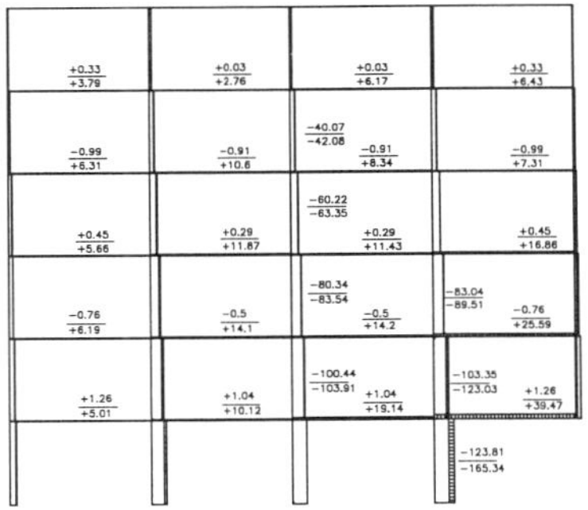

fig.11c. N.F. (t.)

*	Original value
*	After damage value
*	After damage value

values. There is an increase in normal force, which appears in beams over the damaged zone, the maximum compressive normal force appearing at the 6th. storey beam with a value of 8.26 t and the maximum tension appears in the 1st. storey beams with a value of 4.95 t.

5.1.2 Damage of the exterior column in the ground storey (M12)

Change in vertical deflection (Fig. 5a)
Points over the damaged column, (column row e) have the maximum vertical deflection with an average value of 2.8 cm which is about 16 times the average of the original values. In column row d, the vertical deflection has increasing values with an average of 1.6 times the average of the original values at the column row. There are negligible changes in the vertical deflection along the column row a, b and c.

Change in bending moment (Fig. 5b)
In sections over the damaged column, the bending moment inverses its direction relative to the original frame to become a positive.bending moment, with increased values being 2.8 and 2.1 times the original values in the 1st. and 6th. storeys respectively. The maximum increase is at the 2nd. storey with 3.3 times the original value. The negative beding moments in beams on the right to the column line d, (span d) increase in value to 4.8 and 3.0 times the original values in the 1st. and 6th. storeys, respectively.

In spans a, b and c, the negative bending moments in beams increase in value to the average valuess of 1.7, 1.8 and 2.4 times the original values, respectively. In the farthest end of beams, bending moments inverse their direction relative to the original frame in the 2nd. storey in spans a and c and in the 2nd. and 3rd. storeys in span b, and have values with an average of 0.15 times the original values.

There is an increase in bending moments which appears in column rows b, c, d and e, to have values with an average of 8.3, 8.6, 9.7 and 8.94 t-m, respectively, and with maximum values of 12.3, 13.76, 15.9 and 18.08 t-m, respectively. Note that all the maximum values appear in the 2nd. storey except in column row d where it appears in the 1st. storey.

Change in normal force (Fig. 5c)
The only increase in normal force appears in column row d. The increase has values of 1.63 and 1.4 times the original values in the 1st. and 6th. storeys, respectively. The rest of the columns have a decrease in normal force values.

There is an increase in normal forces which appears in beams on spans b, c and d, the maximum compressive normal force appearing in beams is in the 1st. storey with a value of 10.8 t, and the maximum tension appears in beams in the 6th. storey with a value of 8.78 t. Note that, the last two maximum values are in beams of span c.

5.2 Effect of the existence of infill panels
The 4-bay, 6-storey frame of fig. 2. is investigated with infill panels provided in different places as shown in fig. 3. under only two cases of damage (central& exterior column in the ground storey). Also, the original vertical deflection, bending moment& normal force in the frame M10, will be used as a reference lines in Figures 6. to 11. and the hatched areas represent the increase in values of the concerned function.

5.2.1 Damage of the central column in the ground storey

Change in vertical deflection

For cases of infill panels only over the damaged column zone (cases A1, A2), Figs 6a. and 7a. indicate that the average of vertical deflection over the damaged column decreases with an average of 70.2% and 79.3%, respectively, of the average of vertical deflection in case of bare frame M11.

The existence of infill panels all over the frame (case A3), Fig.8a., has the most effective influence on reducing the vertical deflection over the damaged column, as it is reduced with an average of 87.0% of the average of the vertical deflection in case of bare frame M11.

Change in bending moment

In beam sections over the damaged column, bending moments are still in their original direction with a decrease in their values. In the case of infill panels around the damaged column zone except in the ground storey, case A1, as shown in Fig. 6b, has a good effect in reducing the negative bending moment relative to the bare frame M11, in beams over the damaged zone where it has an average of about 1.87 times the original value in the 1st storey beam (spans b and c).

The existence of infill panels in the ground storey around the damaged column and all over the frame (cases A2, A3), Figs. 7b and 8b, represent the most significant effect on reducing the negative bending moments in beams all over the frame relative to the bare frame M11. Where the negative bending moment in the 1st storey beam over the damaged column, which has the maximum value, has values of 1.34 and 1.26 times the original values respectively in each case and there is a slight increase in spans a and e. Also a good effect is noticed in reducing bending moment in columns relative to the case of bare frame M11.

Change in normal force

For the case of infill panels around the damaged column except in the ground storey, case A1, Fig. 6c indicates that the normal forces have values which do not differ much from the values of the bare frame M11. Moreover, there is a high increase in normal forces in beams, specially beams in the 1st storey over the damaged column with a maximum tension of 45.16 t.

The existence of infill panels in the ground storey around the damaged column and all over the frame (cases A2 and A3), has the most effective influence on reducing the normal forces in columns relative to the bare frame M11 and from Figs. 7c and 8c, we can notice that normal force in the 1st storey column (rows b and d) increases to have a value of 1.22 times the original values in case A2 while, there is a decrease in these rows in case A3. In column rows a and e, there is a slight increase in normal force in case A2, while there is an increase in value of about 1.43 times the original value in case A3. Also, there is an increase in normal forces in beams, specially in the 1st storey with maximum tension of 22.1 and 12.79 t, in the two cases A2 and A3, respectively.

5.2.2. Damaged of the exterior column in the ground storey

Change in vertical deflection
The existence of infill panels in the ground storey and all over span d, cases C1 and C2, Figs. 9a. and 10a, has a good effect on reducing the vertical deflection over the damaged column with an average of 66.12% of the value in case of bare frame M12.

The existence of infill panels beside the damaged zone and every where in the frame, case C3, Fig. 11a, has the most effective influence on reducing the vertical deflection either over the damaged column or elsewhere. The vertical deflection over the damaged column is reduced with an average of 85.9% of the average value in case of bare frame M12.

Change in bending moment
In all the studied infill cases, bending moments in beam sections just over the damaged column remain in their original directions with a decrease in values.

The existence of infill panels only over the damaged zone (cases C1, C2), Figs. 9b and 10b, have a small effect in reducing bending moments all over the frame beams. Note that in beams, bending moments in the farthest ends of the damaged column, specially in the first 2-storey beams, inverse their direction relative to the original one with values more than those in case of bare frame M12. Also, we can notice that, there is an increase in bending moments in columns, specially column rows b, c and d.

The existence of infill panels beside the damaged zone and every where in the frame, case C3, Fig. 11b, has the most effective influence on reducing the bending moments all over the frame members (beams and columns), compared with the case of infill around and over the damaged column (case C2).

Change in normal force
For cases C1, C2 and C3, Figs. 9a, 10a and 11a indicate that the normal forces in column row d decrease relative to that in the case of bare frame M12, and the normal forces in the ground storey (column row d) have values of 1.4 and 1.41 times the original value in that column, respectively, in each case.

There is an increase in normal force in beams and we notice that tension appears in all cases C1, C2 and C3, with maximum values of 41.01, 36.09 and 39.47 t, respectively. Note that the maximum tension appears in the 1st storey beams span d in all cases.

6 Conclusions and recommendations

The following conclusions can be drawn out from the present study. In sections just over the damaged column, where the maximum vertical deflection exists, the vertical deflection increases to about 7 times the average of the original value in the case of central column damage, and about 16 times the average of the original values in the case of exterior column damage.

In beam sections just over the damaged column, bending moments inverse their direction relative to the original one. The negative bending moments in beams over the damaged zone have values which are about 4.1 times the original values. The negative bending moments in the beams beside the damaged zone have values about 1.75 times the average of the original values. There is an increase in bending moments in columns, specially in the neighbouring column rows to the damaged one in case of central column damages, and in all column rows in cases of exterior column damages.

In the case of central column damage, the existence of infill panels symmetrically over the damaged column has a significant effect in reducing vertical deflections as it reduces them to about 75% of the vertical deflection in the similar bare frame case. In case of exterior column damage, the existence of infill panels only over the damaged column reduces vertical deflections to about 66% of the vertical deflection in the similar bare frame case.

In sections just over the damaged column, bending moments keep their original direction with a high decrease in their values. The existence of infill panels in the ground storey around the column has a good effect on reducing bending moments all over the frame in case of central column damage. In case of exterior column damage, the infill panels have a limited effect on reducing negative bending moments in beams.

The existence of infill panels in the frame have a limited effect in reducing both vertical deflection or bending moments in case of central column damage compared with the case of infill panels only over the damaged column, while, it has a very good effect in the case of exterior column damage both in reducing vertical deflection and redistribution of bending moments.

7 References

1. Mc Guire, W. (1974) Prevention of progressive Collapse, Proceeding of the Regional Conference on tall buildings, Institute of Technology, Bankok, Thailand, June .
2. Leyendecker, E.V., and, Burnett, E.F.P. (1976) The incidence of abnormal loading in residential buildings, Building Science, Series 89, National Bureau of standards, U.S. Government Printing office, Washington, D.C. September.
3. Bruce Ellingwood, and, E.V. Leyendecker (1978) Approaches for design against progressive collapse, ASCE, Vol. 104,No. ST3, March
4. Bruce Ellingwood, E.V. Leyndecker and James T.P (1983) Probability of failure from abnormal load, ASCE, Vol. 109, No.4, April.
5. Holmes, M. (1961) Steel frames with brickwork and concrete infilling, Proc. Inst. Civ. Eng., Vol. 19, P.P. 473-478.
6. Stafford-Smith, B. and Carter, C. (1969) A method analysis for infilled frames, Proc. Inst. Civ. Eng., Vol. 44, P.P. 31-48.
7. El-Behairy, S. and Abdel-Rahman, A. (1984) Stability of multi-storey infilled frames against failure of columns, Bulletin of faculty of Eng., Ain shams Univ.
8. El-Sayed M.H and El-hozayen (1992) Analysis of closed frames with infill of non-linear behavior Ph.D., Cairo Univ.
9. Rockey, K.C, Evans, H.R. and Nethercot, D.A. (1983) The finite element method, second edition, 1983.
10. Habiballah, A. and Wilson, E.L. (1989) A series of computer programs for the static and dynamic finite element analysis of structures, SAP90", Computer and Structures. Inc., Barkely, California.
11. Disoky, E.H. (1995) Analysis of damaged R.C. frames, M.Sc., Cairo

SEISMIC TESTS ON A FULL-SCALE MONUMENT MODEL
Seismic tests on monument model

A.V. PINTO, G. VERZELETTI and F.J. MOLINA
Safety in Structural Mechanics Unit, ISIS, JRC, EC, Ispra, Italy
A. S. GAGO
Instituto Superior Tecnico, Lisboa, Portugal

Abstract
Pseudo-dynamic and cyclic tests on a full-scale model of part of the cloisters of the São Vicente de Fora Monastery, in Lisbon, are reported. After a first set of pseudo-dynamic and quasi-static cyclic tests performed on the original model, where local damages were observed, the model was retrofitted with four internal continuous bond anchors. A final campaign of quasi-static cyclic tests was carried out on the retrofitted model in order to investigate the effects of this retrofitting technique. The tests were carried out at the ELSA laboratory, JRC, and aimed at the characterization of the non-linear behaviour of limestone-block structures under earthquake loading. Moreover, the assessment of the effectiveness of the retrofitting system was also a major objective of the test campaign. Local and overall stability of the stone-blocks, including columns and arches were assessed for large displacement amplitudes. Comparisons between the test results before and after retrofitting allowed the applicability of these retrofitting solutions and techniques for monumental structures to be investigated.
Keywords: Seismic, tests, model, full-scale, monument structures, retrofitting.

1 Introduction

Public bodies in charge of the maintenance and preservation of cultural heritage are more and more expressing their willingness to assess the seismic vulnerability of important monumental structures and to develop suitable strengthening and retrofitting techniques. The joint research project, *the COSISMO project*, on seismic analysis/assessment of monuments is a first attempt to contribute to the required progress in the field. The Portuguese General-Directorate for National Buildings and Monuments (DGEMN) and the National Laboratory Civil Engineering (LNEC), in Lisbon, and the Joint Research Centre of the European Commission (JRC) set up a research programme including the following main tasks:

Abnormal Loading on Structures edited by K. S. Virdi, R. S. Matthews, J. L. Clarke and F. K. Garas.
Published in 2000 by E & FN Spon, 11 New Fetter Lane, London EC4P 4EE, UK. ISBN 0 419 25960 0

1. Dynamic characterisation of a representative monumental structure, the São Vicente de Fora monastery in Lisbon, by *in situ* testing and numerical modelling;
2. Laboratory testing of a representative model of part of the structure, which will enable the calibration and/or development of non-linear numerical models to be used for predicting the earthquake response of such structures;
3. Development and calibration of non-linear and equivalent linear models appropriate for high intensity shaking;
4. Assessment of the seismic vulnerability of the Monastery, using the developed and calibrated models and appropriate seismic hazard characteristics;
5. Investigation of the applicability of some retrofitting solutions and techniques for monumental structures.

This paper focuses on the laboratory tests on a full-scale façade model carried out at the ELSA Laboratory under the framework of the above-mentioned project. In particular, special attention is devoted to the recently performed tests on the retrofitted model.

2 Structure and test model

The S. Vicente monastery (Fig. 1) represents, from the architectural/engineering point of view, a typical monument of Lisbon, where limestone block masonry columns and arches, forming a resistant structure, are harmoniously combined with stone masonry bearing walls and ceramic dome floors/roofs.

The monument survived the catastrophic November 1st, 1755 earthquake; however, some of the effects of the strong ground shaking are still visible today. The 2 metres thick south external wall of the monastery became curved (mid-span dislocation of about 40 cm); the same happened to the west end-side external wall; and the East end-side edifice, where the Pantheons are presently located, collapsed. A quite detailed description of the damage inflicted to the monastery during the 1755 earthquake is available. Prediction of the damage to the monastery using the present modelling capabilities calibrated on the basis of the experimental results obtained by 'in situ' tests and by laboratory tests is therefore a great challenge.

Fig. 1. S. Vicente de Fora: General view (left), Cloisters internal view (right)

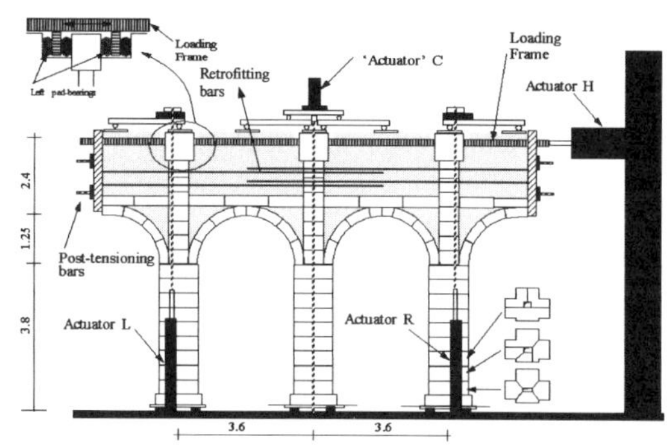

Fig. 2. Façade full-scale model in the ELSA reaction wall laboratory (left) and test set-up - schematic (right)

2.1 Test model

The test model (Fig. 2) was built using materials and construction techniques (stone blocks arrangement) similar to the prototype cloister facade. It is a plane structure with three stone block columns, two complete arches and two external half arches. The upper part of the model is made of stone masonry. Mortar joints three millimetres thick were ensured during the construction. The model was defined taking into account the following two main aspects: it should represent typical monumental structures and it should reproduce the construction techniques (realistic, in terms of materials, scale and stone arrangements).

2.2 The 'retrofitting' technique

The model was retrofitted with four internal continuous bond anchors, two at each level with 2 metres overlapping (Fig. 2). The anchors were placed in horizontal holes drilled from each end side of the model and the grouting was carried out in two phases. First, anchorage of the anchor was guaranteed. Then, a pre-compression of 20 kN was imposed in each steel bar and finally the holes were grouted.

3 Testing of the full-scale model

The question of using a scaled model has been raised many times by the team involved in this project and it was agreed to work with a full-scale partial test-model. It was additionally required to be able to apply simple and realistic boundary conditions to the model. After several numerical simulations with existing models [6], it was decided to adopt the model given in Fig. 2. Even the boundary conditions for a long façade (periodic structure), which include equal displacement of the two lateral end-sides, zero rotation and equal vertical displacements, would have rendered the test set-up very complex. On the basis of the numerical simulations, it was concluded that post-tensioning of the two opposite end-sides of the model would lead to suitable test conditions. Moreover, such testing conditions may represent two distinct parts of a

long façade: one internal (internal column and the two internal half-arches) with period boundary conditions; and the two lateral parts of the model (external columns) representing the external parts of the edifice.
In addition to the lateral boundary conditions, the upper part of the façade (the height of the stone-masonry wall) was also investigated; obviously, it depends on the kind of existing opening (door, window or no opening). In the S. Vicente Monastery, there are windows at the second floor level of the cloisters, which justifies a model with a medium height wall.

3.1 Test set-up, loading devices and measuring system

Having defined the physical model and the required boundary conditions a second phase should be undertaken, which is the definition of realistic loading conditions. Two loading types are considered: the vertical loading due to the remaining upper part of the monument; and the horizontal loading resulting from earthquake excitations.

Concerning earthquake loading two important conditions should be guaranteed namely a rather uniform distribution of forces and a deformation pattern similar to the one resulting from the long façade. Therefore, the overturning moment due to horizontal forces applied at the top of the model (see Fig. 2) must be compensated by means of the vertical servo-actuators located close to the columns of the model. In order to simplify the testing apparatus, a constant vertical force at the central actuator was further assumed. Therefore, in addition to the vertical forces due the missing upper part of the monument, the two vertical external actuators must impose a deformation pattern dictated by the 'shear-like' deformation. The control of the two external actuators is performed by imposing the following two (displacement and force) conditions:

$$d_L = d_R \quad \text{and} \quad F_L + F_R = F_V \tag{1}$$

where d denotes the vertical displacement at the top of the column, the subscripts L and R stand for Left actuator and Right actuator respectively and F_V is the total constant vertical force representing the weight of the 'missing upper part'. This additional load at the top of each column was estimated as 440 kN.

As anticipated above, the additional vertical loads represent the weight of the upper part of the monument, not reproduced in the physical model. Hence, distribution of these loads should reflect the real conditions. For a homogeneous structure a uniform force distribution would be appropriate, but, in our case, a very stiff column-arch structure (limestone blocks) is combined with masonry walls (stone masonry with poor mortars) leading to non-uniform distribution of loads. Numerical simulations with linear and non-linear models indicated that a distribution of forces, in the column and masonry parts represents quite accurately the real situation. Values of 400 and 100 kN were estimated for column and masonry parts, respectively.

The application of horizontal forces, resulting from earthquake excitations was made through an 'original' loading system, which provides uniform distribution of forces. As shown schematically in Fig. 2 (detail at top-left-side), the horizontal forces are equally distributed because they are transmitted from the loading frame to the model through water-pad bearings, which are interconnected. In fact, at each transversal beam of the loading frame, the Left-side pad bearings (L bearing in Fig. 2 – detail) communicate. Therefore, the same pressure develops and the force is proportional to the area of the pad bearing. When the horizontal loading is applied in the left to right sense, the same happens to the right side bearings, R, in Fig. 2). Such a system guarantees a pre-defined distribution of horizontal forces and avoids unrealistic local deformations at the zones of application of loads. In spite of the long and 'flexible' steel-loading frame, a specified distribution of forces can be imposed.

Concerning the measuring system designed for this test, the following types of measurement were adopted:

1. at the base of the columns, underneath the steel base plate, load cells were placed in order to obtain the evolution of vertical forces and the column-base flexural moments;
2. rotations of the column stone-blocks were measured along the height; the horizontal deformation pattern of the columns (internal and one external) can be also obtained from the set of horizontal displacement traducers placed at several levels;
3. deformation of the arches can be derived from the 3 transducers located at the contact zones and from global displacements of the arch stones;
4. vertical and horizontal relative displacements at the key-stone of the arches were also recorded;
5. deformation of the masonry part of the model (upper part) can be derived from the displacement transducers placed in this zone;
6. the forces in each pre-stressing horizontal bar were monitored; in addition, application of forces (horizontal and vertical actuators) were continuously recorded as well as the controlling displacements.

4 Testing programme and test results

Several tests were envisaged for this model apart from the initial dynamic characterization tests in order to obtain frequencies and mode shapes and evaluate damping for very low deformation levels (microns) and the initial stiffness tests. First, two pseudo-dynamic tests were performed for earthquake intensities corresponding to a low and a medium value of return period and then the model was subjected to cyclic tests with pre-defined displacement histories. After these tests the model was retrofitted and more cyclic tests were performed. The cyclic tests with pre-defined displacement histories performed on the original and retrofitted models allowed the influence of boundary conditions on the results, namely the post-tensioning forces in the horizontal ties, and the effectiveness of the retrofitting technique to be investigated.

4.1 Pseudo-dynamic tests

From the dynamic characterization tests and the stiffness tests, initial stiffness was obtained, which, in conjunction with the required initial frequency of the model, dictated the mass to be used in the pseudo-dynamic tests. It is noted that for the pseudo-dynamic tests a one-degree of freedom system (1DOF) was considered. It is well known that the pseudo-dynamic test method allows such an uncoupled definition of the stiffness and mass characteristics, which is not possible in a truly dynamic test. Taking advantage of this feature, the frequency of the system was set to 4Hz, because,

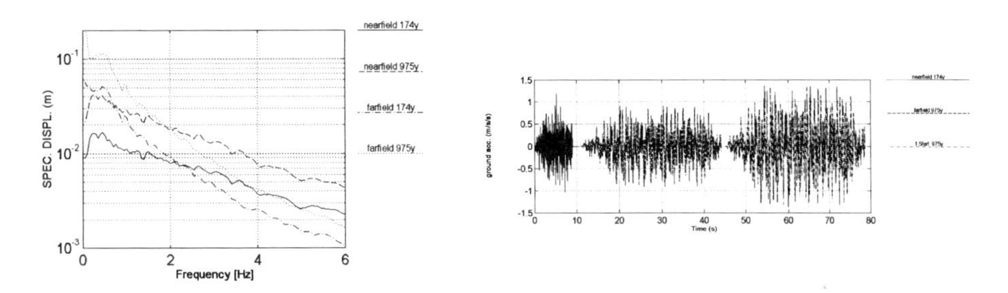

Fig. 3. Accelerograms used and Spectral displacements for far-field and near-field earthquake samples.

the experimental and analytical studies identified such a frequency value [3]. In fact, the results of the in situ dynamic tests of the Monastery [2], and of the numerical dynamic linear analyses found that the frequency corresponding to the first mode of vibration involving global deformation of the cloisters (façade) is approximately 4Hz. Consequently, a mass value of m = 400 tons was adopted for the equivalent 1 DOF system.

Two pseudo-dynamic tests were initially envisaged, corresponding to moderate and high earthquake intensities. Moreover, two types of earthquake loading were used, because two earthquake scenarios should be considered for Lisbon (far-field and near-field) with rather different spectral energy content. As shown in Fig. 3, which plots the displacement response spectra (5% damping) for the two earthquake types and for two return periods (174 and 975 years), the near-field accelerogram response spectrum contains energy in the higher frequency ranges (frequencies higher than 2 Hz), while the contrary happens for low frequencies. Based on these spectral differences and on the pre-test numerical simulations, the following strategy for the pseudo-dynamic tests was adopted. The low-level seismic test was carried out with a near-field type accelerogram (174 years return period) and the high level test was performed with the far-field 975 years return period accelerogram. The accelerograms used in the tests are shown in Fig. 3. The near-field signal, for the low-level test, is a 10 seconds duration while the high-level tests were performed for 30 seconds duration accelerograms. After the high level test, the input signal was multiplied by a factor of 1.5 and a new pseudo-dynamic test was carried out.

The results from the pseudo-dynamic (PSD) tests, in terms of global force displacement diagrams, are given in Fig. 4 for the low-level (LL) and high-level (HL) tests. It must be underlined that the LL test was carried out with a pre-compression force of about 50kN, while the HL test was carried out with a much higher compression force (350kN). During the first PSD test, the LL test, which reached a top displacement less than 8 mm, the moderate value of the compression force lead to the cracking of the model between the column and the masonry. Due to this, the model lost part of its structural 'framing' capacity and the horizontal resisting force dropped down. This case may be considered as representative of a structure without ties located at the extremity of an edifice. In order to avoid premature collapse without exploring other deformation mechanisms, the high level tests were performed with much higher compression forces. Hence, the pre-compression action through the

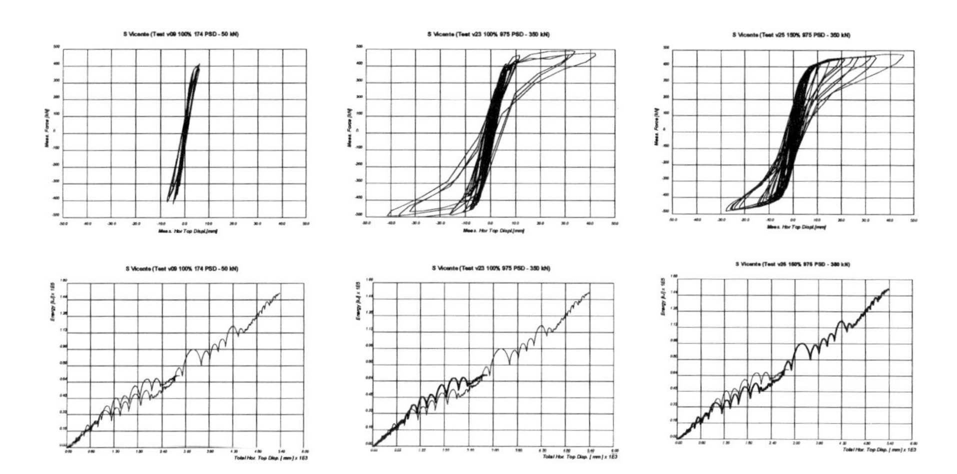

Fig. 4. Force-Displacement and Energy-Displacement (cumulative displacement) diagrams for the low-level (*Column 1*), high-level (*Column 2*) and 1.5x high level (*Column 3*) earthquake tests

steel bars must impose a higher strength of the upper part of the model in order to explore the 'deformation capacity' of the columns.

It is apparent from the diagrams presented in Fig. 4 that for the high-level earthquake tests the structure reaches its strength capacity and maintains it even for high deformation levels. The butterfly-type diagram is typical of these structures (a very high stiffness for low deformation levels and a 'sudden' decrease for important deformations); however, as will be discussed later on, the equivalent damping is quite important for the high level tests. It should be noted that the second high-level test (1.5 times the initial accelerogram (1.5HL)) led to diagrams and mechanisms similar to the HL test and to top displacements proportional to the input intensity. Peak values of top displacement of 43 mm and 64 mm respectively were recorded for the HL and the 1.5HL earthquake tests. There is another important aspect reflected in the energy diagrams given in Fig. 4. In fact it is apparent that the dissipated energy is, for all seismic tests, directly proportional to the cumulative displacement, which confirms that energy dissipation results from friction.

As already mentioned, one of the aims of the research project was to identify suitable and realistic parameters to be used in equivalent linear analyses. To this end, the earthquake test results were analysed using time-domain identification methods [4] [5], and the outcome of this study is shown in Fig. 5. The left column shows the displacement response for the three earthquake tests and the corresponding evolution of eigen-frequencies and equivalent damping. The right column shows the relationship between the displacement amplitude and the eigen-frequency and between the displacement amplitude and the equivalent damping. It is apparent that a linear relationship holds between displacement amplitude and equivalent frequency. Moreover, values no lower than 5% were computed for the equivalent damping for very small displacement amplitudes and the equivalent damping reaches quite high values (7–12%) for medium amplitude levels.

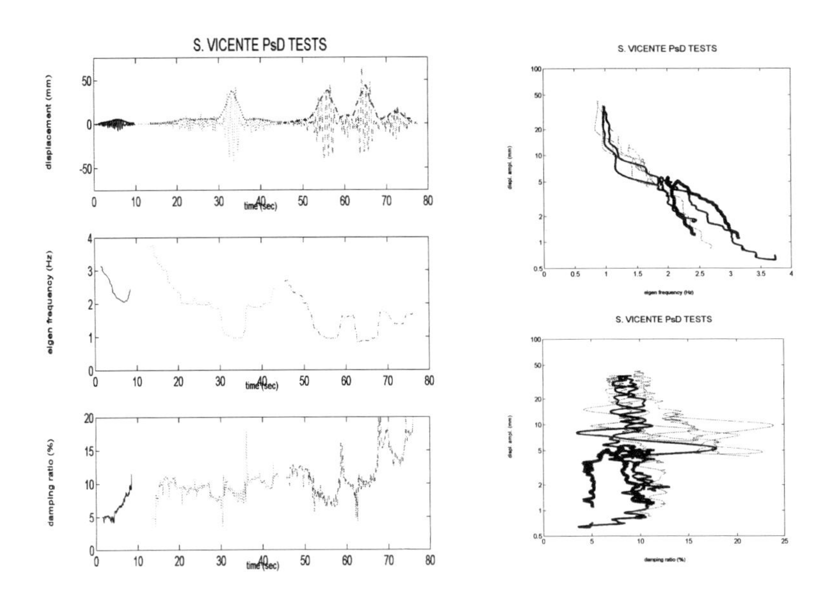

Fig. 5. History of the response to the three imposed earthquakes (left), correlation between displacement amplitude and natural frequency as obtained from the experimental response to the three earthquakes (top right) and correlation between displacement amplitude and equivalent viscous damping ratio as obtained from the experimental response to the three earthquakes (bottom right).

4.2 Quasi-static cyclic tests (original and retrofitted models)

Cyclic quasi-static tests were performed on the original and on the retrofitted model in order to obtain data to calibrate analytical models, to study the effect of the pre-compression forces on the behaviour of the façade and to study the effectiveness of the retrofitting bars. Two cyclic tests were performed on the original model and three were performed on the retrofitted one. The imposed increasing-displacement histories had two constant amplitude cycles for each level and ranged from 8 to 100 mm.

The first cyclic test on the retrofitted model was carried out with the pre-compression force of the last cyclic test carried out on the original model (175 kN). A displacement controlled time history was imposed with amplitudes ranging from 8 to 30 mm. Then the pre-compression force was reduced to 45 kN and a displacement controlled time history was imposed with amplitudes from 30 to 100 mm. Finally, no pre-compression was applied and a new displacement history was imposed, this time with amplitudes ranging from 8 mm to 60 mm.

Force-displacement diagrams for the cyclic tests are given in Fig. 6. The form of the diagrams is apparently the same for both cyclic tests and also for the high-level earthquake test (see Fig. 4). However, a slight difference exists between the test with high compression forces and with the medium compression forces. Both situations develop equivalent strengths for the maximum amplitude (100mm). The main difference between the two comparable cyclic tests (only the pre-compression forces are different) is the smoother transition between the two stiffnesses (closing and

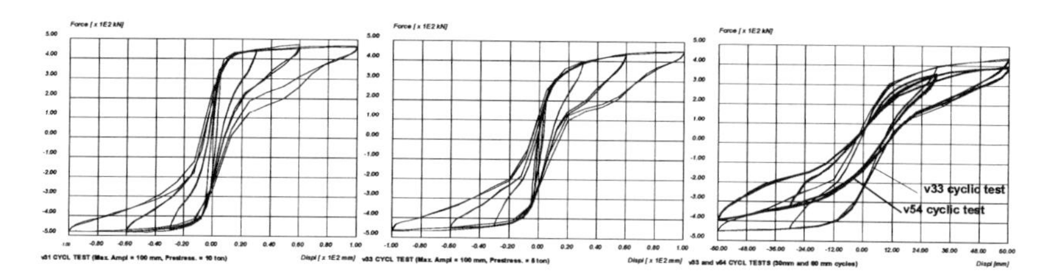

Fig. 6. Force displacement diagrams for the cyclic displacement controlled tests on the original model with high pre-compression forces (*left*), with low pre-compression forces (central) and comparison between the original low compression forces and the retrofitted cases for displacements up to 60 mm (*right*).

opening of the column-block joints). Therefore, one may conclude that 'a minimum' pre-compression level should be applied in order to maintain stability of the upper part of the façade and to improve deformation capacity of these structures. Furthermore, it was verified that compression forces higher than a minimum limit do not improve significantly the cyclic performance of the structure, for the deformation amplitudes experienced.

The results from the tests on the retrofitted model (with continuous-bond anchors), in terms of force-displacement, are given and compared, with the case with low compression forces, in Fig. 6 (right). Similar diagrams were obtained for both cases. However, the retrofitted case showed very important differences in terms of elongation of the upper part of the model. This is apparent in Fig. 7a), which presents the evolution of the elongation of the upper part of the model for the retrofitted and the other comparable cases. Furthermore, the initial length is almost completely recovered in the original models whilst a final permanent deformation is observed in the continuous-bond anchors case. It is shown in Fig. 7c) and 7d) that the opening takes place mainly at the masonry-column interface for all cases; i.e., the deformation patterns and mechanism are similar for all cases. Therefore, the stiffness and strength of the anchors passing through the interface zones control this opening. The continuous-bond anchor (one 20mm diameter bar) and the passing bars (two 36 mm diameter bars) can develop quite different stiffness and strengths, which depend on the steel section, the bond length and the steel yield strength. It is supposed that strong non-linear deformations were experienced in the continuous bond anchor. However, the ultimate deformation capacity was not reached during the tests.

As shown in figure 7b), the energy dissipation is comparable for all cases. In addition, the dissipation mechanisms are similar. Therefore, one may conclude that both solutions are effective. The open issue is to develop suitable models for the design (dimensioning) of the anchors.

5 Final remarks

The tests on the model of the S. Vicente façade of the cloisters have shown the deformation capacity of the column-arch system commonly used in many monumental

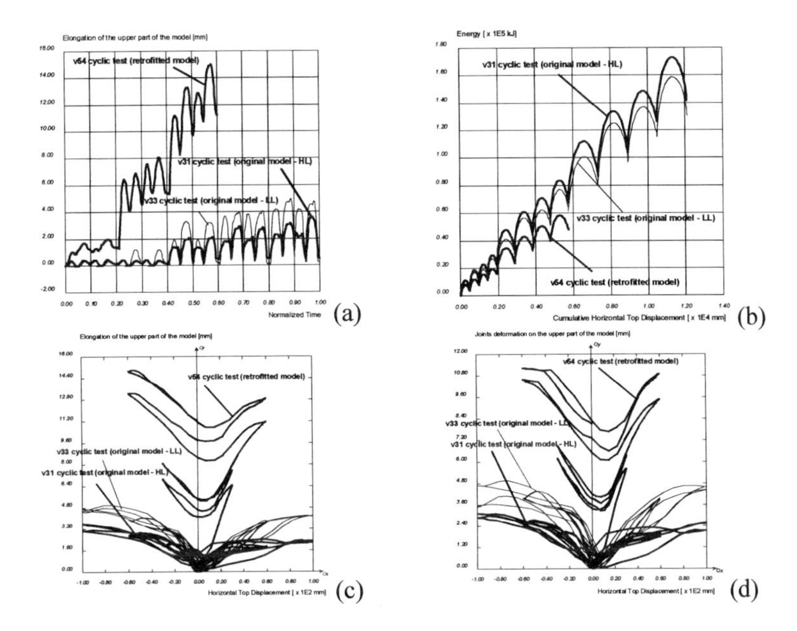

Fig. 7. a) evolution of the elongation of the upper part of the model; b) energy dissipation; c) evolution of the elongation of the upper part of the model; d) evolution of the total joint openings in the upper part of the model.

structures. Drifts of about 2% were imposed without loss of the load carrying capacity. Furthermore, the model has shown important dissipation characteristics due to cracking and friction. Cracking appeared at the block–masonry interfaces in the arch zones and at the interface between the upper part of the columns and the masonry (see Fig. 8). Such dissipation characteristics for important deformation levels exist thanks to the tie bars (pre-compression forces), which allow the required confining of the upper masonry part of the façade model to develop.

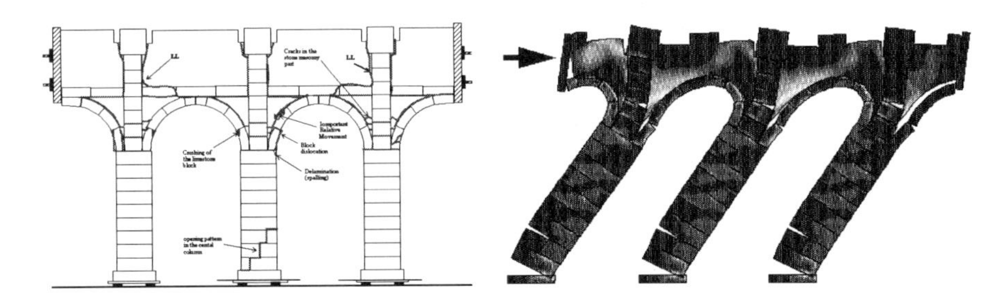

Fig. 8. Deformation pattern and damages (*top*)- analytical result (*bottom*)

'Only' local damages were observed during the tests, namely slight dislocation of column and arch stone-blocks (15 mm maximum value), crushing and delamination of stone blocks at the most stressed contact zones, large cracks in the masonry between contiguous arch-bases and passing through the upper columns and failure (spalling) of a few limestone cover plates. Good deformation and dissipation characteristics for this type of structures are expected if there is a rational distribution of ties at the floor levels. Design and practical application of these tie systems using new analysis tools and construction techniques are under investigation. The test campaign carried out on the retrofitted model showed that the continuous-bond anchors have a rather good performance comparable with that of the pre-compression ties. Moreover, a 'better distribution' of the cracking in the upper part of the model was apparent. In fact, distributed cracks appeared in the stone-masonry wall and the cracks in the masonry-column interface. These tests have shown the applicability and effectiveness of this type of retrofitting in terms of deformation capacity and strength of the model. Another important issue was the performance of the system in the anchors overlapping zone (2m overlapping). It was observed that no damage or loss of bond appeared in this overlapping zone. Therefore, one may assume that such a system can be applied drilling holes with rather small diameters from both sides of the construction, which is an important advantage. There remains however the question of reversibility of the retrofitting solution, which should be taken into account.

6 Acknowledgements

The Portuguese General Directorate for National Buildings and Monuments (DGEMN) financed the work herein presented. The DGEMN and the National Laboratory for Civil Engineering (LNEC), in Lisbon, and the Joint Research Centre of the European Commission (JRC, Ispra, Italy) jointly developed the project COSISMO. The contribution given by Professor C. Sousa Oliveira, IST-Lisbon, to the project definition and development is gratefully acknowledged. Also, the co-operation established with Dr. C.T. Vaz and Dr. E.C. Carvalho from the LNEC in the model construction and testing is underlined. Gratitude is also expressed to P. Pegon from the JRC, for the decisive contribution, through numerical studies, to the definition and preparation of the model and the laboratory tests including the study of the retrofitted model.

The author A. S. Gago acknowledges FCT, Portugal, for the support under the programme PRAXIS XXI (Ph.D. grant 18328/98).

7 References

1. Pinto, A.V., Gago, A. S., Verzeletti, G. and Molina F. J. (1999) Tests on the S. Vicente de Fora Model – Assessment and Retrofitting. *Proceedings of the Workshop on Seismic Performance of Built Heritage in Small Historic Centres - Assisi 99*, Assisi.
2. Campos-Costa, A., Sousa, M. L. N. and Martins, A. (1997) Ensaios Dinâmicos 'in situ' da Portaria de S. Vicente de Fora. *Report LNEC 90/97-C3ES*. LNEC, Lisbon.

3. Dyngeland, T., Vaz, C.T. and Pinto. A.V. (1998) Linear Dynamic Analyses of the S. Vicente de Fora Monastery in Lisbon. *Proceedings of the Workshop on Seismic Performance of Monuments - Monument 98*, Lisbon.
4. Molina, F. J. and Pegon, P. (1998) Identification of the Damping Properties of the Walls of the SAFE Program. *Ispra: European Commission*, Technical Note No. I.98.35.
5. Molina, F. J., Pegon, P. and Verzeletti, G. (1999) Time-domain identification from seismic pseudo dynamic test results on civil engineering specimens. *2nd International Conference on Identification in Engineering Systems*, University of Wales, Swansea.
6. Pegon, P. and Pinto, A.V. (1998) Numerical Modelling in Support of Experimental Model Definition – The S. Vicente de Fora Model. *Proceedings of the Workshop on Seismic Performance of Monuments - Monument 98*, Lisbon.

SEISMIC ANALYSES OF A R/C BUILDING - STUDY OF A RETROFITTING SOLUTION

Seismic retrofitting of R/C buildings

A.V. PINTO and H. VARUM
ELSA, SSMU, ISIS, Joint Research Centre (EC), Ispra, Italy
E. COELHO and E.C. CARVALHO
C3ES - National Laboratory of Civil Engineering, Lisbon, Portugal

Abstract

The preliminary experimental results from the tests on a 4-storey R/C frame structure are presented and discussed. The full-scale model is representative of the common practice of 40~50 years ago in most southern European countries. Special attention is devoted to the study of a retrofitting solution based on bracing and rubber dissipaters, which is intended to increase stiffness and damping and consequently reduce the earthquake deformation demands.
Keywords: Dissipating devices, earthquake tests, global modelling, reinforced concrete buildings, seismic retrofitting.

1 Introduction

The recent earthquakes have dramatically demonstrated that research in earthquake engineering must be directed to the assessment and strengthening of existing structures lacking appropriate seismic resisting characteristics. The very recent 'European earthquakes' (e.g. Italy - 1997, Turkey - August 1999 and Greece - September 1999) confirm and highlight that also Europe may suffer from the vulnerability of its existing building stock.

There is an increasing effort devoted to the issue; however, the great difficulty of the problem is also recognised. In fact it involves several parties namely, the earthquake engineering (EE) community, policy makers and building owners who must work together for a successful solution. To the EE community should be assigned the following tasks: development of effective retrofitting solutions and techniques, and development of codified re-design methods and rules allowing their widespread application by the technical community.

Abnormal Loading on Structures edited by K. S. Virdi, R. S. Matthews, J. L. Clarke and F. K. Garas.
Published in 2000 by E & FN Spon, 11 New Fetter Lane, London EC4P 4EE, UK. ISBN 0 419 25960 0

Along these lines, a European project, the ICONS project, financed by the TMR programme of the Commission, was recently set-up. Under the ICONS-Topic 2 - Assessment, Strengthening and Repair research programme it is foreseen to test pseudo-dynamically two full-scale reinforced concrete frames [1], which are supposed to be representative of the design and construction practice of 40~50 years ago in most southern European, Mediterranean countries. Design of these frames was performed at LNEC [2] under the framework of the ICONS project and the tests will be carried out at the ELSA laboratory of the Joint Research Centre financed by the TMR-Programme, Access to Large-scale Facilities.

Aiming at a preliminary assessment of the structure, and to evaluate the effectiveness of different retrofitting solutions, several ICONS participants are performing non-linear analyses and are also studying different retrofitting solutions. In addition to the non-linear analysis of the frames, the effectiveness of a retrofitting solution based on bracing with rubber dissipaters is assessed. A preliminary analytical assessment of the frame capacity was made by Griffith [3] who also made a simplified design of the bracing system. The final design of the bracing system was made by Taucer [4] without taking into account the infill panels.

This paper summarises the results from the non-linear analyses of the structure considering several cases, namely: the bare frame (Frame), the infilled frame (Frame + Inf) and the retrofitted frame (Frame + Inf + Ret). Also included is the analysis of the retrofitted frame without infill panels, because the design of the retrofitting system was performed ignoring the infills. These numerical results were labelled (Frame + Ret).

Section 2 gives details on the structure, materials, loads and retrofitting solution. The modelling aspects (models, assumptions, etc.) and corresponding parameters are presented in section 3. Section 4 focuses on the non-linear analyses and corresponding results. Section 5 briefly presents and discusses the experimental results from the bare frame tests, discusses the issue of modelling (refinement, parameters, etc.) and compares the numerical results with the available experimental ones. Finally, section 6 summarises the main conclusions of the study.

2 Structure, materials, loads and retrofitting solution

2.1 Structure geometry and material properties

The dimensions of the building and section details are shown in Fig. 1. It can be seen in the elevation and plan drawings that the storey heights are 2.7 m and there are two 5 m span bays and one 2.5 m span bay. Brick masonry infill (200 mm thick) is contained within each bay. The left-hand bay infill contains a window (1.2 x 1.1 m) at each of the 4 levels. The central bay contains a doorway (2.0 x 1.9 m) at ground level and window openings (2.0 x 1.1 m) in each of the upper 3 levels of the building. The right-hand (2.5 m span) bay contains solid infill (i.e., without openings). It should be noted that the longitudinal reinforcing steel was smooth round bars, not the deformed steel bars used for reinforcement today. All beams in the direction of loading are 250 mm wide and 500 mm deep. The transverse beams are 200 mm wide and 500 mm deep. The concrete slab thickness is 150 mm. The column splice joint detail and the column stirrup detail should be noted in particular. Their likely poor seismic performance will be discussed later.

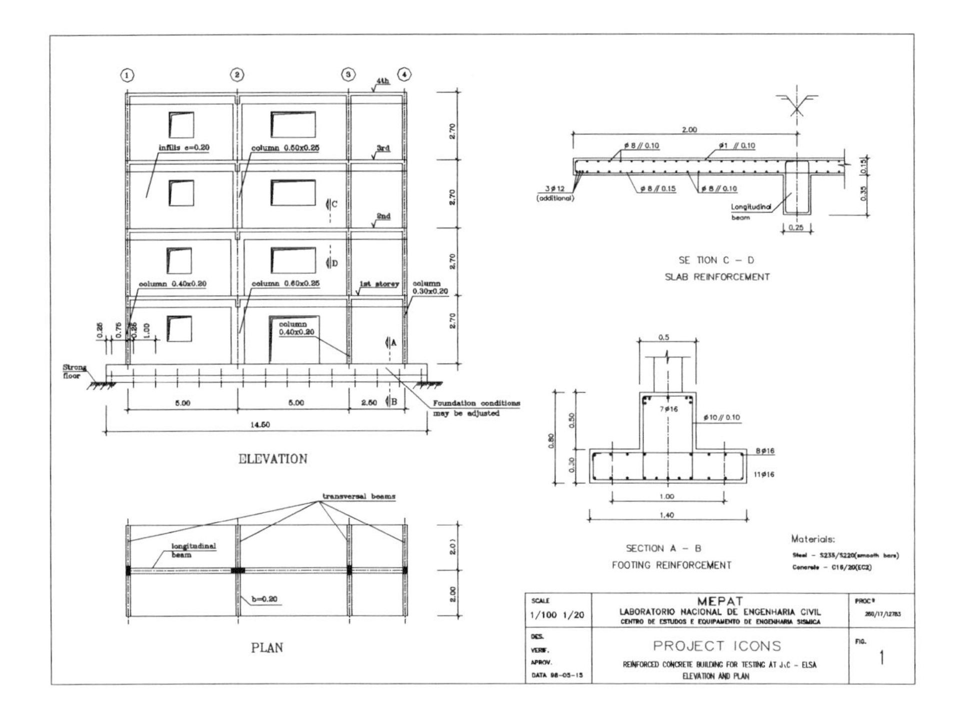

Fig. 1, Plan and elevation views of concrete frame plus masonry infill building

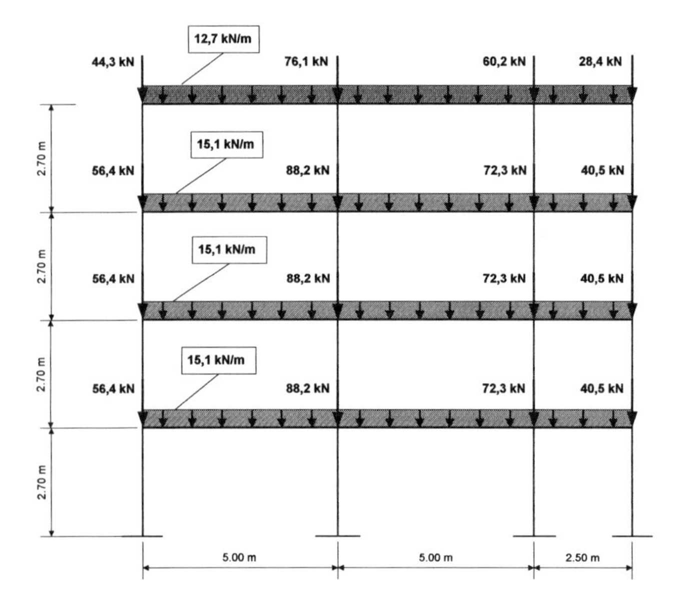

Fig. 2, Scheme of vertical loads for non-linear analyses

Preliminary calculations have been carried out in order to establish which failure mechanisms are most likely to occur under seismic loading. In order to do this, the mean values for the respective material strengths shown in Table 1 have been used.

2.2 Loads, masses and input motions

Vertical loads - For the analyses, vertical distributed loads on beams and concentrated loads on the column nodes were considered, in order to simulate the dead load other than the self-weight of the frame (live-load, weight of partitions, finishing). Fig. 2 gives the details of the loads considered [2]. The accelerograms considered in the non-linear analysis were derived from hazard consistent response spectra corresponding to several return periods. Accelerograms with 15 seconds were assumed. The storey masses considered were: 40.0 tons for the last floor and 44.6 tons for the others. A Rayleigh damping of 2% for the first and second modes was considered.

2.3 Retrofitting solution

It is expected that the 4-storey RC frame under analysis will perform unsatisfactorily for the earthquake motions corresponding to those assumed in the present design codes. Several deficiencies were identified in the structure, such as, inadequate dissipation/collapse mechanisms, inadequate detailing of members and joints. In order to improve the seismic performance of such a structure, a retrofitting intervention is required. There are three basic solutions to increase seismic performance of the structure, namely: to isolate the structure, to increase its deformation capacity and to increase its stiffness, strength and damping characteristics.

Table 1. Material properties

Material	Relevant Properties (*mean values*)			
Steel (FeB22k)	$f_{sy} = 235MPa$	$f_{su} = 365MPa$	$\varepsilon_{su} = 29.9\%$	
	$E_s = 200x10^3\,MPa$			
Steel (tests results)	$f_{sy} = 337MPa$	$f_{su} = 455MPa$	$\varepsilon_{su} = 25.0\%$	
Concrete (C16/20)	$f_{cu} = 24\,MPa$	$\varepsilon_{cu} = 0.2\%$	$f_{tu} = 1.9\,MPa$	$E_c = 20x10^3\,MPa$

The retrofitting herein studied is based on the last solution. It is a bracing system with rubber dissipation devices, which will increase stiffness and damping of the system and consequently reduce the deformation demands.

Two alternative layouts were studied for the bracing: one located in the central bay (K-bracing) (see Fig. 3), which leads to better distribution of the storey forces but interferes with the openings (door and windows) and the other (X-bracing) located in the shorter external bay. These two alternative solutions led to similar results.

The design of the bracing system, including the dissipation devices, was performed assuming [3] that 1% drift (27 mm inter-storey drift) corresponds to the ultimate limit state for the frame under analysis. Furthermore, it was assumed that, for these deformation levels, the effects of the infill panels are negligible. Further, it was assumed that the peak base shear strength of the frame, for the 1% drift, is 150kN and the effective stiffness (secant stiffness) of the equivalent SDOF system with the mass located at 2/3rd of the total height of the building leads to a Period (Ts = 1.8 sec).

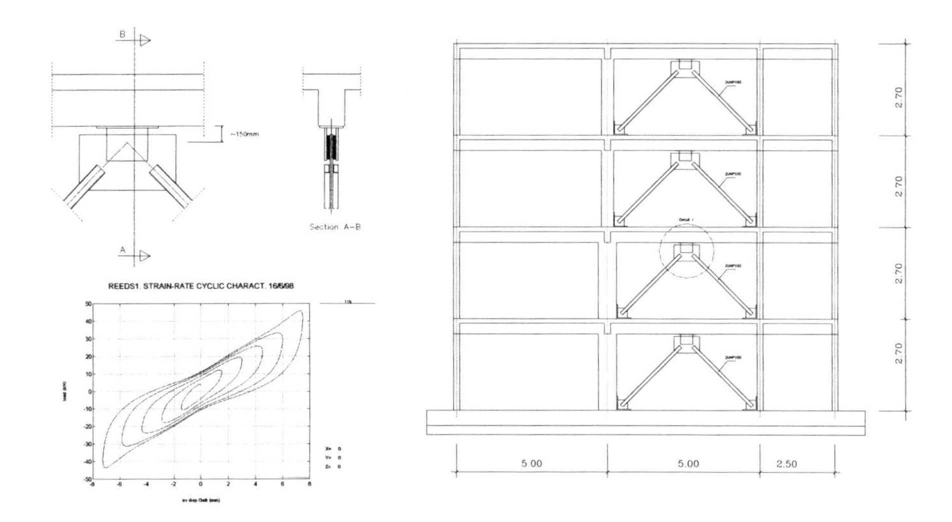

Fig. 3, Bracing system: Device details and general layout and typical diagram for a device tested at ELSA under the project REEDS

The design displacement spectra for the different damping ratios were derived from a basic one for 5% damping (assumed to increase linearly from 0, for T=0 seconds, to 200 mm for T=2 seconds, and being constant for higher periods) using the following 'correction factors' (SQRT(5/ζ)).

For a 50-years non-exceeding probability of 10%, a device is required at each storey with the characteristics given in Table 2.

Table 2. Characteristics of the Energy dissipation devices (one device)

Int.-Storey Drift 1%	**Location Storeys**	**DLF**	**F_u (kN)**	**D_u (mm)**	**F_y**	**K_1**	**Comment**
10% Non-exceeding probability	0-2	0.35	80	25	$F_u/3$	$K_0/10$	1 device per storey (Fig. 3)
	2-4		50	25			

Energy dissipation device loss factor - DLF; DLF = tang δ; $\delta = \sin^{-1}(2W/(\pi\Delta W)$; W - area surrounded by the hysteresis loop; ΔW - half of the area of the rectangle that inscribes the hysteresis loop (= $2F_{max}.D_{max}$)

*Note: Devices are able to accommodate displacements and forces up to 140% of their nominal capacity (**F_u,D_u**)*

3 Modelling

The structure (reinforced concrete frame) has been modelled by beam elements with non-linear behaviour at the potential hinge zones (vicinity of the frame joints) and linear elements in the internal parts of the structural elements. Furthermore, an elastic element was also considered to simulate the joint thickness. The non-linear elements are represented by a fibre model with uniaxial constitutive laws for concrete and steel. To simulate the slab contribution, 1.0 m was considered for the effective flange width.

The infill panels were simulated with bidiagonal struts and the bracing system with dissipaters were represented with bar elements (bracing) and a non-linear spring element for the dissipater.

The length of the non-linear fibre element was estimated on the basis of empirical formulae and taking into account that this element is a Timoshenko element with constant curvature (one integration point only). Assuming a common empirical expression for the effective plastic hinge length and that the curvature in the plastic hinge zone has a parabolic distribution, the equivalent length hinge-element, L^*_p, calculated for the same chord rotation, depends on the ductility. However, it tends asymptotically to half of the empirical value of the plastic hinge length.

3.1 Concrete model

In compression, a parabolic curve is assumed from the initial unloaded stage up to the peak stress values, with initial tangent modulus equal to the concrete Young's modulus. The softening branch is described by a straight line, whose slope depends on the degree of confinement. Under tensile stresses, the behaviour is described by a linear elastic branch with a subsequent softening branch, which accounts for tension stiffening effects. The cyclic behaviour of concrete has been firstly described by a crude model representing the main feature of the concrete behaviour under cyclic loading. In a second stage, the model has been improved in order to account for secondary effects such as crack closing and to avoid eventual numerical difficulties in the algorithms. Analytical formulae and a detailed description of this model can be found elsewhere [5].

3.2 Steel model

This model includes typical curves for monotonic and cyclic loading. The monotonic curve is characterised by an initial linear branch followed by a plateau and a hardening branch up to failure. The cyclic behaviour is described by an explicit formulation proposed by Giuffré and Pinto and implemented by Menegotto and Pinto [5].

3.3 Masonry (infill) model

The model for infill panels is the strut model proposed by Combescure [6]. It is a general multi-linear model that accounts for cracking, compression failure and strength degradation due to either monotonic or cyclic loading as well as for the pinching effects due crack-closing. The model assumes no tensile resistance and the behaviour in monotonic compression is described by a multi-linear curve including a primary linear elastic behaviour, a second branch approximating the cracking process and two final branches representing two phases of the masonry behaviour, which can be considered as plastic behaviour (crushing of the masonry panel) with positive and subsequently negative strain hardening. Cyclic behaviour is characterised by a linear unloading-reloading law without plastic displacement in the primary branches. The hysteretic behaviour, after having reached the plastic point, is also governed by a multi-linear curve with specific rules to account for plastic deformations, crack closing strength degradation.

Table 3. Mechanical properties - Mean values used in the analyses

E_p GPa)	G_p (GPa)	f_{tp} (kPa)	C_R	ν	μ
1.28	0.24	200	0.9	0.05	2.5

Identification of the strut model parameters was performed by empirical expressions suggested in [7]. The values showed in Table 3 were considered in the analyses, where: E_p - Young modulus; G_p - Shear modulus; f_{tp} - reference tensile strength; C_R - factor of quality of masonry work; ν - post-yield slope of envelope curve; and, μ - ductility factor.

3.4 Dissipater model

The dissipaters were simulated by a bilinear model (see Fig. 3). The steel model introduced above was used to represent the constitutive uni-axial law of the dissipater setting the model parameters according to the relevant requirements, namely a sharp transition between the linear and the 'post-yielding curves and the tangent of the asymptotic curve defining the post-yielding range.

4 Non-linear analyses

Static pushover analyses were initially performed, in order to identify the global behaviour of the structure and to compare relative strengths (frame and frame+infills) and corresponding evolution with the imposed deformations. Non-linear analyses were performed for several earthquake intensities. Some results (drift profiles and vulnerability functions) from the non-linear analyses are hereafter illustrated.

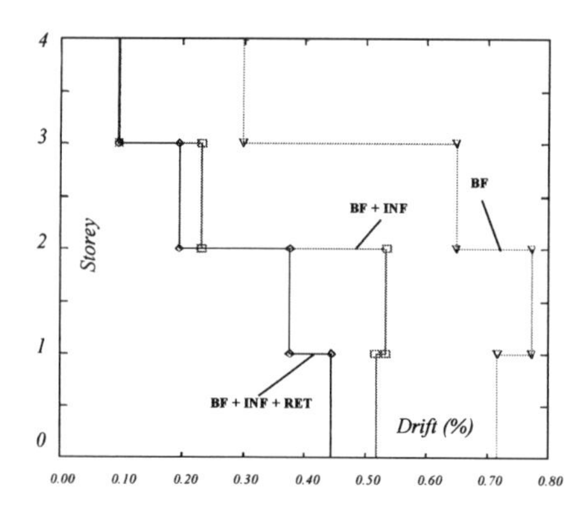

Fig. 4, Maximum drift profiles for bare, infilled and retrofitted frames (975yrp)

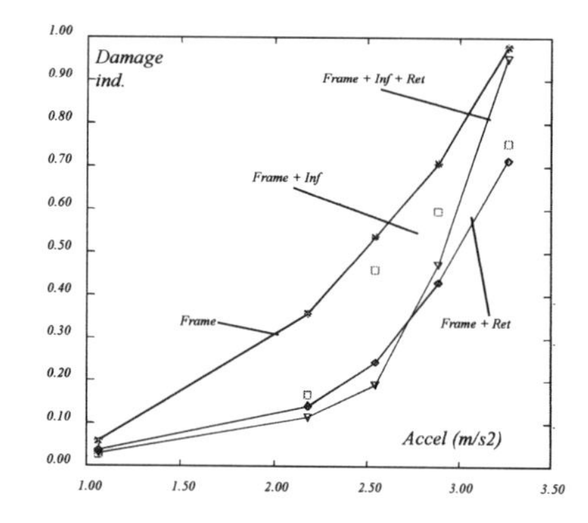

Fig. 5, Global damage on the frame structure (evolution with input intensity)

Table 4. Frequencies calculated for bare and infilled frames

Frequencies (Hz)	1st Mode	2nd Mode	3rd Mode	4th Mode
Bare frame	1.47	4.32	7.04	9.55
Infilled frame	3.85	10.90	14.13	16.78

5 First test results against numerical results

5.1 Results from the tests on the bare frame specimen

A series of pseudo-dynamic (PsD) tests on the reinforced concrete full-scale frame model is currently being carried out at the ELSA laboratory [1]. It is programmed to test both the infilled and the bare frame and to assess experimentally the effectiveness of various retrofitting solutions and techniques. The tests on the bare frame were just performed and a few results are hereafter presented.

The bare frame specimen (full-scale 4 storey R/C frame - without masonry infill) was subjected to one earthquake corresponding to 475 years return period (475-yrp) and subsequently a second PsD test with a 975-yrp was carried out. The results from these tests are given in Fig. 8-9, in terms of maximum inter-storey drift profiles for positive and negative deformations and shear-drift diagrams for the 3rd storey.

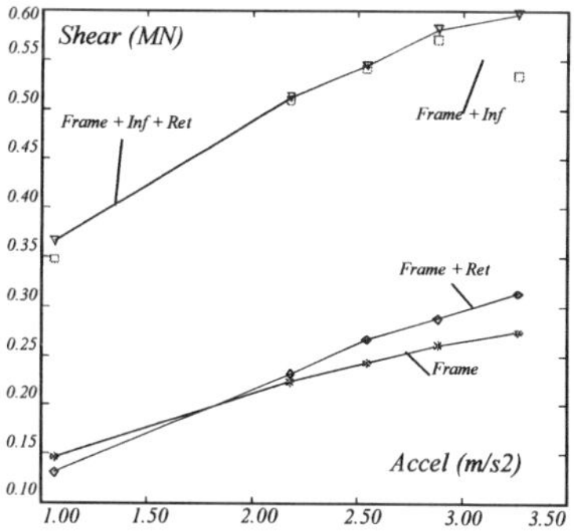

Fig. 6, Maximum base-shear (evolution with input intensity)

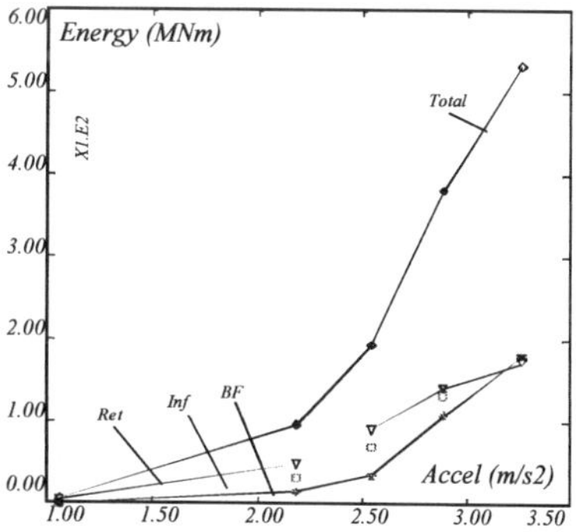

Fig. 7, Energy dissipation (evolution with input intensity)

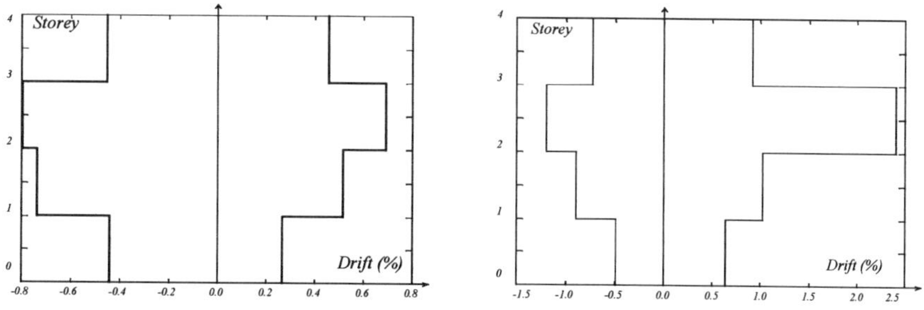

Fig. 8, Maximum Drift Profiles for 475 yrp (left) and 975 yrp (right)

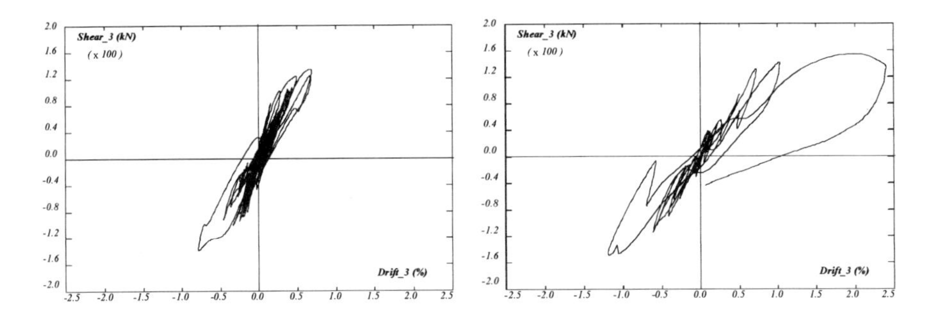

Fig. 9, 3rd Storey Shear/Drift diagram for 475 yrp (left) and 975 yrp (right)

It is apparent that the deformation demands concentrate in the 3rd and 4th storeys for the 475-yrp earthquake test and collapse of the 3rd storey was almost reached for the 975-yrp test. This test was stopped at 7.5 seconds in order to allow repair and retrofitting and to assess their effectiveness in the subsequent tests.

From these tests on the bare frame it is possible to confirm the storey mechanism, which was expected to develop during the earthquake response. In fact, the structure represents common design practice of ~40 years ago when seismic loading was roughly considered or not even taken into account. From the shear-drift diagrams for the 475-yrp test it is apparent that a rather limited non-linear behaviour (storey ductility of about 2 at the 3rd storey) and quite limited damage occurred during the test. Slight cracking at column extremities, as well as in the girders (at the slabs - for negative moments) could be observed and no spalling of cover concrete occurred. The 975-yrp test was subsequently performed and was stopped at 7.5 seconds because failure of the 3rd storey was imminent. In fact, clear hinging of the strong column of the 3rd storey at the base, top and also at the bars termination zone (700 mm from the base of the column) developed with severe damage (yielding, spalling and yielding of the stirrups at the bars termination one). Disclosure of the 90 degrees bent stirrups was not observed but it would certainly have occurred if the test had been continued.

The results have only been recently available and a more detailed analysis is required. However, it is already possible to confirm the high vulnerability of these structures. In fact, it was demonstrated that, in spite of the very limited damages for the 475-yrp earthquake, the demands for a slightly higher intensity earthquake (1.3 times the reference earthquake, in terms of peak acceleration) led to imminent storey failure and consequent collapse of the structure. Development and validation of effective (also economical) retrofitting solutions and techniques for this type of structures is therefore urged. The second part of the testing campaign will devoted to these issues.

5.2 Numerical modelling - refinement and model parameters

One of the important objectives of the numerical benchmark on the response of the structure is to find out the most suitable numerical models to predict the seismic response of this kind of structures and to identify the sensitivity of the models to their characteristic parameters. It is also expected that such structures will experience shear failure, failure at the beam column joints and phenomena such as slippage of rebars (steel rounded bars) and strain penetration. Therefore, one should use appropriate models to take into account most of the above mentioned phenomena.

The JRC used a fibre model [5] considering a rectangular cross-section for the columns and a T-beam to represent the girders because such a model allows both bending and shear to be considered, which is likely to control failure in the central stocky column. However, the following aspects were not taken appropriately into account: the inter-storey height was uniformly considered with 2.7 m but, as the beam element supporting the cross-section should be located at the cross-section centroid, the first story height must be shortened. Therefore, the first storey stiffness and strength were underestimated. Additionally, the slab participation was also almost neglected. This point is particularly relevant for the refined modelling considered because the effects of the slab reinforcement can be significant. In fact, as the equal displacement condition for the storey nodes is not imposed, the girder is allowed to deform axially and the section stiffness drops suddenly after cracking. On the contrary, this drop does not happen in the columns. Consequently the relative strength of the columns and girders may differ strongly from the reality.

The results from post test non-linear analyses (maximum drift profiles) taking into account the aspects discussed above are shown in Fig. 10 together with the results from the experimental tests and the numerical results obtained at LNEC. These non-linear analyses were performed with a Takeda-type model and the parameters for the multi-linear constitutive laws were obtained assuming full-cracked sections. Furthermore, bilinear models were considered for the envelope curve (pointing directly from origin to yielding). Therefore, the 2% damping considered by LNEC seem to be insufficient to take into account the cracking affects. The higher flexibility of the LNEC model is apparent in Fig. 10. However, drift profiles (pattern) are rather well in agreement with test results.

Therefore, it is concluded that much care should be taken in the modelling of these structures. Furthermore, the use of refined models may lead to unrealistic results if the model parameters are not correctly chosen. It is also clear that the sensitivity of the response to such model parameters increases with the complexity of the models.

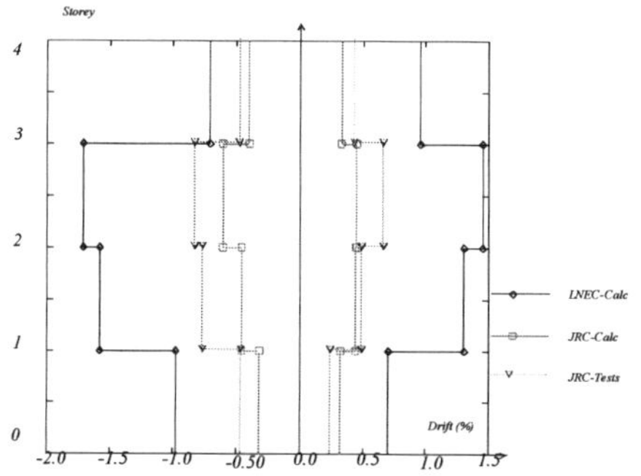

Fig. 10, Drift profiles (numerical and experimental)

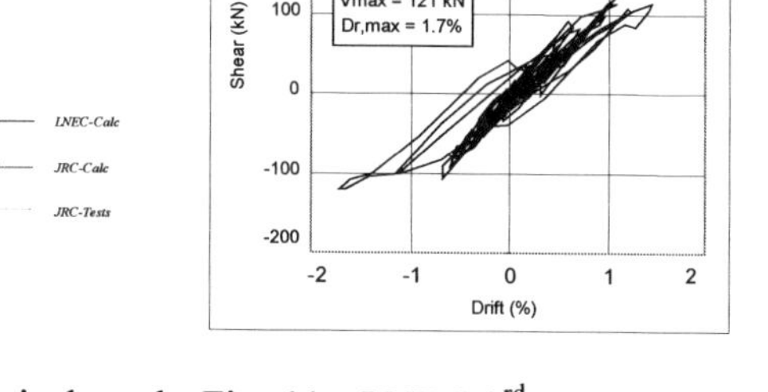

Fig. 11, LNEC 3rd Storey Shear/Drift diagram

6 Conclusions

The results from the analyses show that the infill panels considerably protect the reinforced concrete frame. The numerical analyses for the retrofitted frame case allow the following conclusions to be made:

- The proposed light retrofitting solution is effective for low, medium and high intensities but not particularly effective for very high intensities, when infill panels exist. This retrofitting system was designed for the bare frame and it is very effective for this case. However, a more accurate design shall take the infill panels into account.
- The system leads only to a small increase of storey shear forces.
- Additional energy dissipation - the energy dissipation is equally shared by the RC frame, the infill panels and the retrofitting devices.

The preliminary results from the bare frame tests demonstrate how vulnerable this type of structures are. In spite of a 'satisfactory performance' for the nominal input motion, the structure exhibits a premature storey collapse mechanism (column hinging at the 3rd storey) for an input motion slightly higher than the nominal one.

7 Acknowledgements

The work presented has been developed under the ICONS TMR-Network research programme and the ELSA experimental tests on the 4-storey RC frames were financed by TMR - Large-scale facilities programme of the European Commission.

The co-operation and contribution of the European institutions and researchers involved in the ICONS research network is gratefully acknowledged.

8 References

1. Pinto, A., Verzeletti, G., Molina, J., and Varum, H. (1999) Pseudo-dynamic tests on non-seismic resisting R/C frames, *JRC-Publication (in print).*
2. Carvalho et al. (1999) Design of a 4-storey R/C frame to be tested at the ELSA laboratory, *LNEC report*, Lisbon.
3. Griffith, M. (1999) Seismic retrofit of R/C frame buildings with masonry infill walls - Literature review and preliminary case study; *JRC-Publication (in print).*
4. Taucer F. (1999) A displacement based design methodology for retrofitting of frame structures using dissipation devices, *JRC-Publication.*
5. Guedes, J. (1997) Seismic behaviour of RC bridges - Modelling, numerical analysis and experimental assessment, *PhD Thesis, FEUP*, University of Porto.
6. Combescure, D.; Pegon, P. (1996) Introduction of Two New Global Models in CASTEM 2000 for Seismic Analysis of Civil Eng. Structures; *JRC-Publication.*
7. Zarnic, R.; Gostic, S. (1998) Non-linear Modelling of Masonry Infilled Frames, *11th ECEE, Balkema*, Rotterdam, ISBN 90 5410 982 3; Paris.

CYCLIC TESTING AND ANALYSIS OF STEEL END-PLATE CONNECTIONS

Cyclic Testing of Steel Connections

A.W. THOMSON and B.M. BRODERICK
Department of Civil, Structural and Environmental Engineering
Trinity College Dublin, Dublin, Ireland

Abstract

A test program in which flush end-plate connections of variable plate thickness are loaded under cyclic displacements of variable amplitude was undertaken as part of a study of the behaviour and modelling of semi-rigid partial-strength connections. This paper presents the details of the testing regime and describes the initial observations of the experimental work. The moment-rotation hysteresis curves are determined and initial conclusions are drawn on the effect of the plate thickness on the cyclic response of the flush end-plate connections. The yield and ultimate rotational capacities are predicted using the Eurocode 3 T-stub method and a second method; these being compared to the experimental results. It is intended that one of these methods will form the basis of a new cyclic response prediction model. This model will allow the seismic behaviour of connections using flush end-plate connections to be predicted from their material and geometrical properties.

Keywords: flush end-plate connections, moment capacity calculation, semi-rigid partial-strength, cyclic response testing

1 Introduction

In recent years, there has been a shift away from the full-strength, heavily-welded connections conventionally favoured in seismic engineering towards semi-rigid moment connections. These moment connections can display many attractive features such as high ductility and energy dissipation characteristics. One such connection is the end-plate connection for which an abundance of design guidance is available [1]. In particular, the flush end-plate connection, which is classified as a partial strength connection, may be designed to yield under strong ground motion. Considering the brittle failure of many connections inspected after the Northridge and Kobe earthquakes, this ability to yield at a specified load in a controlled manner offers

Abnormal Loading on Structures edited by K. S. Virdi, R. S. Matthews, J. L. Clarke and F. K. Garas.
Published in 2000 by E & FN Spon, 11 New Fetter Lane, London EC4P 4EE, UK. ISBN 0 419 25960 0

obvious advantages. However, to ensure that the correct yielding mode is obtained, careful selection of design details such as end-plate thickness, bolt size and bolt spacing is required. While the existing guidance for wind-moment design includes these as considerations for ductility requirements, it is unlikely that the ductility provided will be adequate for seismic engineering applications where a much higher ductility demand is experienced.

The use of semi-rigid partial-strength connections offers a number of advantages over full-strength heavily-welded typologies. The connections are simpler to construct and structures incorporating them usually have a lower initial stiffness, resulting in a longer natural period and the advantage of lower seismic design forces. The fact that the connections are partial-strength means that the yield point may be controlled in the design process, allowing ductility demand to be predicted during different loading conditions. This in turn leads to a better understanding of the frame response, and resulting economy. These partial-strength connections can also increase the energy dissipation potential of the frames. The wind-moment connections advocated for use in the United Kingdom design codes [1] are also easier to construct on site and are less susceptible to lack-of-fit and tolerance problems.

Although a great deal of work has been carried out on semi-rigid connections, much of this has centred around angle connections and extended end-plate connections. Adey et al [2] and Ghobarah et al [3][4] carried out tests to determine the cyclic characteristics of extended end-plate connections, as did Tsai and Popov [5], while Troup et al [6] developed a finite element model to predict their response. Engelhardt and Husain [7] investigated the cyclic response of angle connections while Elghazouli [8] carried out cyclic and pseudodynamic tests to determine the ductility of the connections. In 1995, Calado and Ferreira [9] developed a cyclic prediction model for top- and seat-angle connections using specialised line elements. However, little work has been carried out into the response of flush end-plate connections. Kennedy and Hafez [10] and Davison et al [11] carried out monotonic tests to determine the rotational stiffness of flush end-plate connections, while Phillips and Packer [12] examined the effect of the plate thickness. Madas [13] developed a mechanical prediction model for seismic loading while Macken [14] carried out experimental investigations to determine the response of flush end-plate connections to cyclic loads.

This paper describes the initial results of a series of tests carried out on flush end-plate connections which are used to determine the rotational capacity of the connections under conditions representative of strong ground motion. Beam-column sub-assemblages are subjected to monotonic and cyclic displacements of increasing amplitude until failure occurs. Three connection details are tested in which the only variable altered is the thickness of the end-plate, and consequent change in yield pattern within the connection. The influence of this change on the dissipative capacity of the connection is observed.

The test results are compared with predictions for yield moment capacity obtained using established methods, such as that developed by Johnson and Law [15], or the newer T-stub method as proposed in Eurocode 3: Annex J [16], to determine the validity of these methods when applied to the cyclic response of flush end-plate connections. It is envisaged that one of these methods will form the basis of a new cyclic response model.

2 Testing specimens and procedure

2.1 Flush end-plate specimens

The test specimens are beam-column sub-assemblages using a flush end-plate. The column section is a one metre length of 203 x 203 x 86 UC and the beam consists of a one metre length of 254 x 102 x 22 UB. The end-plate plates are 277 mm x 200 mm with thicknesses of 8 mm and 12 mm. The bolt holes are spaced at 45 mm from the centre of the beam web and 60 mm from the centre of the beam flanges. The bolts used are all Grade 8.8 M20 bolts. An initial torque of 300 Nm was applied to each bolt to ensure that all of the tests are performed under identical and controlled conditions.

The two different end-plate thicknesses were selected to allow the first two failure modes detailed in Eurocode 3 [16] to be evaluated under cyclic loading. Mode 1 failure occurs when the end-plate forms plastic hinges at the bolt line and at the beam flange as shown in Fig. 1(a). It is representative of the behaviour of a connection with a relatively thin end-plate and large bolts.

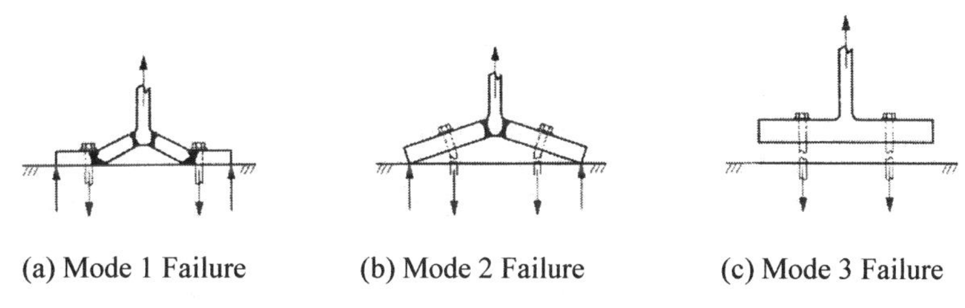

(a) Mode 1 Failure (b) Mode 2 Failure (c) Mode 3 Failure

Fig. 1. Failure Modes as defined in Eurocode 3: Annex J (1993)

Mode 2 failure is defined as the formation of plastic hinges at the beam flange followed by yielding of the bolts as shown in Fig. 1(b). The third failure mode is defined as yielding of the bolts while the end-plate remains elastic (Fig. 1(c)). This failure mode results in a brittle failure which is not recommended in seismic zones.

2.2 Testing procedure

The testing procedure used for cyclic tests on beam-column sub-assemblages is a variation of the ECCS recommended procedures [17]. These recommend that for every cyclic test carried out, two monotonic tests are also performed to determine the elastic limits of the structural steel elements in the tension (+ve) and compression (-ve) ranges. As all of the connections tested were symmetrical flush end-plates, a single monotonic test was sufficient as the yield point could be assumed to be the same in both directions. The cyclic waveform characteristics are calculated as follows:

- One cycle in the $\delta_y/4$ interval;
- One cycle in the $\delta_y/2$ interval;
- One cycle in the $3\delta_y/4$ interval;
- One cycle in the δ_y interval;
- Three cycles in the $2\delta_y$ interval;
- Three cycles in the $(2+2n)\delta_y$ interval (n = 1, 2, 3, ...).

where δ_y is the elastic displacement limit of the structural connection. It should be noted that additional waves of alternative amplitudes, or additional cycles at the above amplitudes, may be added as required. This results in a waveform such as that shown in Fig. 2.

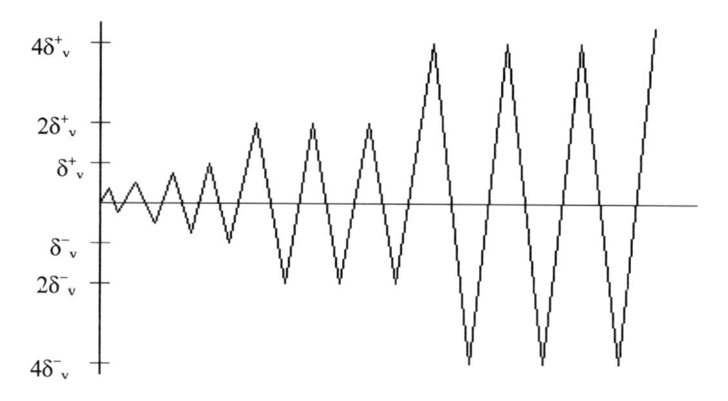

Fig. 2. Cyclic Waveform consistent with ECCS Recommended Procedures (1986)

2.3 Test set-up and control

The specimens were placed in a loading rig and attached to a servo-hydraulic actuator using a specially constructed hinge. The actuator used for this series of tests is capable of delivering a load of 100 kN with a stroke distance of ±50 mm. In order to facilitate loading of the specimens the column section was placed in a horizontal position while the beam section was positioned vertically. The specimens are placed on two I-beam pieces with steel bars welded on top. These steel bars ensure that the reaction forces in the column section are point loads in the column flanges. The columns are held in place vertically by threaded bars running through the I-pieces and a box section as can be seen in Fig. 3. Steel bars are also welded to the box section to ensure the loads imposed on the top of the sections are also point loads. This system allows the column to rotate at its ends. An adjustable length was used to secure the column section in the test rig. This consists of a threaded bar inside a sheath. A torque can be applied to the threaded bar in order to simulate axial loading on the column. On the opposite end of the column section, a reaction block was attached to the loading.

The actuator is pinned at both ends as shown in Fig. 4. The actuator is a computer controlled servo-hydraulic system with a built in LVDT and a load cell attached to the piston. The full test set up is shown in Fig. 4 where the labels are as follows:

1 universal beam section; 2. universal column sections; 3. actuator hydraulic system; 4. actuator arm with detachable load cell; 5. adjustable arm for holding specimen in place; 6. upper box section of vertical holding system; 7. lower I-beam section of vertical holding system; 8. reaction block; 9. hinge connecting specimen to actuator; 10. hinge connecting actuator to loading rig; 11. loading rig.

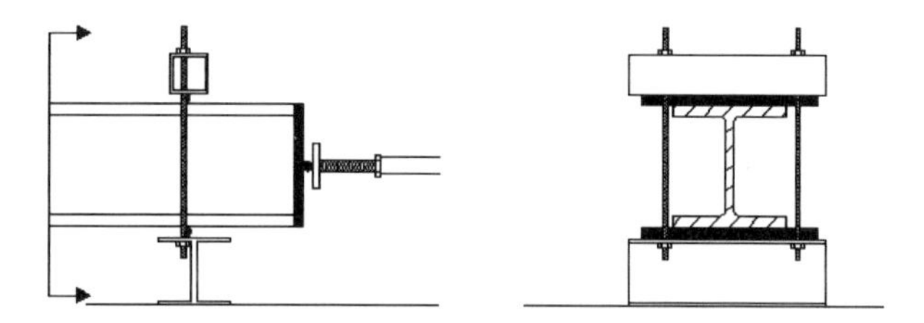

Fig. 3. System for holding column sections in place on testing frame

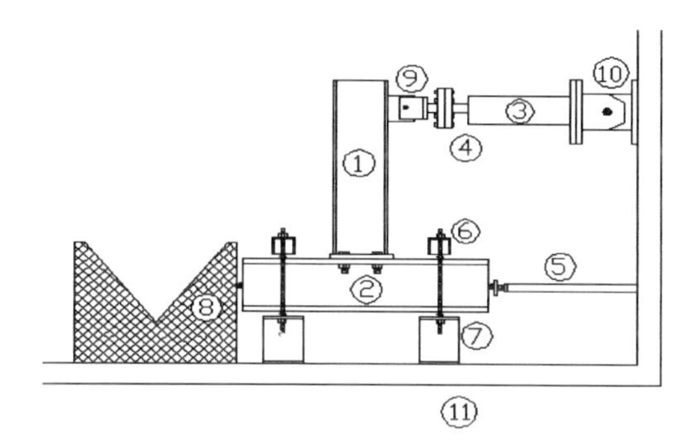

Fig. 4. Test Specimen Set-up in Cyclic Tests

3 Analysis techniques

The ability of two existing connection analysis models to predict the experimental results is examined. The first method is based on a model by Johnson and Law [15] for the analysis of composite connections with end-plates. The second method forms the basis of the design rules for semi-rigid connections set out in Eurocode 3: Annex J [16]. The methods are outlined below and the results presented in Section 4.

3.1 After Johnson and Law (1981)

This method was originally devised to calculate the stiffness of connections in composite frames. These composite connections consisted of flush end-plates with continuous concrete slabs. As the method determines separate coefficients for the concrete, column section, end-plate and the bolts in the joint, it is possible to determine the initial stiffness for the flush end-plate connection by assuming that the concrete stiffness is equal to zero. This prediction is based on the connection geometry and the material properties. However, the authors acknowledge that the method may result in a conservative yield rotation estimate.

The model assumes that the tensile force in the beam flange is related to the deformation of the connection, δ, by an overall joint coefficient, C, as shown in Eqn. 1.

$$C = \frac{T}{\delta} \qquad (1)$$

The full expression for the joint coefficient is:

$$C = \frac{1}{0.5\left(\frac{1}{C_b} + \frac{1}{C_c}\right)\lambda + \frac{1}{C_e}} \tag{2}$$

where C_b is the stiffness contribution from the bolts, C_c is the stiffness contribution from the column flange, C_e is the stiffness contribution from the end-plate and λ is a coefficient representing the eccentricity of the bolt line from the upper beam flange. The following assumptions are made:

- When determining C_b, bolts are treated as axially loaded members
- Any deformation in the column web is negligible; and
- Stiffeners and bolt holes have no effect on the compression zone of the joint.

Because this method only calculates the initial stiffness of the steel connection, it is assumed that the post-yield stiffness could be approximated using the strain hardening coefficient for steel. This resulted in the following formula:

$$K_2 = \mu K_1 \tag{3}$$

where K_1 is the initial connection stiffness, K_2 is the secondary stiffness and μ is the strain hardening coefficient for steel.

This stiffness was then used to calculate the ultimate capacity of the specimen based on the rotational capacity determined in Eurocode 3: Annex J [16].

3.3 After Eurocode 3: Annex J

This method was proposed in Eurocode 3 to allow the stiffness of semi-rigid bolted end-plate connections to be determined. The method calculates both the rotational stiffness and rotational capacity of the connections. Unlike the method described above, it calculates secant stiffness as well as initial elastic stiffness. However, the ultimate capacity of the connections may be underestimated in some cases depending of the exact geometry of the connection type. The initial stiffness of the connections may approximated by:

$$S_{j,ini} = \frac{Eh^2 t_{wc}}{\sum \frac{\mu_i}{k_i}\left(\frac{F_i}{P_i}\right)^2} \tag{5}$$

where $S_{j,ini}$ is the initial joint stiffness, E is Young's Modulus for steel; h is the lever arm from the bolt row to the compression centre, t_{wc} is the thickness of the column web; μ_i is a modification factor for component i; k_i is the stiffness of component i; F_i is the force in component i due to moment M; and P_i is the capacity of component i. The methodology given in Eurocode 3 also allows for the determination of the rotational capacity of the connections as follows:

$$\phi_{cd} = \frac{10.6 - 4\beta_{cr}}{1.3h_1} \tag{6}$$

where ϕ_{cd} is the rotational capacity of the connection, h_l is the distance from the bolt-line below the tension flange of the beam to the centre of the compression zone and β_{cr} is a factor determined by the failure mode of the connection and the geometrical properties.

4 Experimental and prediction results

Results from the experiments and both prediction models are summarised in Tables 1 and 2. Both specimen types exceeded the rotational capacity predicted by the method presented in Eurocode 3: Annex J [16]. In reality, it was not possible to determine the ultimate rotational capacity of some of the specimens, as this was greater than the full travel length of the servo-hydraulic actuator. This is most relevant for the 8mm thick end-plate tests.

The prediction model by Johnson and Law [15] gives a conservative value for the initial stiffness of both specimens, resulting in erroneous yield rotation predictions. Further, as this method employs a bi-linear curve, the ultimate moment is difficult to quantify. The predictions based on the Eurocode 3: Annex J [16] are reasonably similar to the experimental values for connections with the 12 mm thick end-plate. However, the predicted behaviour for the 8 mm end-plate is significantly different to that noted during the experimental testing. This may be attributable to material variability in the end-plate, although this possibility has not been fully investigated.

Table 1. Experimental and prediction model results for 8mm thick end-plate

	Initial Stiffness	Yield Rotation (mrad)	Yield Moment (kNm)	Ultimate Rotation (mrad)	Ultimate Moment (kNm)
EC 3: Annex J	1.03×10^6	13.65	14	39.05	21
Johnson and Law	3.98×10^6	5.28	21	39.05	22.63
Experimental Results	5.1×10^6	6.9	35.1	67.3	54.2

Table 2. Experimental and prediction model results for 12mm thick end-plate

	Initial Stiffness	Yield Rotation (mrad)	Yield Moment (kNm)	Ultimate Rotation (mrad)	Ultimate Moment (kNm)
EC 3: Annex J	2.83×10^6	9.4	26.6	31.18	40
Johnson and Law	11.3×10^6	3.5	40	31.18	43.08
Experimental Results	8.51×10^6	5.7	48.4	40.2	69.7

Both the models under-predicted the yield moment of the connection by nearly a factor of two. A highly ductile response was observed; the maximum moment reached 67.3 kNm and the connection continued to reach approximately the same at higher amplitude cycles. A rotational ductility factor of 9 was measured before the actuator reached its full travel distance. It can be seen in Fig. 5 that on both the tension (+ve) and compression (-ve) sides of the curve, the actuator limits were reached. However, on both sides of the hysteresis curve, the ultimate moment capacity approaches a limiting value reasonably quickly.

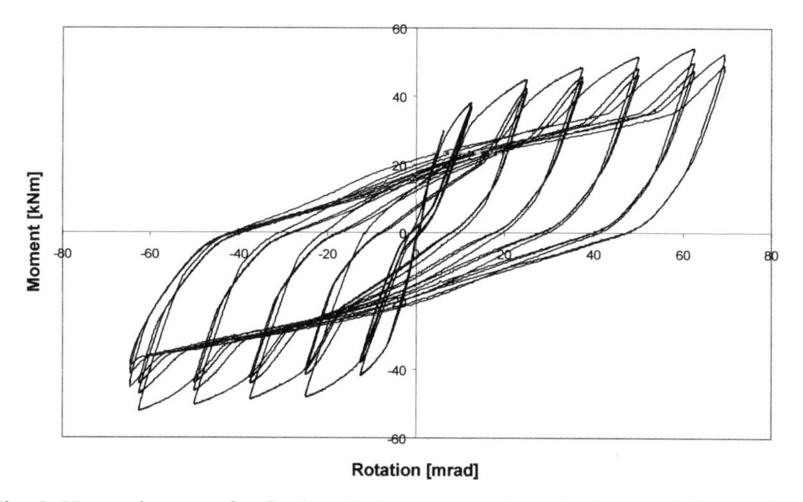

Fig. 5. Hystersis curve for flush end-plate connection with 8mm thick end-plate

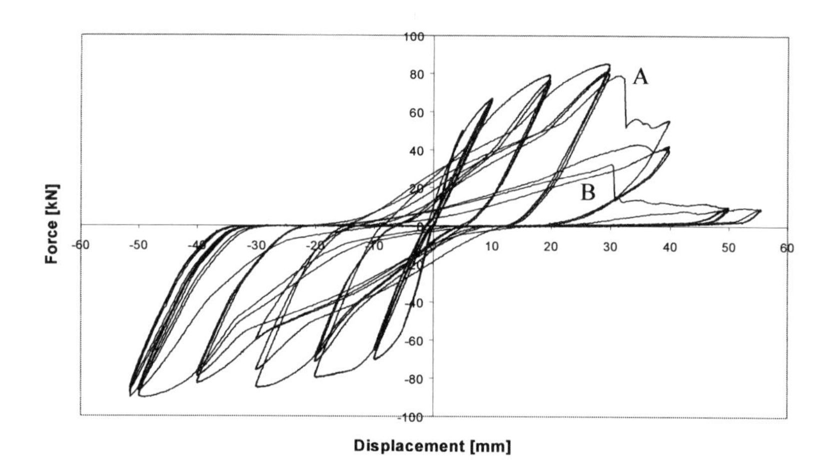

Fig. 6. Hysteresis curve for 12 mm thick end-plate connection

The cyclic response for the 12 mm thick end-plate resulted in a symmetrical moment-rotation curve for the first 40 mrads of the imposed displacement waveform. As expected, a mode 2 failure was observed. However, both bolts in the bolt row did not fail at the same point, resulting in the response seen in Fig. 6. At point A, the first of the two bolts failed which resulted in a drop in the moment resistance from 62 kNm to 44 kNm. At this point, the connection was able to maintain a relatively constant maximum moment for the remaining two cycles in the group. However, when the next group of three cycles was begun, the second of the two bolts failed, shown by point B, and the moment resistance of the connection fell to almost zero.

The connection continued to resist rotation on the other side, without bolt failure, reaching a maximum moment resistance of 70.6 kNm. However, the connection may be considered to have reached the ultimate capacity when the first of the two bolts fails.

5 Conclusions

The results of two cyclic tests on flush end-plate connections have been presented. Both of the tested connections exceeded their predicted yield and ultimate moment resistance values. This can be attributed to conservatism on the part of the prediction models combined with material over-strength. It was seen from the comparison of the experimental test results and the behaviour predicted by two numerical methods that a significant difference existed between them. The values that were predicted using the method published by Johnson and Law resulted in conservative estimates for both the initial stiffness and the yield rotation, although the ultimate moment capacities of the different connection types were reasonably accurate. In comparison, the predicted Eurocode 3: Annex J values resulted in lower estimates for the both connection end-plates. The predicted yield moment for both of the connections also was lower than the experimental results showed. Therefore, both of the models examined above resulted in conservative values when compared to experimental results. However, both connections behaved according to the failure mode predicted by Eurocode 3.

It is proposed to conduct further cyclic tests for comparison with the analytical models. This further test series will also include connections whose failure mode is classified as Mode 3 by Eurocode 3. These connections will consistent of the same beam and column members, connected using a 20 mm thick end-plate which will cause the bolts to fail in a brittle fashion. It is anticipated that this set of tests will clarify the previous results as well as provide further information about the connection deformation under cyclic loads. It is anticipated that these tests will facilitate the development of a component based model for the cyclic behaviour of semi-rigid flush end-plate connections.

6 References

1. British Constructional Steelwork Association (1995) *Joints in Steel Construction: Moment Connections*, Steel Construction Institute.

2. Adey, B.T., Grondin, G.Y. and Cheng J.J.R. (1998) Extended end plate moment connections under cyclic loading, *Journal of Constructional Steel Research*, Volume 46, No. 1-3, Paper 133.
3. Ghobarah, A., Korol, R.M. and Osman, A. (1990) Behaviour of extended end-plate connections under cyclic loading, *Engineering Structures*, Vol.12, pp.15-27.
4. Ghobarah, A., Korol, R.M. and Osman, A. (1992) Cyclic Behaviour of Extended End-plate Joints, *Journal of Structural Engineering*, Volume 118, No. 5, pp. 1333-1353.
5. Tsai, K.C. and Popov, E.P. (1990) Cyclic Behaviour of End-Plate Moment Connections, *ASCE, Journal of Structural Engineering*, Vol. 116 (11), pp. 2917 - 2930.
6. Troup, S., Xiao, R.Y. and Moy, S.S.J. (1998) Numerical Modelling of Bolted Steel Connections, *Journal of Constructional Steel Research*, Vol. 46, No. 1 - 3, Paper 362.
7. Engelhardt, M.D. and Husain, A.S. (1993) Cyclic Loading Performance of Welded Flange-Bolted Web Connections, *ASCE, Journal of Structural Engineering*, Vol. 119(12).
8. Elghazouli, A.Y. (1996) Ductility of Frames with Semi-Rigid Connections, *Eleventh World Conference on Earthquake Engineering*, Paper No. 1126.
9. Calado, L. and Ferreira, J. (1995) A Numerical Model for Predicting the Cyclic Behaviour of Steel Beam-to-Column Joints, *10th European Conference on Earthquake Engineering*, Vol. 3, pp. 1721 - 1725.
10. Kennedy, D.J.L. and Hafez, M.A. (1984) A Study of End-Plate Connections for Steel Beams, *Canadian Journal of Civil Engineering*, Vol. 11, Pt 2, pp. 139-149.
11. Davison, J.B., Kirby, P.A. and Nethercot, D.A. (1987) Rotational Stiffness Characteristics of Steel Beam-to-Column Connections, *Journal of Constructional Steel Research*, Vol. 8, pp. 17-54.
12. Phillips, J. and Packer, J.A. (1981) The Effect of Plate Thickness on Flush End-plate Connections, *In Joints in Structural Steelwork (edited by Howlett et al)*, pp. 6.77 - 6.92
13. Madas, P.J. (1993) Advanced Modelling of Composite Frames Subjected to Earthquake Loading, *Ph.D. Thesis*, University of London.
14. Macken, C. (1997) Testing of Steel Moment Connections, *M.Sc. Thesis*, Trinity College Dublin.
15. Johnson, R.P. and Law, C.L.C (1981) Semi-Rigid Joints for Composite Frames, *In Joints in Structural Steelwork (ed. by Howlett et al)*, pp. 3.3-3.19.
16. Comité Européan de Normalisation (1992) *Eurocode 3: Design of Steel Structures: Part 1.1 general rules and rules for buildings*, ENV 1993-1-1
17. ECCS Technical Committee 1 - Structural Safety and Loadings (1986) *Recommended Testing Procedure for Assessing the Behaviour of Structural Steel Elements under Cyclic Loads*, European Convention for Constructional Steelwork, 1st Edition.

NUMERICAL SIMULATION AND MODELLING OF ORTHOTROPIC STEEL BRIDGE DECKS

Modelling orthotropic bridge decks

D.L. KEOGH and A.T. DEMPSEY
Department of Civil Engineering, University College Dublin, Ireland

Abstract

This paper describes the modelling of orthotropic steel bridges. The factors leading to the choice of model type are discussed, and particular emphasis is given to modelling of the orthotropic plate deck. An experimental load testing programme which was carried out on an orthotropic steel bridge in France is described. This involved driving two pre-weighed trucks across the bridge, a number of times, at different velocities. A three dimensional combined grillage and finite element model of the bridge was generated and analysed statically for the experimental truck loading. Comparisons are made between the predicted response of the bridge to the statically applied loading and the observed experimental response. The effect of truck velocity on bridge dynamics is highlighted and the paper concludes with some interesting observations relating dynamic amplification factors to truck velocity.

Keywords: Bridge modelling, dynamic analysis, finite element analysis, instrumentation, natural frequency, orthotropic bridge decks, steel bridges.

1 Introduction

Orthotropic steel bridge decks are characterised by the form of the plate making up the running surface of the bridge. The stiffness of the steel plate in the longitudinal direction of the bridge is enhanced by adding stiffeners. The transverse stiffness is not so increased. Consequently, this plate behaves in an orthotropic manner. Fig. 1 shows a portion of a typical orthotropic plate where the longitudinal stiffeners are made from a single piece of steel plate, rolled formed and welded to the top plate.

Bridge decks of this type are often used on relatively light, long span bridges. Consequently, the ratio of live load to dead load is greater than for other types of highway bridges. This, in combination with the occurrence of numerous welded

Abnormal Loading on Structures edited by K. S. Virdi, R. S. Matthews, J. L. Clarke and F. K. Garas.
Published in 2000 by E & FN Spon, 11 New Fetter Lane, London EC4P 4EE, UK. ISBN 0 419 25960 0

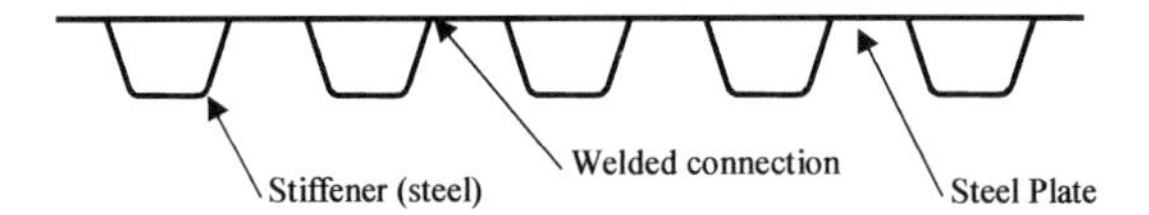

Fig. 1. Composition of a typical orthotropic plate

connections on the structure (especially between the stiffeners and the deck plate), renders the bridge highly sensitive to both normal and abnormal traffic loading. Knowledge of how the bridge reacts to these loads, and especially for their extreme values, is important in both design and maintenance of these bridges, which tend to be located at strategic points in the road network.

2 Modelling of orthotropic bridge decks

Modelling of orthotropic bridge decks can be a complex procedure. The development of a suitable model should consider both static load predictions and dynamic predictions. The recommendations given here are most suited to bridges with main longitudinal girders, continuous over a number of spans, and transverse beams connecting them. The orthotropic plate deck is attached to both the longitudinal girders and the transverse beams. The recommendations given here generally relate to static and dynamic models, but where specific differences exist this is noted.

2.1 Model type

The selection of the model type can have a large bearing on the amount of effort required for both the modelling and the interpretation of results. One of the first decisions to be made is whether to use a two or three dimensional model. Many orthotropic bridges lend themselves to three dimensional models as the orthotropic plate is at a different (vertical) level to the longitudinal and transverse beams. Fig. 2 (a) shows a three dimensional model of a portion of an orthotropic bridge. The longitudinal and transverse beams are located at one level, namely that of their centroids, and the plate is modelled by finite elements or a grillage of beams located above them. The two parts are connected by vertical beam members. These members should be assigned as 'rigid' or given very high section properties so that they do not deform. In such a model, if the longitudinal and transverse beams have centroids at different levels, they too can be on different levels and joined by rigid vertical members.

Fig. 2 (b) shows a two dimensional model of the same bridge. In this, the longitudinal beams are represented by the thick lines while the plate is represented by the members (or elements) between them. The transverse beams and the transverse stiffness of the plate are represented by the transverse members or elements. In this model, it is necessary to consider what portion of the plate acts as the top flange to the longitudinal and transverse beams. This should be determined in accordance with 'shear lag' calculations [1]. Care should be taken not to account for the stiffness of the plate twice, i.e., as a top flange to a beam and as part of the plate.

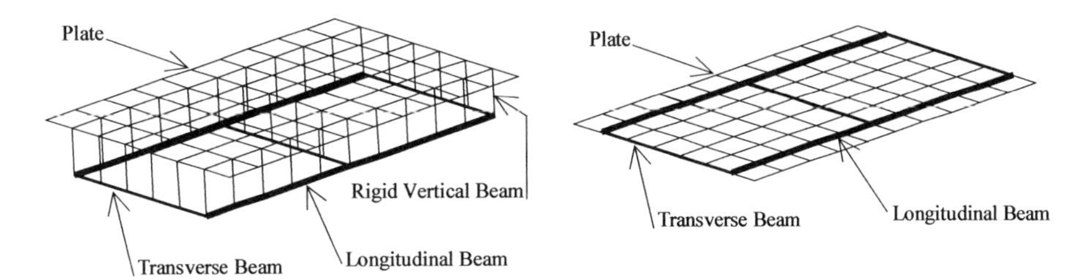

Fig. 2 (a). Three Dimensional Model

Fig. 2 (b). Two Dimensional Model

The two and three dimensional models each have their own advantages and disadvantages. Traditionally, the two dimensional model was preferred as it was more economical on computer resources, but this is no longer an issue. Three dimensional models can be solved by today's computers in very little extra time than their two dimensional equivalents. In addition to this, many of the commercial programs have graphical interfaces which allow the generation of a three dimensional model almost as easily as a two dimensional one. The choice of model type will depend largely on the amount of effort required to determine the member properties and to interpret their output. The properties of the members in the three dimensional model are often easier to determine, as each part of the bridge, beams, plate etc., are individually discretised. For the two dimensional model, a member may represent, for example, a longitudinal beam and a portion of plate. Not only has the quantity of plate to be decided upon, but the resulting property is more difficult to determine.

Interpretation of results is an important consideration when formulating a model. For example, the three dimensional model shown in Fig. 2 (a) may present particular difficulties when attempting to determine the moment carried by the longitudinal beam. In the model, these moments will appear not only as moments in the beam and plate but also as opposing axial forces (with a lever arm between them) in the beam and plate. In a two dimensional model the situation is simpler as the program will simply give a moment. When it comes to the design stage, the three dimensional model may be more relevant as it will tell the user what force or moment exists in each element of the structure, but many designers prefer the two dimensional approach and decide for themselves what portion of the structure to design to resist the moment.

With dynamic modelling it is often the natural frequencies which are of concern. If the user is not concerned with design moments at this stage, a three dimensional model may be the most appropriate. Such a model may lend itself to an easier description of both the stiffness and the weight distribution of the bridge. The remaining recommendations concentrate on three dimensional modelling as this is considered most appropriate for static and dynamic models of orthotropic decks and as much guidance is already available in the literature on two dimensional modelling.

2.2 Longitudinal and transverse beams

In general, the three dimensional model should resemble the real bridge as much as possible. This will mean locating longitudinal and transverse members at the same level as the centroids of the beams which they represent. If the flanges of the beams vary in thickness so much as to significantly change the level of their centroids, consideration should be given to allowing for this in the model by placing different portions of the members at different levels. If the bridge is skewed and skew transverse beams occur, such as at the ends of the bridge or above intermediate supports, these should be allowed for in the model in the same way. Where members occur on different levels in the model, these should be connected at their ends by 'rigid' vertical members. Care should be taken to provide sufficient rigid vertical beams such that members on different levels behave compositely and do not bend individually between vertical beams. If the program being used does not allow the definition of 'rigid' members, properties should be chosen for these members such as to render them effectively rigid. This may be achieved by checking deformations in these members, or by repeatedly increasing their stiffness in the model until a stage is reached where further increase has no effect on the model. Care should be taken as too high a stiffness in these members may introduce numerical instability into the solution of the model.

2.3 Modelling of Orthotropic Plate

The orthotropic plate is perhaps the most challenging part of the bridge to model correctly. It is referred to as orthotropic as it possesses different stiffness in the two orthogonal directions, ie, longitudinally and transversely. This is achieved by the provision of some form of plate stiffener in the longitudinal direction only. The plate is referred to as 'geometrically orthotropic' as it is the geometry that gives it different stiffnesses in the longitudinal and transverse directions. It is made of one homogenous material, namely steel, which gives it the same material properties in all directions. Most commercial finite element programs do not incorporate elements which allow for modelling of this type of plate directly. Alternatively, they use 'materially orthotropic' elements. These assume the same stiffness in both orthogonal directions by adopting a single thickness for the plate. The orthotropy is accounted for by allowing the specification of different moduli of elasticity in the two directions. As the stiffness of the plate is a function of the product of modulus of elasticity and moment of inertia, EI, by adjusting the value of E in the two directions the difference in moments of inertia in the real plate can be allowed for.

For dynamic modelling, it is important that the weight of the model be accurately allowed for as well as the stiffness. For this reason, it is convenient to choose the depth of the elements so as to give the correct cross sectional area of the plate (and hence correct weight). The modulus of elasticity in both the longitudinal and transverse directions can then be adjusted to give the correct EI values (and hence correct stiffness). Further information on this technique is given in [2].

In order to check the validity of the chosen model of the orthotropic plate, especially with regard to predicting natural frequencies, a small portion of plate can be modelled in isolation as a geometrically orthotropic assemblage of isotropic elements. Fig. 3 shows a portion of such a model. The dimensions of the plate should correspond to those of the actual bridge. The boundary conditions of the portion of plate may be difficult to determine. A sensible approach may be to consider both extreme cases of simply supported and fully fixed at the edges in turn. Consideration may also be given to investigating both single spanning and two way spanning cases. A materially orthotropic finite element model of the plate can be analysed with the

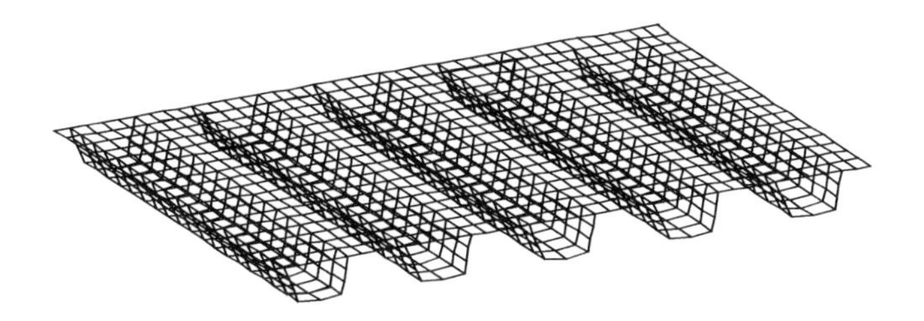

Fig. 3. Portion of geometrically orthotropic finite element model

same boundary conditions. Comparisons can be made between the natural frequencies, deflections, flexural stresses etc. predicted by the two models.

Great care should be taken to ensure the correct weight of the bridge is used in the model. The weight of road surfacing, footpaths and other additions such as parapets should be estimated as accurately as possible as they may account for up to 30% or more of the dead weight of the bridge. This is a particular feature of long span steel orthotropic bridges. It may be particularly difficult to estimate this as the thickness of surfacing often changes during the lifetime of the bridge. A final check could be made by comparing the weight of the model with the weight obtained from such sources as bills of quantities or as determined by its designers or constructors in the case of existing bridges. A check of this type may only be of benefit for highlighting gross errors.

3 Experimentation of Autreville Bridge

An orthotropic steel deck bridge, known as Autreville Bridge, located on the A31 motorway between Nancy and Metz in Eastern France, was instrumented. The bridge consists of three spans, 74.5m, 92.5m and 64.75m. There are four lanes and two emergency lanes, which are carried by a steel plate of approximately 30.5m width. It has longitudinal stiffeners, which are trapezoidal in shape at 600mm centres. The plate is supported by transverse cross beams at between 3.8 and 4.6m centres. These in turn span between two, 3.8m deep longitudinal I-beam girders. Details of the bridge can be seen in Fig. 4 and Fig. 5.

Fig. 4. Autreville bridge in eastern France

Fig. 5 Details of transverse cross beams and orthotropic plate

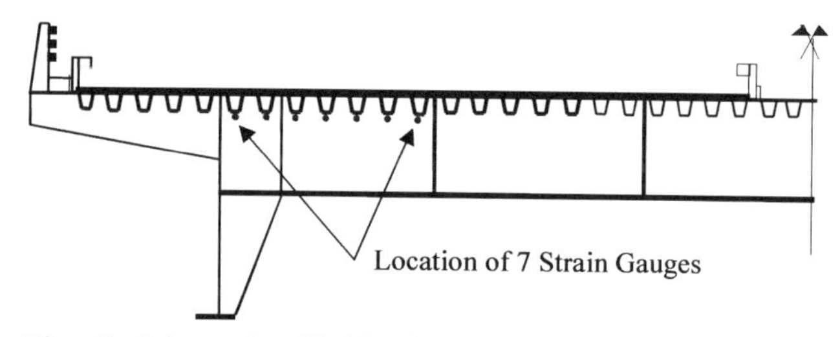

Fig. 6. Schematic of bridge instrumentation

The bridge was instrumented with seven strain gauges located on the underside of seven adjacent stiffeners. They were located half way between two transverse beams in one of the end spans, towards the end of the bridge. A data acquisition system was used to record the strain in all gauges at a frequency of 200 Hz. Two pre-weighed trucks were driven across the bridge four times each at each of 5, 10, 20, 30, 40, 50, 60 and 80 km/hr. One of the trucks had two axles and the other five axles.

Table 1. Truck configurations for load test

	Axle Weights (kN)					Axle Spacings (m)			
	A_1	A_2	A_3	A_4	A_5	AS_{12}	AS_{23}	AS_{34}	AS_{45}
Truck 1	55	129	-----	-----	-----	4.8	-----	-----	-----
Truck 2	48	87	64.7	64.7	64.7	3.5	3.4	1.38	1.38

4 Modelling of Autreville bridge

A three dimensional combined grillage and finite element model of the Autreville bridge was created in accordance with the recommendations of Section 2. The orthotropic plate was modelled using materially orthotropic finite elements and all of the other members were modelled using beam elements. The STRAP [3] structural analysis package was used for this. Fig. 7. shows the model geometry. The finite element mesh was refined locally in the region of the instrumented section to

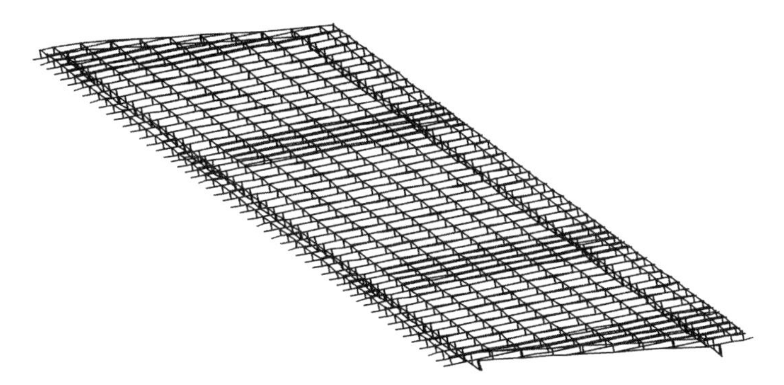

Fig. 7. Three dimensional combined grillage and finite element model

allow for more accurate predictions as the truck loads passed over. This refinement is not shown in the figure.

In order to check the validity of the materially orthotropic model of the plate, a small portion was considered in isolation and modelled as a geometrically orthotropic assemblage of isotropic elements. This was carried out in line with the recommendations of Section 2.3 and Fig. 3. The response of this was compared with that of an equivalent materially orthotropic model. Good correlation was found and it was concluded that the materially orthotropic model was satisfactory.

5 Comparison of results from experiment and model

The average strain from the seven gauges on adjacent stiffeners was determined from the experiment as each of the two trucks passed over the region of the instrumentation. This was determined for each of the seven velocities from 10 to 80 km/hr and converted to stress by multiplying by the modulus of elasticity of steel. This average stress was also determined from the predictions of the bridge model. Fig. 8 shows the average stress from both the experiment and model predictions for the moving 2 axle truck. The experimental values are only shown for the 10 and 80 km/hr runs as these represent the extreme values of stress, i.e. all of the other values fall between these lines. Fig. 9 shows the same quantities for the 5 axle truck. It should be noted that the model predictions of the moving trucks represent a large number of static loadcases and do not take account of the bridge or truck dynamics.

Fig. 8, for the two axle truck, shows very good comparison between the predicted and observed experimental bridge response. This indicates the level of accuracy obtainable from the type of model used. Fig. 9 also compares well but it is felt that the additional complexity leads to increased dynamic effects in the real bridge and slightly less accurate predictions from the model. The latter could be overcome by either refining the finite element mesh or by increasing the number of static loadcases used to describe the moving load.

Another interesting conclusion from these figures is that the bridge response, in this case stress, reduces with increasing truck velocity. If the model predictions are taken as being correct for zero velocity then this also agrees with this finding. This

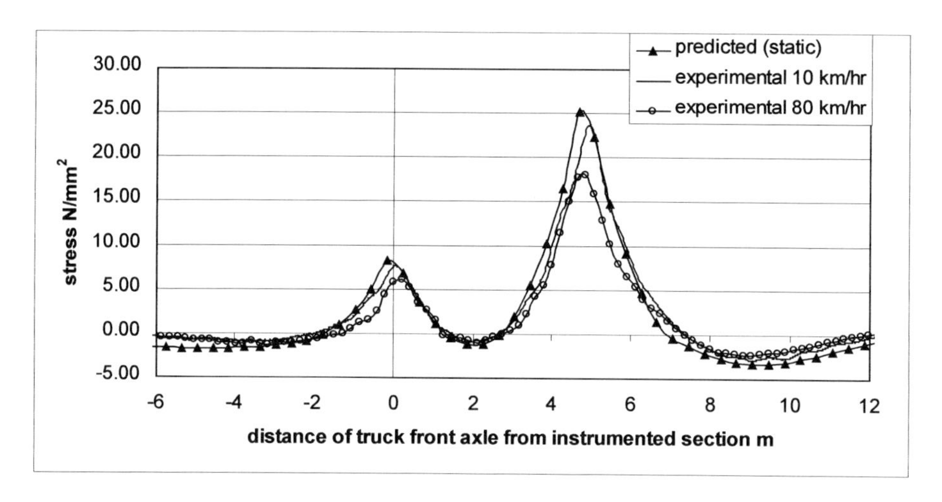

Fig. 8. Average stress at instrumented section as 2 axle truck passes above.

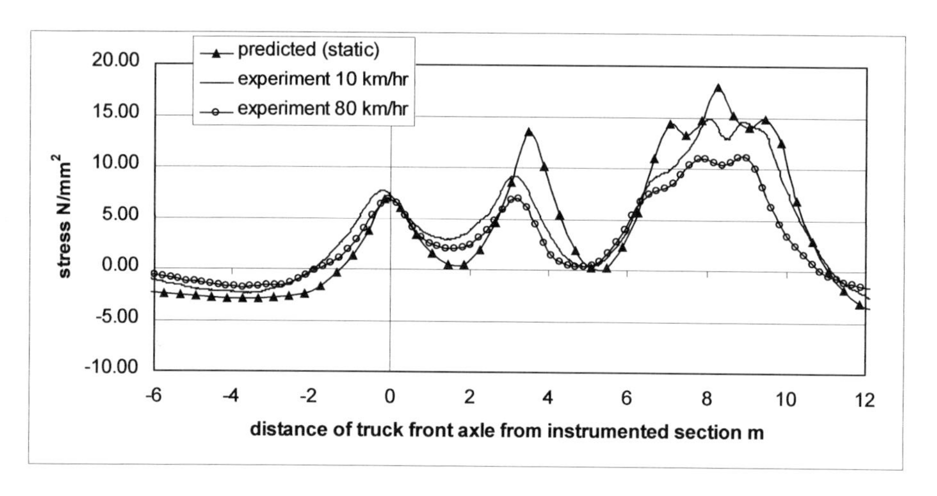

Fig. 9. Average stress at instrumented section as 5 axle truck passes above.

finding has implications when considering the dynamic amplification factors (DAF) given in many bridge design codes. From this experiment, it appears that there is a net dynamic de-amplification. It is felt that this is because the orthotropic plate has a short span between transverse beams, but the bridge has low fundamental natural frequency.

Fig. 10 illustrates this point for a simple one dimensional dynamic bridge model. This figure shows the DAFs for a continuously supported, one dimensional, bridge model. The bridge beam is supported every 4.62m to resemble the support given to the plate by the transverse beams. Five different beams were considered in the analysis, with first natural frequencies of 0.5, 1, 7, 14, and 29 Hz. A moving force

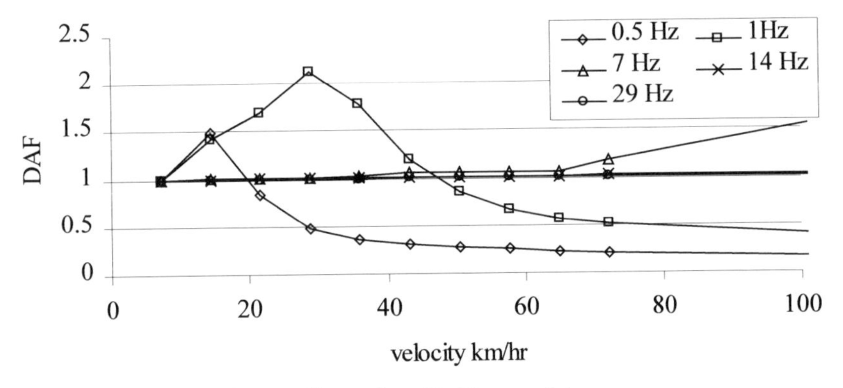

Fig. 10. DAFs for one dimensional bridge model

was passed across the bridge at velocities ranging from quasi-static (5 km/hr) to 100 km/hr and the DAFs were determined for each velocity for each of the beams. It is interesting to note that the phenomenon of reducing DAFs with increasing velocity was observed for beams with low first natural frequencies as was observed in the experiment. However, to accurately predict the DAFs for the bridge, a two or three dimensional model would be required. The one dimensional model only confirms the phenomenon but agrees with the findings of Fryba [4].

Fig. 11 shows the relationship of the DAF to truck velocity. This was derived from the experimental data for the two and five axle trucks. It is simply the dynamic response (strain) of the bridge divided by the static response. In this case, the static response was taken as that from the truck driving across the bridge at 5 km/hr.

This figure indicates that the dynamic response of this bridge increases with increasing velocity up to a velocity of about 10 km/hr. After this, the response of the bridge decreases with increasing velocity and drops below the static (5 km/hr) response at about 45 km/hr. The response of the bridge to a truck travelling at 80 km/hr is about 20% less than that of traffic moving very slowly across the bridge.

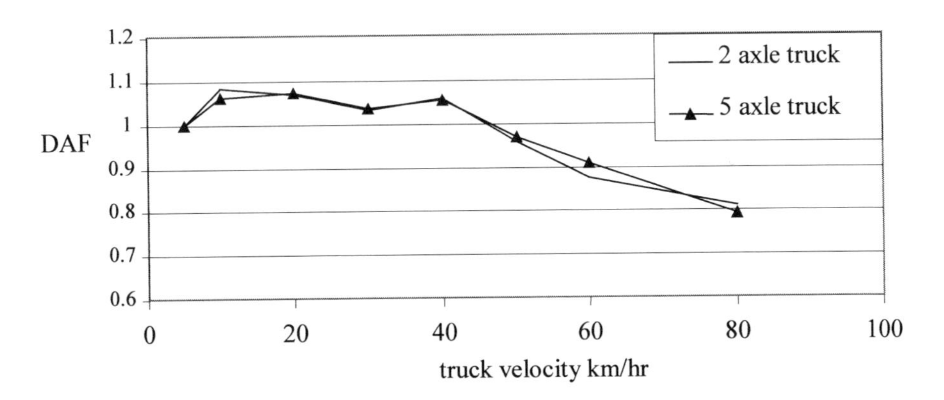

Fig. 11. Dynamic amplification factor for various truck velocities

The findings of this paper were somewhat unexpected in that the response of the bridge due to trucks moving at normal highway speeds is less that that at slower speeds. This means that the bridge effectively 'sees' less than the static load. This is in effect a beneficial abnormal load. It is the authors' opinion that the dynamic behaviour of the bridge is responsible for the reduced dynamic amplification factors. Further investigation is required to confirm this, and to this end, dynamic modelling of the bridge is currently in progress.

6 Conclusions

The modelling of orthotropic bridges was discussed and recommendations were given for setting up such a model. Particular emphasis was given to modelling of the orthotropic plate. An experimental programme on a bridge in France was discussed and results were given for stresses on the underside of the plate as two different trucks drove across the bridge. The experiment was repeated a number of times at different velocities. A three dimensional combined grillage and finite element model of the bridge was used to predict the response due to the truck loadings. Good correlation was found between observed and predicted response thus confirming the validity of the model. When the relationship between bridge response and truck velocity was investigated, it was found that the response diminished as the velocity increased over a velocity of about 10 km/hr. Dynamic amplification factors were compared to truck velocities and again it was found that these reduced with increasing truck velocity and dropped below unity for velocities over about 45 km/hr. It was concluded that these effects were due to the dynamic behaviour of the bridge. This was also observed using a simple one dimensional bridge model and with the theory of Fryba [4]. More advanced two and three dimensional dynamic modelling of this bridge is currently underway.

7 Acknowledgements

The authors would like to thank Mr B Jacob, Mr J Carracilli and the technical staff of Laboratoire Central des Ponts et Chaussées for the instrumentation and experiments conducted on the Autreville Bridge.

8 References

1. Hambly, E. C. (1991) *Bridge Deck Behaviour*, E & FN Spon, London.
2. O'Brien, E. J. and Keogh, D. L. (1999) *Bridge Deck Analysis,* E & FN Spon, London.
3. Atir (1998) *STRAP, Structural Analysis Programs User's Manual Version 7.30,* ATIR Engineering Software Development ltd., Tel Aviv.
4. Fyrba, L. (1972), *Vibration of Solids and Structures under Moving Loads*, Noordhoff International Publishing, Groningen, The Netherlands.

DYNAMIC RESPONSE OF STEEL DUCT PANELS TO BLAST LOADING

Dynamic response of steel plates

W. GROGAN
Culley Associates Inc., San Francisco, USA

Abstract

This paper provides some insight into the response of rectangular steel ducting when subjected to high velocity internal pressure waves. Panel deformation characteristics were determined from a finite element simulation of typical impulse loading. A small scale model was tested in the laboratory under static internal pressure conditions, and behaviour of the structure was compared with finite element models of the same scale.

General agreement between the results of physical testing and finite element modelling indicated the validity of the finite element techniques employed. Dynamic behaviour was examined by extending the finite element model and, on the basis of the observed response characteristics, design guidance is proposed.

Blast, dynamic, experimental, steel plates, numerical model, large displacement, plastic

1 Introduction

The phenomenon of dust explosions in confined spaces is well documented, but the mechanism is not sufficiently understood to prevent or adequately control such events. In general terms, fine dust particles within a confining structure can become electrically charged. The charge may be grounded by a spark, and the resulting chain reaction sends a pressure wave along the structure, which absorbs energy by deforming.

The confining structures in this study (Primary Air ducts in coal-fired power stations - typically 1.5m x 1.2m) delivered dry air to the coal crushing mills i.e. rotating drums in which steel spheres crush coal fuel. A spark induced event, which occurred approximately every three years, forced a pressure wave along the duct system. Energy was absorbed by steel panel deformation and, in rare cases, cracked panel edge welds. This paper describes the steel panel responses and gives proposed design guidance.

Abnormal Loading on Structures edited by K. S. Virdi, R. S. Matthews, J. L. Clarke and F. K. Garas.
Published in 2000 by E & FN Spon, 11 New Fetter Lane, London EC4P 4EE, UK. ISBN 0 419 25960 0

2 Plate bending theories

The behaviour of a rectangular steel plate under given loading conditions is the result of bi-directional interaction which cannot be accurately defined by a simple system of equations [1]. Theoretical conditions of edge fixity cannot generally be achieved, while plate response can be significantly influenced by slight yielding of fixed edges [2]. The magnitudes of plate stress and deflection are typically derived using rigorous analytical techniques such as the Finite Difference and Boundary Element methods. Leissa et al. [3] have determined relative merits for several numerical methods. Szilard [4] has discussed both classical and numerical methods, tabulating specific solutions.

2.1 Plate classification

Classifying a plate as either 'thick' or 'thin' is an attempt to describe its predominant mode of behaviour under load. The property in question is the ratio of side length to thickness, generally referred to as the plate 'slenderness'. A ratio of 40 has been suggested [5] as the boundary between thin and thick plate classification, although caution should be exercised when using such a simplified approach. This value refers to dominant characteristics, but does not represent a step change in plate behaviour.

2.2 Small/large deflections

Plate deflections in the order of $t_p/2$ are described as being 'small displacement' in nature [2]. The stresses developed through the plate depth are predominantly from bending action. When plate deflection increases beyond $t_p/2$ membrane action starts to develop [5], introducing direct (or diaphragm) stresses in addition to bending stress, and the behaviour is described as 'large displacement'. Several studies have explored large displacement behaviour for specific geometric and loading arrangements, e.g. [6] [7] [8] [9] [10], expressing response terms in the form of dimensionless coefficients.

2.3 Elastic/plastic stresses

Plasticity of steel plates is defined as "...the property of sustaining appreciable (visible to the eye) permanent deformations without rupture..." [5]. In the case of thin plates, appreciable diaphragm stresses may develop before plastic yield develops, but the latter will tend to offset the increased stiffness produced by the former.

3 Static loading test, J1

A small scale model section of Primary Air ducting was tested in the laboratory under conditions of incremental static internal pressure (Fig. 1a). The model consisted of a three bay unit enclosed by end plates. Each bay was separated by a stiffening ring. The centre bay is considered to have been isolated from end plate restraint effects, and therefore representative of a typical ducting bay. Properties are summarised in Table 1.

3.1 Instrumentation

Deflection and strain measurements were made for the central bay, to determine key panel response data. Plate response was considered to be symmetrical about centreline axes, and strains were therefore measured for one quadrant (Fig. 1b).

Table 1. Physical and Material Properties of Testpiece J1

Element type	Dimensions		Material properties	
	Thickness (mm)	Width (mm)	Young's modulus (kN/mm^2)	Yield stress (N/mm^2)
Folded plate	1.0	-	178	180
Stiffener ring	8.5	8.0	170	307

3.1.1 Deflection measurement

Key panel deflection data was obtained using mechanical dial gauges located at the centre of each mid-bay panel (Fig. 2). Such deflections are a good indicator of panel deformation characteristics and can be used as a measure of plate stiffness, including the effects of in-plane force and plastic yield.

3.1.2 Strain measurement

Strains were measured for one of the longer central panels using 45° rosette gauges (Fig. 1). The majority of strain gauges were deployed at the outer panel surface, to achieve a reasonable contour map of strain. Three rosettes were also attached to the inside surface at key locations, facilitating the measurement of bending and direct stresses. These gauges were protected from moisture with nitrile rubber solution.

3.2 Execution

Internal pressure was applied using a hand operated water pump, with applied pressure values read from a dial with 2 p.s.i. accuracy. Pressure was increased in steps of 2 p.s.i. to a value of 34 p.s.i., by which time the plates were observed to have undergone significant plastic distortion.

At each loading step the material was allowed to stabilise, with negligible changes in strain or deflection observed for a period of one minute, before recording strain/deflection data. The pressure was then increased to 40 p.s.i., in order to observe strains and deflections in this highly plastic condition, before removing the load.

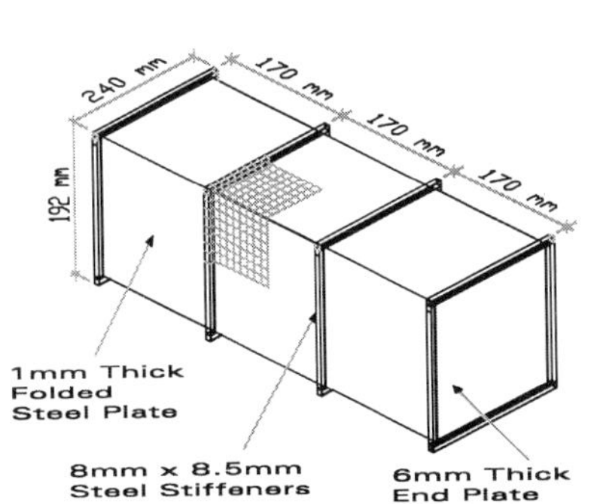

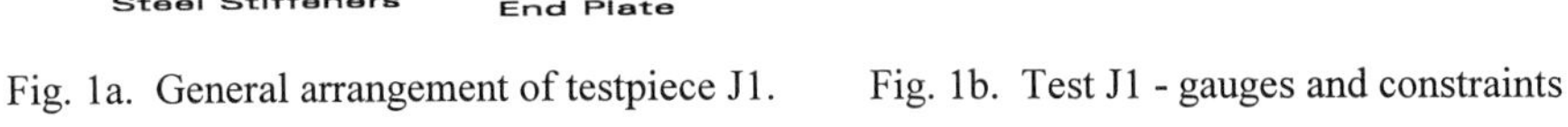

Fig. 1a. General arrangement of testpiece J1.

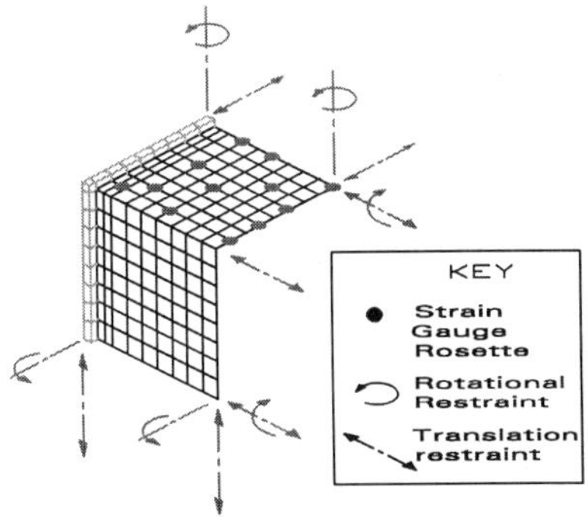

Fig. 1b. Test J1 - gauges and constraints

3.3 Results

Significant distortion of the testpiece was observed, with only minor leakage of the pressurised water at the upper end of the loading range. Most of the strain gauges remained intact giving reliable readings throughout, including the inner surface gauges.

3.3.1 Panel Deflections

Mid-panel deflections are shown in Fig. 3. Large displacement theory indicates that as the magnitude of panel deflection increases, direct stress will act to reduce the gradient of the deflection/load curve [2].

Initially the dominant source of displacement was elastic panel strain, resulting in moderately increased plate stiffness as direct force developed. Subsequent plate yield (Fig. 4) will have relieved edge restraint [5], off-setting the influence of direct stress. In addition, the stiffeners which constrained the mid-bay panels underwent inelastic deformation, contributing to the apparent stiffness reduction.

3.3.2 Stresses

Following the development of large displacements, direct stresses were found to be dominant at the plate centre, whereas both direct and bending stress were significant at plate boundaries. The latter observation demonstrates that the adjacent, orthogonal panels acted as constraining boundary elements, providing rotational restraint.

Strain readings at gauged locations were examined at each load value, and compared with the Hencky/von Mises yield criterion, to map the spread of plasticity. The results for the outer surface (long panel) are shown in Fig. 4.

Fig. 2. Test model J1 after loading.

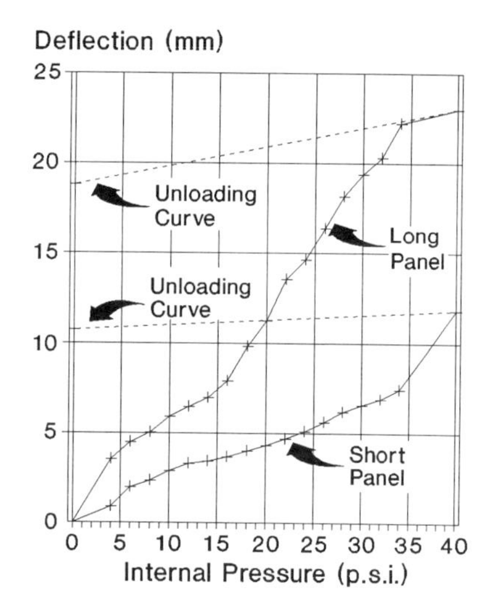

Fig. 3. Deflection v. pressure at centre of middle bay panels – testpiece J1.

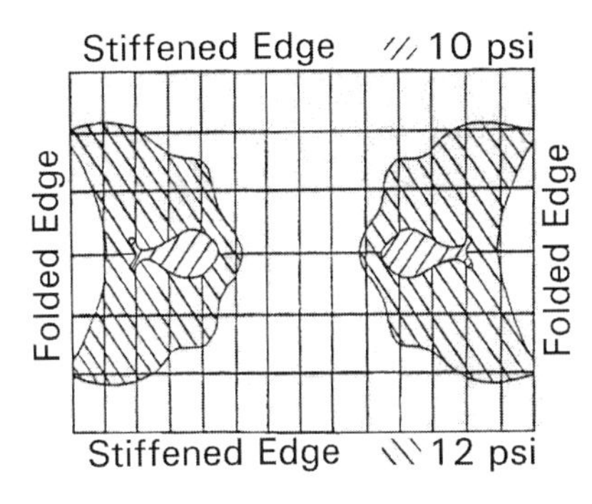

(a) Yield locations at 10 & 12 psi.

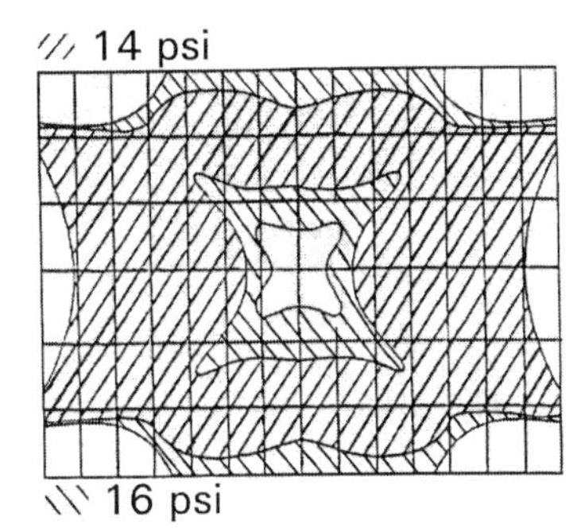

(b) Yield locations at 14 & 16 psi

Fig. 4. Spread of plastic yield along outer surface for long (gauged) panel of testpiece J1.

4 Finite element simulations of laboratory test : models L1, L2 & L3

In order to explore the behaviour of large cross-section steel duct sections both during and following the application of a dynamic impulse, numerical (finite element) models were developed using the PAFEC program [11]. This approach was necessary in order to stay within the stated economic and time constraints. The first step was to model behaviour of the laboratory testpiece, under static load conditions, for comparison.

4.1 Boundary conditions

Using the assumptions of symmetry outlined in section 3.1 above, the model was simplified by isolating a portion of the duct model, and by constraining degrees of freedom at the boundaries as indicated in Fig. 1b.

The influence of a corner radius at the panel junction was investigated by creating two similar models, one with an approximated corner radius at the panel intersection (model L1) and the other without (model L2). See Figs. 5 and 6 respectively.

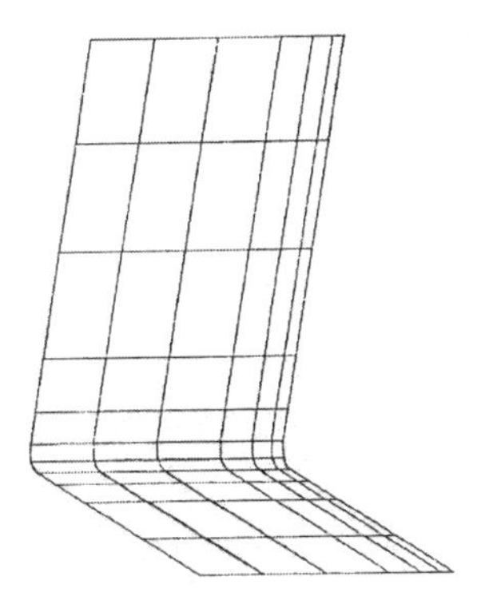

Fig. 5. Finite element model L1.

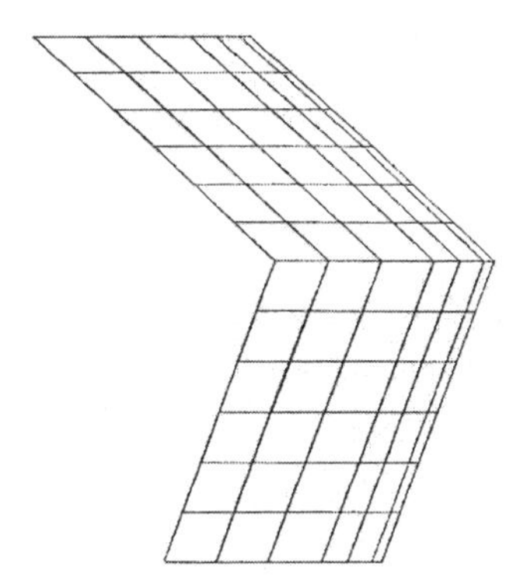

Fig. 6. Finite element model L2.

4.2 Element mesh

Element mesh grids were created using the PAFBLOCKS [11] generation module. Mesh densities were progressively increased towards the stiffened panel edges, where the expected panel curvature due to bending suggested a need for more detail. The narrow boundary element also served to accommodate stiffener modelling.

Mesh refinement was also carried out for model L1, at the panel intersection, to facilitate detailed modelling of plate interaction at the corner (Fig. 5).

4.3 Elements

The elements used to represent the steel panels were the 'semi-loof' plate elements (designation 43210), in which direct and bending stresses are accounted for during large displacements. An attempt to model the boundary ring stiffeners with beam elements for large displacement models produced calculation errors at the element interface, and subsequent failure of the model. This problem was resolved by using a moderately increased plate element thickness at the stiffener locations (2mm) and increasing the elastic modulus to produce the required stiffness along these strips. Physical properties were chosen to reflect laboratory test model J1 (see Table 1).

4.4 Analytical procedures

Both models L1 and L2 were analysed with large displacements theory using the 'SNAKES' sub-routine in which physical properties are re-calculated following each load step to incorporate deformation effects. A maximum pressure of 10.15psi was achieved by the model after which singularities developed, and increased loads could not be achieved. A further model, L3, with the same arrangement of elements as model L2 was subjected to a small displacements analysis for comparison.

4.5 Results

A key measure of the relative behaviour of steel plates is obtained by comparing the load/deflection curves for panel mid-points. The characteristic curves in Figure 7 indicate the behaviour of numerical models L1, L2, L3 and test model J1.

Comparing the curves for models L1 and L2 (large displacements) with those for model L3 (small displacements) the restraining effects of diaphragm force are evident. It is also clear that the higher stiffness in large displacements modelling is more pronounced for the longer panel, confirming that direct stress is more dominant for plates with higher slenderness. The similar plate stiffnesses for test J1 and for numerical models L1 and L2 indicate a correct overall modelling approach, while the benefits of a more detailed mesh at the corner radius were not significant.

5 Full size finite element model, F1, under static loading

The scale of finite element model L1 was increased to reflect the dimensions of the full size structure. Other aspects of the model (mesh density, elements, boundary conditions) were unchanged, and the model was subjected to static loads using large displacements theory. The deflection/load curve for the long panel is given in Fig 8, together with theoretical plate bending curves.

5.1 Comparison with plate bending theories

Equations for both small and large deflection theory [2] have been applied to the long panel in model F1, and the resulting curves are plotted in Fig. 8. Model deflection was, as expected, less than predicted using small displacement theory. However, predicted values of deflection using large displacements theory were 50% lower than for model. The discrepancy may be accounted for by the influence of flexible boundaries in one direction on the finite element model, and approximations embedded in the theoretical formulations.

6 Dynamic finite element model, F2

An elastic dynamic analysis using the Newmark method, and excluding the effects of damping, was performed on a full size model of the rectangular ducting. Damping was considered to have minimal influence during the critical period both during and immediately following load application, and was not incorporated into the model.

6.1 Model definition

Element type and boundary restraints were similar to those used for model F1. In view of the complexity of dynamic modelling, and the limited benefits of including a corner radius (see section 4.5), both the radius and boundary mesh refinement were excluded. The dynamic procedure was not prone to problems encountered in large displacements analysis when using 'beam' element stiffeners. Beam properties were used for typical fabrication details i.e. 102x51 RSC stiffeners between duct sections and 50x50x6 RSA backing material along the junction between orthogonal plates.

6.2 Loading characteristics

Measurements of a typical blast loading episode were provided by the project sponsors, as shown in Fig. 9. The response of the structure is divided into two distinct phases. Phase I covers the period of load application (initial 0.005 seconds).

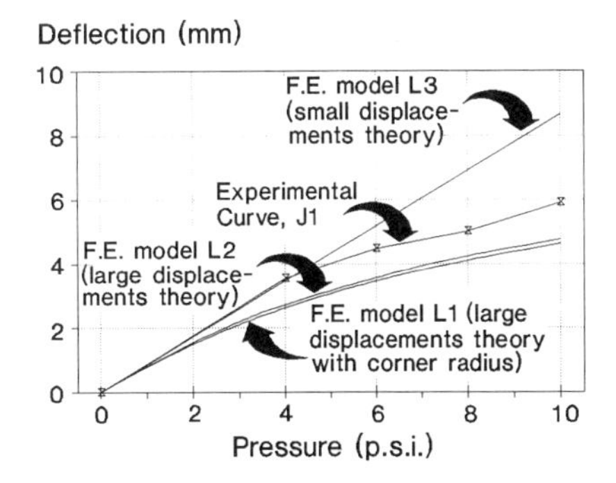

Fig. 7. Deflection v. pressure at centre of long panel for finite element models L1, L2, L3 and testpiece J1.

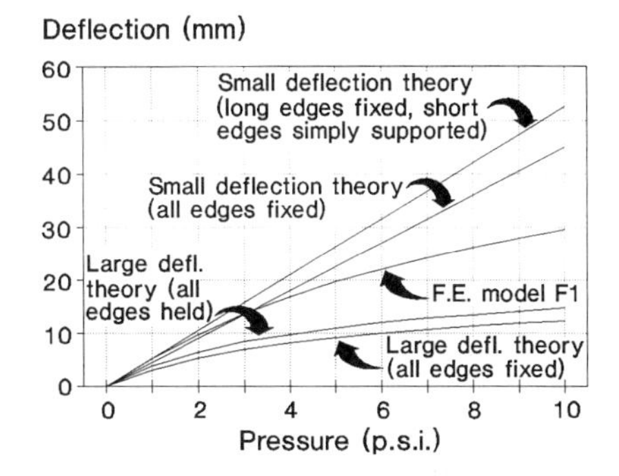

Fig. 8. Deflection v. pressure at centre of long panel for model F1- finite element and theoretical values.

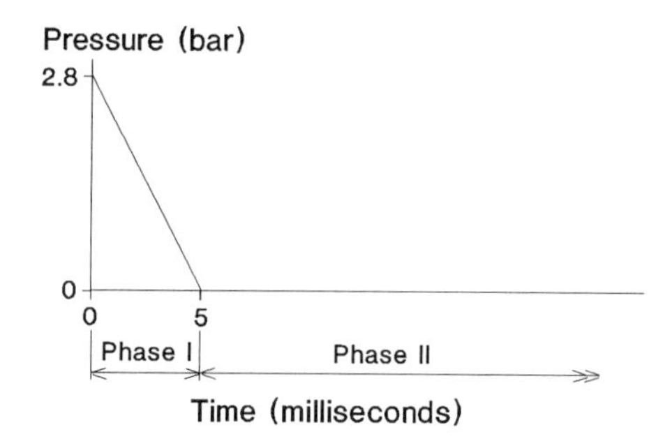

Fig. 9. Dynamic load characteristic applied to full size finite element model F2.

During Phase II the structure deflects under free vibration. A Newmark Beta parameter of 0.25 was chosen, which assumes a linear acceleration profile [12] [13] [14] between the solution time intervals.

6.3 Mode shapes and frequencies

Deflected shapes and frequencies of vibration for the first 24 dynamic modes were calculated by the program. The applied load pattern (Fig. 9) represents one quarter of a 50 hertz cyclic load phase, which is approximately double the value of the fundamental mode (28 hertz) and half of the fourth mode (96 hertz). The first four modes (Fig. 10) are considered to be dominant influences on the structural response.

6.4 Dynamic response

The initial structure response, referenced at the centre of the long panel, is shown in Fig. 11. Peak deflection occurred at the end of the first quarter cycle, approximately 10 milliseconds after the start of load application. Subsequent peak responses are attenuated due to restraining interaction between the two orthogonal plates.

Deflection profiles across the long plate centre-line, between corner intersections, are shown in Fig. 12. Dominant features are dictated by modes 1 and 2. Deflected shapes of modes 3 and 4 are also evident at time intervals 2, 5 and particularly 6.

6.5 Stresses

Critical stresses were observed along the plate boundaries, adjacent to the ring stiffeners and to the orthogonal panels. The most striking observation was the relative maximum boundary bending stresses compared with static analysis values.
For model F1, peak bending stress values were approximately equal at the centre of all edges. However, the results of dynamic analysis (model F2) indicated that peak stress at the centre of the stiffened edges was approximately 50% higher than along the corner edges. Also, the stiffened edge peak stress for model F2 was achieved at maximum response, whereas peak values at the corner edges occurred prior to maximum panel deflection. Clearly, dynamic deflections (Fig. 12) and stress distributions were different in character to those produced by static pressure loading.

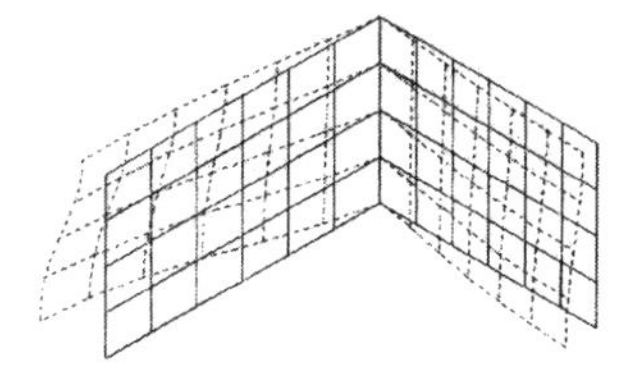

(a) Mode shape 1 with frequency of 28 Hz.

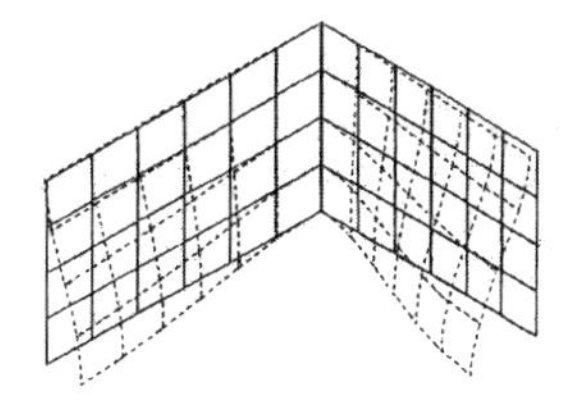

(b) Mode shape 2 with frequency of 40 Hz.

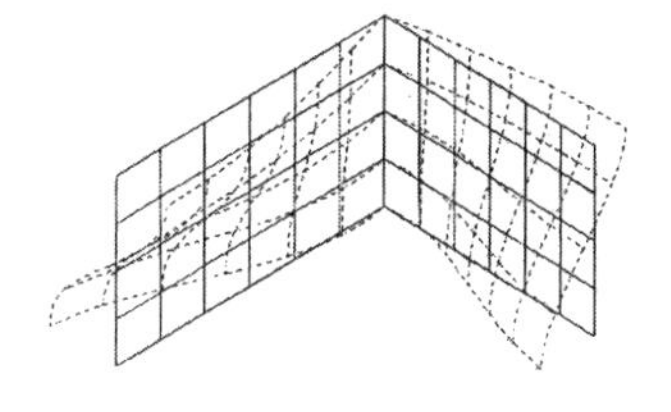

(c) Mode shape 3 with frequency of 66 Hz.

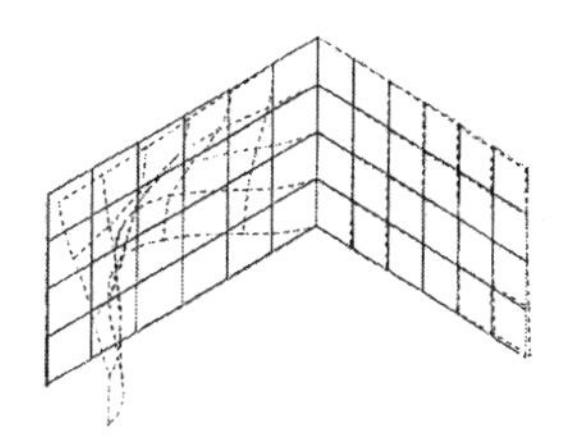

(d) Mode shape 4 with frequency of 96 Hz.

Fig. 10. First four mode shapes and frequencies for dynamic finite element model F2.

7 Conclusions

Laboratory test J1 demonstrated the interaction between bending stress, direct stress and plasticity for panels with the given slenderness under static pressure. Measurement of strain on opposing sides of one panel indicated the extent of plate bending which resulted from boundary restraint. The results of test J1 were used to validate finite element modelling arrangements, which were extended to incorporate response to dynamic loading. The results of dynamic modelling (model F2) highlighted a

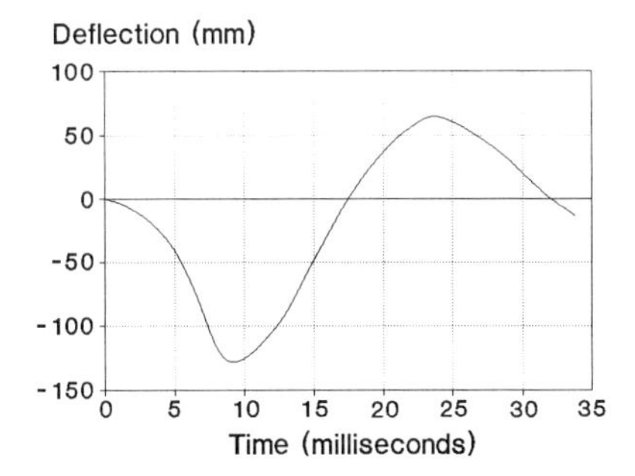

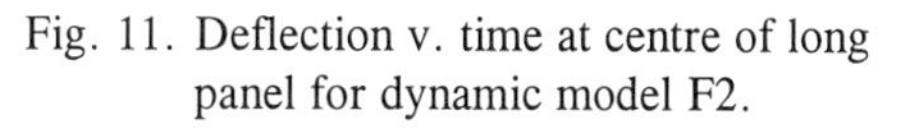

Fig. 11. Deflection v. time at centre of long panel for dynamic model F2.

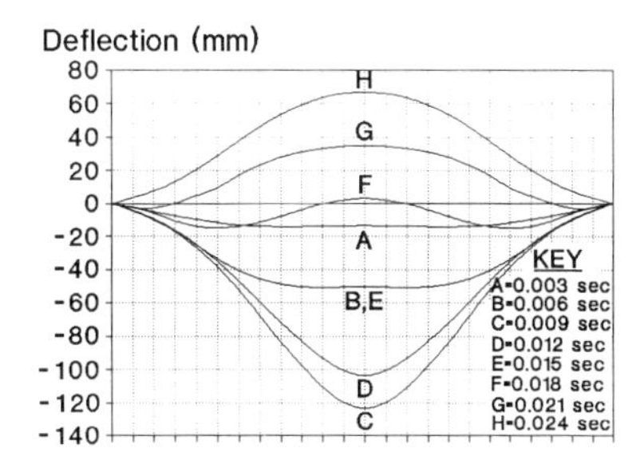

Fig. 12. Deflection profiles across centreline of long panel (F2) parallel to stiffrs.

departure from static load response characteristics (model F1) introduced by the combination of various modes and by interaction between orthogonal panels.

Comparison between strain profiles obtained from static and dynamic analyses (models F1 and F2) suggested that the stiffened edge was relatively highly loaded in the latter case compared with the former, suggesting that stiffened edges across the width of ducting panels, where welded connections are typically made, are more susceptible to failure under these loading conditions. It is suggested that such connections be located at between one fifth and one quarter span locations, where stress levels were observed to be of a consistently low magnitude.

8 References

1. Timoshenko, S.P. (1959) *Theory of Plates and Shells*, McGraw-Hill, New York.
2. Roark, R.J. and Young, W.C. (1975) *Formulas for Stress and Strain*, McGraw-Hill, New York/London.
3. Leissa, A., Clausen, W., Hulbert, L. and Hopper, A. (1969) A Comparison of Approximate Methods for the Solution of Plate Bending Problems. *AIAA J.*, Vol. 7, No. 5.
4. Szilard, R. (1974) *Theory and Analysis of Plates; Classical and Numerical Methods*, Prentice-Hall.
5. Save, M.A. and Massonnet, C.E. (1972) *Plastic Analysis and Design of Plates, Shells and Disks*, North-Holland Publishing Company.
6. Levy, S. (1942) Bending of Rectangular Plates with Large Defelections. *Natl. Adv. Comm. Aeron.*, Tech. Note 846.
7. Levy, S. (1942) Square Plate with Clamped Edges under Normal Pressure Producing Large Defelections. *Natl. Adv. Comm. Aeron.*, Tech. Note 847.
8. Levy, S. and Greenman, S. (1942) Bending with Large Deflection of a Clamped Rectangular Plate with Length-width Ratio of 1.5 under Normal Pressure. *Natl. Adv. Comm. Aeron.*, Tech. Note 853.
9. Wang, C.T. (1948) Nonlinear Large Defelection Boundary-Value Problems of Rectangular Plates. *Natl. Adv. Comm. Aeron.*, Tech. Note 1425.
10. Wang, C.T. (1948) Bending of Rectangular Plates with Large Defelections. *Natl. Adv. Comm. Aeron.*, Tech. Note 1462.
11. P.A.F.E.C. (Program for Automatic Finite Element Calculations), Levels 5&6.
12. Clough, R.W., Penzien, J. (1975) *Dynamics of Structures*, McGraw-Hill, New York.
13. Newmark, N.M (1971) *Fundamentals of Earthquake Engineering*, Prentice-Hall.
14. Paz, M. (1985) *Structural Dynamics : theory and computation*, Van Nostrand Reinhold, New York.

SIMULATION OF LATERAL LOADING ON COKE OVEN WALLS DURING RAM PUSHING

Lateral loading on coke oven walls

Y.O. ONIFADE
Piest & Co. Ltd., Kharkov, Ukraine

Abstract

A mathematical model is presented that aims to describe the behaviour of coke mass during "normal" and "sticker" pushes to the coke oven wall. The model is capable of simulating the distribution of ram pressure in the coke mass and the resultant lateral loading on the refractory wall for given oven length and height. Coke under high temperature (500-1200 ^{0}C) is assumed to be a visco-elasto-plastic body deforming within the confines of the oven walls. Lateral pressure on the wall is expressed as a fraction of ram pressure on the coke mass through proportionality coefficient. Comparison of results obtained with statistical estimate of ram pressure for industrial coke oven batteries shows a satisfactory agreement.

Keywords: Ahler's beam, coke ovens, ram pushing, refractory masonry, visco-elasto-plasticity.

1 Introduction

Coke oven battery is a refractory masonry consisting of 65-100 ovens. Each oven is made up of two parallel walls located 0.45m from each other. Geometrical parameters of existing industrial ovens are: length L=16.0m, height R=4.3m, 5.0m, 5.5m, 6.0m and 7.0m with wall thicknesses r=0.736m, 0.85m, 0.91m and 0.99m respectively. The battery is roofed by a massive refractory masonry of thickness δ between 0.9m and 1.1m, depending on the oven height.

Ovens are loaded with coal, which is then heated in the ovens for 14 to 16 hours at 500-1200^0 C. This is known as the coking period. The practical situation of coke industries indicates a life span of 25 years for ovens of height R=4.3m and 12 years for R=7.0m before the development of progressive failure of the oven walls. This paper, therefore, limits its investigations to ovens of heights R=4.3m and 7.0m.

Abnormal Loading on Structures edited by K. S. Virdi, R. S. Matthews, J. L. Clarke and F. K. Garas. Published in 2000 by E & FN Spon, 11 New Fetter Lane, London EC4P 4EE, UK. ISBN 0 419 25960 0

2 Review of pushing phenomenon

After the coking period, the product – coke – is discharged from the coke ovens into waiting wagons by special pushers exerting ram force on the coke mass. "Normal" pushes are obtained with high rank coals producing well contracted coke which doesn't come into contact with oven walls during pushing. Coke, in this case, is assumed to be an elastic body developing negligible deformation during pushing. On the other hand, coke product may cause "sticker" pushes with coke readily expanding outwards to the walls under pressure from the ram. Coke packs against the oven walls leading to significant increase in the value of residual pressure on the walls and failure of the masonry at 1.0 – 2.0m from the machine side of the oven.

3 Existing design methods

Generally coke oven refractory walls are designed using Alher's beam model [1] [2] [16] [17]. According to this model, one meter wide beams fixed at both ends are taken longitudinally at mid-height and vertically at mid-span, and assumed to be bending under uniformly distributed load from coking mass. Another approach [10][11], incorporates the effect of temperature and transverse anchorage on the refractory masonry. All existing design methods ignore the effect of load transmitted to oven walls during pushing.

4 Simplification of problem

The oven is idealised into a three-layer composite structure with two bearing layers (oven walls) sandwiching a relatively weaker layer of coke mass. Most importantly, the ram force P is applied to the infill (coke mass) leading to the emergence of normal stress in the coke mass.

5 Approach to be followed

To determine the proportional pressure transmitted to the oven walls during ram pushing, a description of the longitudinal distribution of the pressure developed in the coke mass as a result of ram pressure is needed. To this end, a set of equations is formulated describing the resistance of a idealised elementary layer of coke mass to pushing from the oven. Frictional forces between coke and oven walls, between the oven floor and coke and as well as the proportionality coefficient are all taken into consideration. The proportionality coefficient is assumed to be a function of plastic strain of the coke mass. Then the behaviour of coke during pushing, the pushing phenomenon and the lateral loading on the walls during pushing can be simulated.

A right-handed Cartesian co-ordinate system is introduced viz: Oxyz (space fixed). Oxz is situated in the surface of the coke at rest, Ox axis is at a right angle to the vertical refractory walls. Ram force from the pusher is applied on the surface of the coke mass and will be referred to as the machine side of the oven. Origin O lies at

$x = \pm h/2$ where h is the width of the oven. The ram pressure is denoted by P and the corresponding normal stress in the coke - σ_y. Forces developed due to coke to wall friction and coke to oven floor friction are F_{xy} and F_{zy} respectively. Coefficient of friction, acceleration of gravity, coke density and proportionality coefficient are k, g, ρ and α, respectively. For a definition sketch see Fig. 1.

The equilibrium equation for an elementary layer of coke between y and y+dy under pushing force $P = \int_A \sigma_y dA$ is

$$\int_A \sigma_y dA - \int_A (\sigma_y + d\sigma_y)dA - \int_{A_1} F_{xy} dA_1 - \int_{A_2} F_{zy} dA_2 - k\rho gRhdy = 0 \quad (1)$$

in which A=Rh, A_1=2Rdy, A_2=hdy and the pressure in the coke mass σ_y is in the form

$$\sigma_y = q(1\text{-}z/R)f(y), \qquad 0 \le z \le R \qquad (2)$$

where f(y) is the function expressing the distribution of σ_y on the y-axis to be derived from Eqn. 1 with boundary condition

$$P_{max} = \int_A \sigma_y|_{y=0} dA \qquad (3)$$

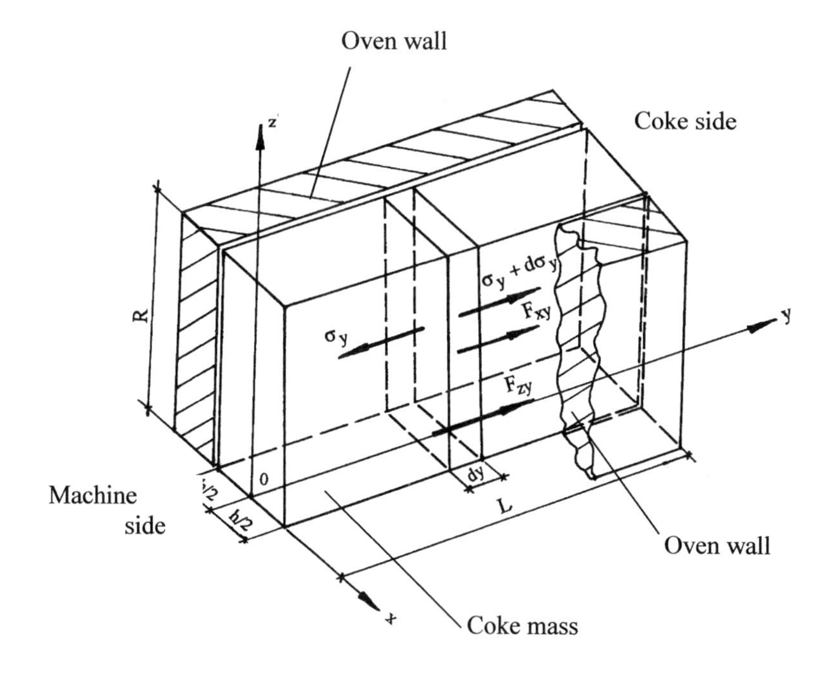

Fig.1 Definitions: A coke oven

i.e. at f(0)=1 the constant q=2P/R.

In [4] a formula for predicting ram force is given as:

$$P - \frac{B}{2L\overline{\alpha}}(1+\frac{b\overline{\alpha}}{h})(\exp^{\frac{L}{\lambda}}-1) \quad (4)$$

where b = height of coke mass; $\lambda = Rh/(2k\overline{\alpha}(R+h))$; $\overline{\alpha} = \nu/(1-\nu)$;
ν - Poisson's ratio; B = ρRhL- weight of the whole coke mass.

Eqn. 4 suggests that coke is an elastic body and it's always dischargeable. The coefficient α in the elastic range is a constant $\alpha = \overline{\alpha}$ [4]. Beyond the elastic region (under plastic flow), α becomes variable and is dependent of the plastic strain ε_y^p of coke:

$$\alpha = 1-(1-\overline{\alpha})\exp(-\varepsilon_y^p/2\beta_1) \quad (5)$$

where β_1 = dimensionless parameter characterising the influence of ε_y^p on α. Then

$$F_{xy} = k\alpha\sigma_y,$$
$$F_{zy} = kq\alpha f(y) \quad (6)$$

From Eqn. 1, we obtain a non-homogenous differential equation

$$f'(y) + f(y)/\lambda + 2kg\rho/q = 0 \quad (7)$$

from which function f(y) in Eqn. 2 can be derived. Here $\lambda = Rh/(2k\alpha(R+h))$.

The stress-strain relationship for coke in the assumed state of stress can be determined through the theory of viscoelastoplasticity [12] where full stress σ_{ij} in tensor form

$$\sigma_{ij} = \overset{*}{\sigma}_{ij} + \sigma_{ij}^a \qquad (i= 1, 2, 3; j= 1, 2, 3) \quad (8)$$

is the sum of $\overset{*}{\sigma}_{ij}$ - effective and σ_{ij}^a – equilibrium stresses. On the other hand, equilibrium stress σ_{ij}^a, similar to [7][8] is

$$\sigma_{ij}^a = \sigma_{ij}^e + \sigma_{ij}^d \quad (9)$$

a sum of σ_{ij}^e – active and σ_{ij}^d – additive stresses. The plastic flow condition is given as

$$\frac{3}{2}S_{ij}^e S_{ij}^e = \psi^2 \quad (10)$$

where $S_{ij}^e = \sigma_{ij}^e - \sigma_0^e\delta_{ij}$ – deviator of active stress, $\sigma_0^e = \sigma_{ii}^e/3$ – sphere stress component, δ_{ij} – Kroneker's symbol, δ_{ij} = 1 when i=j, δ_{ij}=0 when i≠j. In the space of principal equilibrium stress, potential surface ψ = constant is a set of cylinders with common axis equally inclined to the axes of the co-ordinates. Vectors of strain velocity are not orthogonal to these surfaces.

In the same state of stress in a five-dimensional space of deviators of equilibrium stress, Eqn. 10 is a hypersphere with its centre displacing under plastic deformation thus:

$$\sigma_{ij}^{d} = g(\varepsilon_i^p)e_{ij}^p \qquad (11)$$

where $S_{ij}^d = \sigma_{ij}^d - \sigma_0^d\delta_{ij}$ - components of the deviator of additive stress, $\sigma_0^d = \sigma_{ii}^d/3$ - component of sphere stress, $\varepsilon_i^p = (\frac{2}{3}e_{ij}^p e_{ij}^p)^{1/2}$ - intensity of plastic strain in which $e_{ij}^p = \varepsilon_{ij}^p - \varepsilon_0^p\delta_{ij}$, $\varepsilon_0^d = \varepsilon_{ii}^d/3$ - component of sphere strain, $g(\varepsilon_i^p)$ - function of the influence of plastic strain on additive stress in a given state of stress.

To obtain flow equation, coke's properties are assumed to be characterized by volume ζ and shear η viscocities. Skorohod [14] proposed shear viscosity η as

$$\eta = \eta_0 \frac{\sigma_i^a}{\sigma_i} \qquad (12)$$

and that η is related to ζ by the expression

$$\zeta = \frac{4}{3}\eta_0 \frac{1-\Theta}{\Theta} \qquad (13)$$

in which Θ - porosity of given material;
σ_i^a – intensity of equilibrium stress and
σ_i – intensity of full stress.

The stress-strain relationship can be expressed as

$$\dot{\varepsilon}_0^p = \sigma_0/\zeta \qquad (14)$$

$$\dot{e}_{ij} = \sigma_{ij}^{*}/\eta \qquad (15)$$

Avoiding some manipulation, from Eqns. 12 – 15, $\dot{\varepsilon}_y^p$ and $\dot{\varepsilon}_x^p$ are expressed as

$$\dot{\varepsilon}_y^p = \frac{3}{4}(\sigma_y + 2\sigma_x)\frac{\Theta}{\eta_0(1-\Theta)} + \frac{(\sigma_y - \sigma_x)(\sigma_y - \sigma_x - 2C)}{2\eta_0 C}, \qquad (16)$$

$$\dot{\varepsilon}_x^p = \Theta\frac{\sigma_y + 2\sigma_x}{4\eta_0(1-\Theta)} - \frac{(\sigma_y - \sigma_x)(\sigma_y - \sigma_x - 2C)}{12\eta_0 C} \qquad (17)$$

which show that with high porosity Θ, the determinant role in the value of $\dot{\varepsilon}_y^p$ and $\dot{\varepsilon}_x^p$ is played by the component of Eqns. 16 and 17 with Θ as a multiple. Indeed, during pushing, coke is compressed and $\Theta \to 0$ leading to predominant influence of volume viscosity on $\dot{\varepsilon}_y^p$ which can then be expressed as

$$\dot{\varepsilon}_y^p = (\sigma_y - \sigma_x)(\sigma_y - \sigma_x - 2C)/2\eta_0 C \qquad (18)$$

and the potential C which varies under shear flow due to deformational hardening

$$C = C_0\exp(\varepsilon_y^p/\beta_2) \qquad (19)$$

where C_0 – pre-exponential parameter; β_2 – dimensionless parameter.

The stresses σ_y and σ_x are related as $\alpha = \sigma_x/\sigma_y$ and under non-equilibrium shear flow i. e. $\sigma_y - \sigma_x >> 2C$ the $\sigma_y(t)$- ε_y^p relationship is

$$\varepsilon_y^p = (\beta_1\beta_2/(\beta_1+\beta_2))\ln(1+\frac{\beta_1+\beta_2}{\beta_1\beta_2}((1-\overline{\alpha})^2)\omega\int_0^t \sigma_y^2 dt \qquad (20)$$

where $\omega^{-1}=2\eta_0 C_0$ is the parameter showing the influence of viscosity parameter η_0 and deformational hardening parameter C on the plastic strain ε_y^p during coke discharge from the oven. The expression for α can then be written as:

$$\alpha = \frac{1-\overline{\alpha}}{(1-((\beta_1-\beta_2)(1-\overline{\alpha})^2/\beta_1\beta_2)\omega\int_0^t \sigma_y^2 dt)(\beta_2/2(\beta_1+\beta_2))} \qquad (21)$$

which shows that $\overline{\alpha} \le \alpha \le 1$ and that α depends on the parameters β_1, β_2 and ω. Eqn. 21 demonstrates the effect of loading regime $\sigma_y(t)$ and plastic strain parameter ω on α. An increase in α turns the parameter λ = Rh/(2αk(R+h)) in Eqn. 7 to a variable and the equation itself nonlinear with respect to σ_y. Eqn. 7 is solved using Rounge-Count's method of the fourth order accuracy.

The design method presented in [10][11] had suggested a plate model for coke oven wall using finite element analysis for plate bending problem. In view of this, a computer program capable of computing lateral load at given point on the oven wall during pushing is formulated. However the results discussed in this paper are limited to the oven floor level (z=0) with more attention paid to the effect of plastic strain of coke on load distribution.

6 Results

Fig.2 shows the pushing force required to discharge the whole coke mass from the ovens of various lengths and heights when oven width $h = 0.45$m, coefficient of friction $k = 0.5$, Poisson's ratio $\nu = 0.03$ coke density $\rho = 7.0$kN/m^3 and dimensionless parameters $\beta_1 = \beta_2 = 0.04$. Two levels of plastic strain are considered: $\omega=10^{-8}$, 10^{-6} kPa^2 Hour. The discharge period $t = 40$ s. The results can be summarised as follows:

a) The combined effect of length and height of oven shows that as the length increases, a given increase in the oven height requires an increase in ram force to push the coke from the oven;

b) Curves 1 and 2 represent pushing forces for coke with low level of plastic strain parameter ($\omega=10^{-8}$ kPa^2 Hour) which are in good agreement with the results obtained in [4]. They illustrate the fact that coke discharge is always possible for given oven length and height. Low level of plastic strain parameter, therefore, corresponds to elastic state of coke mass;

c) On the other hand, curves 1' and 2' show pushing forces for given oven length and height necessary for coke discharge under high level of plasticity. A very marked increase in the ram force is observed with the emergence of "contact" zones between coke and oven walls. An increase in ω results into a sharp rise in the value of P in the "contact" zone beyond which its effect is not felt thereby making coke discharge impossible. This corresponds to "sticker" pushes usually experienced in practice.

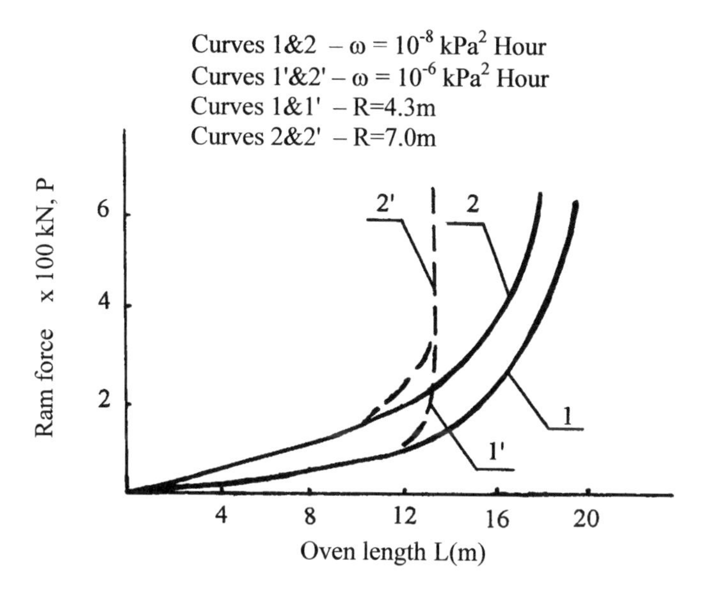

Fig. 2 Ram force – oven length for different coke oven heights.

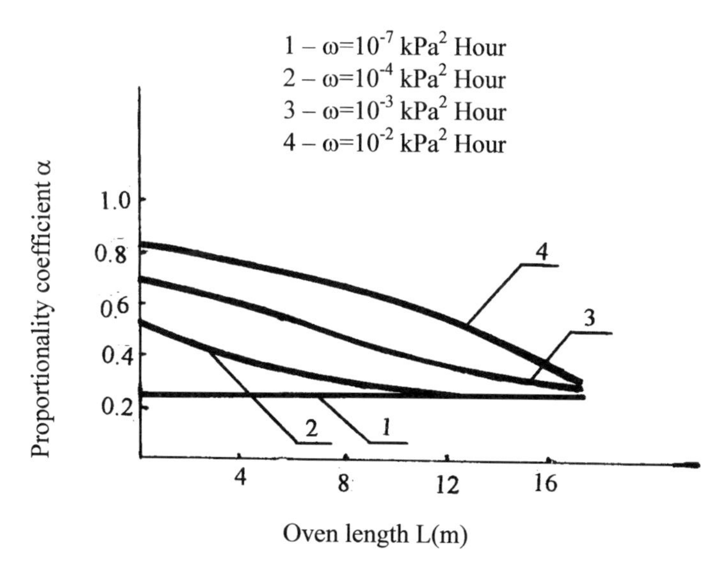

Fig.3 Proportionality coefficient – oven length for different plastic strain parameter.

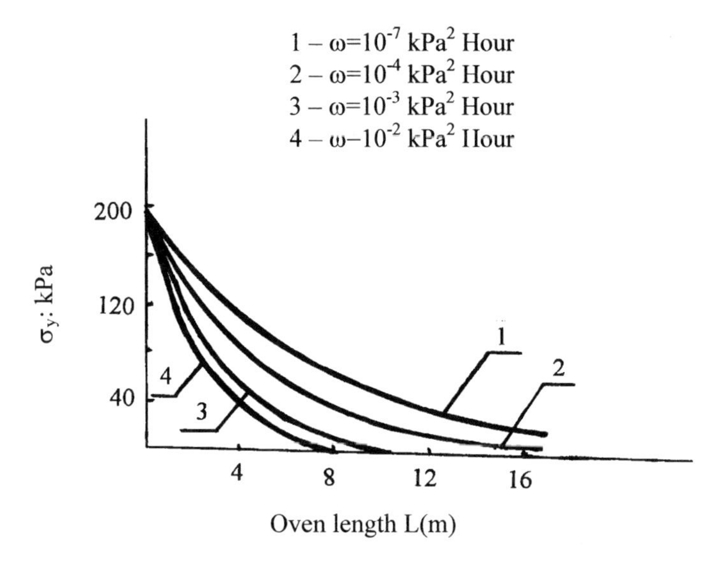

Fig.4 Effect of varying ω on σ_y for given oven length L at the oven floor level (z=0) for oven height R=7.0m

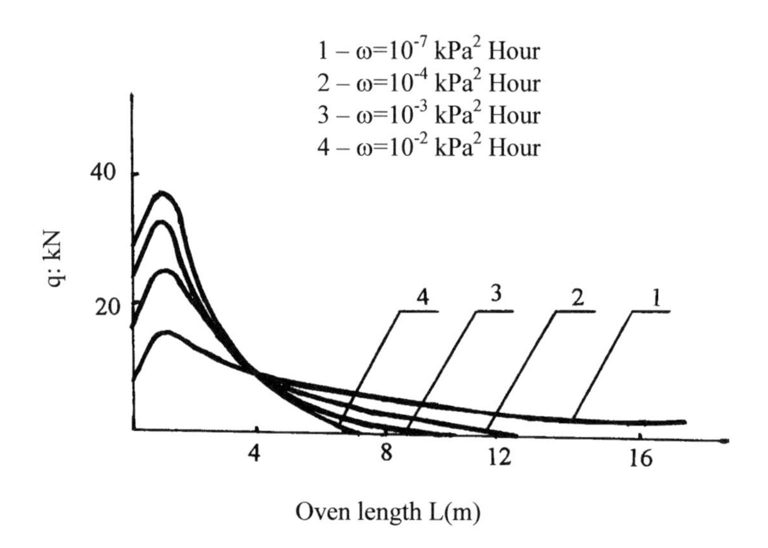

Fig.5 Lateral load q distribution on oven wall at oven floor level (z=0) for R=7.0m for varying plastic strain parameter.

Further investigation were made for oven height $R = 7.0$m, Poisson's ratio $\nu = 0.2$, coefficient of fraction $k = 0.12$, ram force $P = 348.6$ kN, coke density $\rho = 7.0$ kN/m^3, discharge period $t = 10$ s, oven width $h = 0.45$m and dimensionless parameter $\beta_1 = \beta_2 = 0.04$. Four levels of ω were considered: $\omega = 10^{-2}$, 10^{-3}, 10^{-4} and 10^{-7} kPa^2 Hour, represented as Curves 1, 2, 3 and 4 in Figs. 3-5, respectively. Observations made are:

e) The α - L relationship for different ω is shown in Fig.3. Curve 1 shows that α behaves as a constant when $\omega = 10^{-7}$ kPa^2 Hour. This is the elastic state of coke. An increase in the value of ω results into α becoming a variable with its value increasing two to four times at the machine side of the oven (Curves 2, 3 and 4);

f) The pressure on coke due to ram force is illustrated in Fig.4. It shows that σ_y is distributed along the whole oven length when $\omega = 10^{-7}$, 10^{-4} kPa^2 Hour (Curves 1 and 2). This reveals the fact that ram force is felt by the whole coke mass in the oven and pushing is normal. With an increase in ω a rapid fall in the value of σ_y is observed for a given increase in the oven length. σ_y diminishes to zero at $y = L/2$;

g) when $\omega = 10^{-7}$ kPa^2 Hour lateral load is evenly distributed on the oven wall with its ultimate value at $L = 1.0$m. This is the normal push. An increase in ω, i.e. "sticker" pushes, significantly influences the load - oven length relationship in that load distribution is confined to $0 \leq y \leq 8.0$ m i.e. first half of the oven length (Fig.5, Curves 2 - 4). The ultimate value of lateral load in this case is more than doubled.

7 Conclusions

1. A mathematical model for predicting lateral load on coke oven walls during coke discharge from the oven is presented. The model is well-suited for simulating both "normal" and "sticker" pushes for industrial coke oven batteries.

2. The accuracy of the model has been demonstrated extensively by its satisfactory agreement with results obtained for ram force P prediction in [3][4].

3. The model has been applied to study the pushing phenomenon. The results are found to be in good accord with practical observations in industrial ovens. Overall, the results show that coke pushing regime depends on the strain state of coke. Based on the theory of visco-elasto-plasticity, expressions for proportionality coefficient and for lateral load on the walls were derived.

4. Computed results show that normal discharge is possible for a given length and height of oven when coke mass's behaviour is elastic. Coke, beyond the elastic range (high level of plastic strain of coke), is found to require high level of ram force for discharge, and, at the same time, developing "contact" zones with the oven walls thereby causing "sticker" pushes.

5. The simulated lateral load is found to be evenly distributed on the oven walls for normal pushes. Load is redistributed and confined to the first half of the oven length during heavy pushes. Ultimate load is situated at $L = 1.0$m and is doubled or tripled for heavy pushes. The average service life of coke ovens are 25 years for $K = 4.3$m and 12 years for $R = 7.0$m with the emergence of vertical cracks on the oven walls at $L = 1.44$m. Load distribution simulated in this work is, therefore, very important for coke oven designers in the analysis of the fatigue life of coke oven batteries.

8 References

1. Ahlers, V. Grenzbelastung einer koksofenwanden. 1. Ernutlung der grenzbelastung einer koksofenwand. Stahl und Eisen.,1959, B.79, No.7, 397-405.
2. Ahlers, W. Grenzbelastung von koksofenwanden. 2.Moglichkeiten zur beeinflusung der Grenzbelastung von koksofenwand. Stahl uad Eisen, 1959, B. 79, No. 7, 622-629.
3. Cork D. Sticker ovens and heating wall damage. Cokemaking International. – 1998. – v. 11. – N.2 – 44-47.
4. Gibson, G. and Gregory, D.H. Blending and proportioning of coal in coking blend. Effects on production operations and quality. Ironmaking Proc., AIME, 1979, v.38, 52-61.
5. Valanis, K.C. A theory of viscoplasticity without a yield surface. Arch. Mech., Stos., 1971, v. 23, 517-551.
6. Valanis, K.C. On the foundations of the theory of viscoplasticity. Arch. Mech., Stos., 1975, v. 27, 857-868.
7. Ishlinskii, A.Y. General theory of plasticity with linear hardening Ukr., J. Math.,1954, v.6, No.3, 314-324.(in Russian).
8. Kadashevich, Y.I. and Novojilov, V.V. Theory of plasticity with reference to microstress. J. Appl. Math. and Mech., 1958, v. 22, No.1, 78-89. (in Russian).
9. Onifade, Y.O. State of stress-strain of structural elements of coke oven batteries in extreme working conditions. Ph.D. dissertation, St.Peterburg Civil Engineering Institute, 1989, 160. (in Russian).
10. Romasko, V.S. and Onifade, Y.O. Statical analysis of coke oven walls under thermotechnical factors; compressive strength of refractory masonry; shear bond strength of refractory mortar. Report Kharkov Civil Eng. Inst., Govt. No. 01860134730, Inv. No. 02880020532. (in Russian).
11. Romasko, V.S and Onifade, Y.O. Analysis of short term and long term deformational properties of refractory masonry on stress-strain of coke oven batteries. Report. Kharkov Civil Eng. Inst. Govt. No 0184007622., Inv. No 02860027607. (in Russian).
12. Romasko, V.S. High-temperature creep for refractory materials with anisotropic hardening. Proc. Ukr. Inst. Strength Problems., J.Strength Problems., 1984, No.6, 19-22, (in Russian).
13. Romasko, V.S. Prediction of creep of visco-elasto-plastic structural elements. Ph. D. dissertation, Kharkov, Ukraine, 1981. (in Russian).
14. Skorohod, V.V. Ecological fundamentals of the theory of caking. Kiev, Naukova Dumka, 1972, 149. (in Russian).
15. Shkodzinskii, O. K. Phenomenological model of visco-elasto-plastic materials. Proc. Ukr. Inst. Strength Problems., J. Strength Problems., 1987, No.11, 53-56. (in Russian).
16. Tshelkov, A.K. Coke oven batteries designer's handbook. Mettallurgia, Moskow, 1965, v.2, 287.(in Russian).
17. Tshelkov, A.K. Coke oven batteries designer's handbook. Mettallurgia, Moskow, 1966, v.5, 454.(in Russian).

FULL-SCALE TEST OF PROTOTYPE SUPERHEATER SECONDARY RESTRAINT SYSTEM

Test of superheater secondary restraint

R.S. MATTHEWS, A.G. CHALMERS and K.D. WALLER
Taywood Engineering, London, UK

Abstract

During regular maintenance inspections at Dungeness B Power Station, concerns were identified regarding the potential failure of a particular weld in the Superheater penetration assemblies which could result in an element being ejected from the pressure vessel wall, leading to:

- Rapid loss of [radioactive] reactor coolant, and
- Damage to adjacent plant items.

Taywood Engineering were responsible for the design and installation of a system of secondary restraints to limit the consequences of this event. Because of the nuclear safety critical nature of the structures, and the demanding design solution required to resist the abnormal loads within the available space, extensive testing was undertaken.

This paper describes the test carried out on a full-scale prototype restraint assembly. It presents the design of the test, including the dynamic loading system, the control and instrumentation system (including triggers and safety interlocks) and the test procedure. A discussion of the key results is given, including a comparison between test behaviour and analysis predictions that indicated a likely mode of behaviour not previously identified and consistent with lower loads.

The test showed the prototype behaviour to be satisfactory and gave confidence in the overall robustness of the design solution. The system tested went on to be successfully installed on Station.

Keywords: test, full-scale, dynamic, gas pressure, secondary restraint, safety, nuclear, prestressed concrete pressure vessel.

Abnormal Loading on Structures edited by K. S. Virdi, R. S. Matthews, J. L. Clarke and F. K. Garas.
Published in 2000 by E & FN Spon, 11 New Fetter Lane, London EC4P 4EE, UK. ISBN 0 419 25960 0

1 Introduction

During routine maintenance inspection at Nuclear Electric's (NE) Dungeness 'B' Power Station, minor cracking was observed in the steam-side welds of the superheaters which penetrate the walls of the two prestressed concrete pressure vessels (PCPVs). This discovery led to concern about the integrity of the gas-side, circumferential welds (weld 'C') which cannot satisfactorily be inspected (Fig 1).

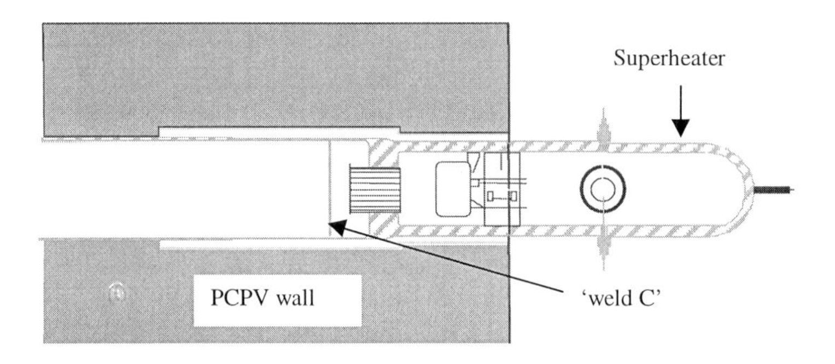

Fig. 1 Cross-section through PCPV wall showing Superheater assembly.

The superheaters form part of the primary circuit pressure boundary. Failure of weld 'C' would leave a 6 tonne outboard portion unrestrained against the gas pressure which, at operating temperature and pressure (650 °C and 3.627 MPa), would be ejected from the vessel with an initial acceleration of up to 40 g. This would lead to:

- Rapid loss of [radioactive] reactor coolant, and
- Damage to adjacent plant items.

Taywood Engineering were engaged by Nuclear Electric to undertake the design, procurement, testing, manufacture and installation of a system of secondary restraints to limit the consequences of the postulated failure and allow for the safe shutdown of the reactor. The restraints were designed to accommodate normal operational movements of the header, and sized and shaped (with a very accurate stiffness) to apply a highly predictable dynamic load forcing function to the restraint. They comprised a special impact load-absorbing helmet connected to large stainless steel plates (with over 170 components each) connected to posts tied back to the PCPV wall by proprietary undercut anchors (Fig 2). It was also required to limit the leakage of circuit gas under fault conditions, and this was achieved by providing a specially developed bellows device to seal the zone where the penetration emerges from the PCPV wall.

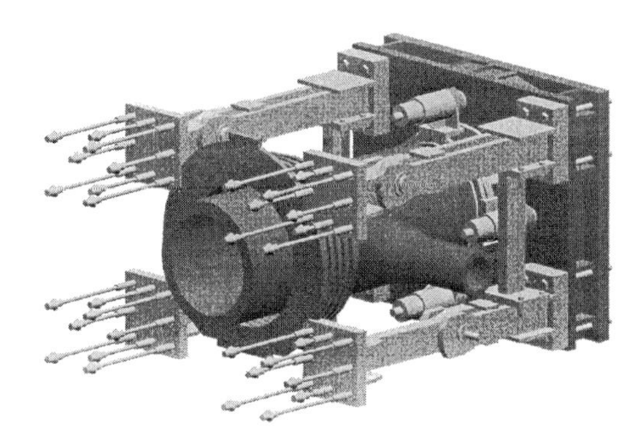

Fig. 2 Restraint assembly including bellows (leak limiter).

Because of the nuclear safety critical nature of the structures, and the demanding design solution required to resist the abnormal loads within the available space, extensive testing was undertaken as follows:

- Concrete mix development to simulate PCPV concrete in the laboratory,
- Individual and group tests to anchor bolts,
- Static and dynamic tests of energy absorbing materials, and
- Full-scale dynamic and sustained load test of prototype restraint assembly.

This paper describes the test carried out on a full-scale prototype restraint assembly.

2 Test design

2.1 Objectives

The objectives of the test were to:

- Establish that the restraint system could withstand the dynamic loads that occur following failure of weld 'C', without causing local failure of the PCPV wall.
- Show that the system could support follow-up steady-state pressure applied for 15 hours, without progressive deterioration or failure.
- Demonstrate the adequacy of the bellows device for limiting gas leakage.
- Provide data for use in validating the design.

Additionally, the full-scale test structure was used to assess the quality procedures and method statements to be used for the site installation.

2.2 Test structure

To meet the above objectives it was necessary to undertake the test at full-scale. A full reproduction was constructed of a part of the pressure vessel wall (it was not necessary to model the curvature), including a superheater penetration. The dimensions were sufficiently large to allow the development of a full concrete failure cone around the fixings (6.5×4.3×3.81 m), with the thickness equal to that of the PCPV wall.

The wall section was constructed using concrete that accurately reproduced the in situ properties of the [30 years old] Dungeness PCPV concrete. The reinforcement in the test structure was similar to that in the PCPVs and the general late-life stress-state was accurately reproduced using post-tensioning bars to induce a mean compressive stress of 2.5 MPa. The wall contained representative inclusions such as cooling water pipes and pre-stressing ducts.

The shutter tube in the wall was the same as in the PCPV, except for the rear section that was thickened to form a pressure chamber for the loading system (see 2.3).

A specimen Superheater penetration (minus steam pipes and fittings etc.) was provided for the test so as to model the correct mass, mass distribution, stiffness and natural frequency etc. For convenience the test superheater penetration was terminated at the rear of the tube plate, forward of the weld 'C' position. The tube plate was sufficiently thick to make the necessary connections to the test loading system.

An entire prototype restraint system was installed using the same constraints and procedures to be used on site.

A cross-section through the test structure is shown in Fig. 3.

2.3 Loading system

The worst case dynamic load occurred at elevated temperature (190 °C). However, it was impractical to undertake the test at this temperature and so the load conditions were applied at ambient temperature. This was conservative, because of the increased stiffness of some components at reduced temperatures (see 6. Discussion of results).

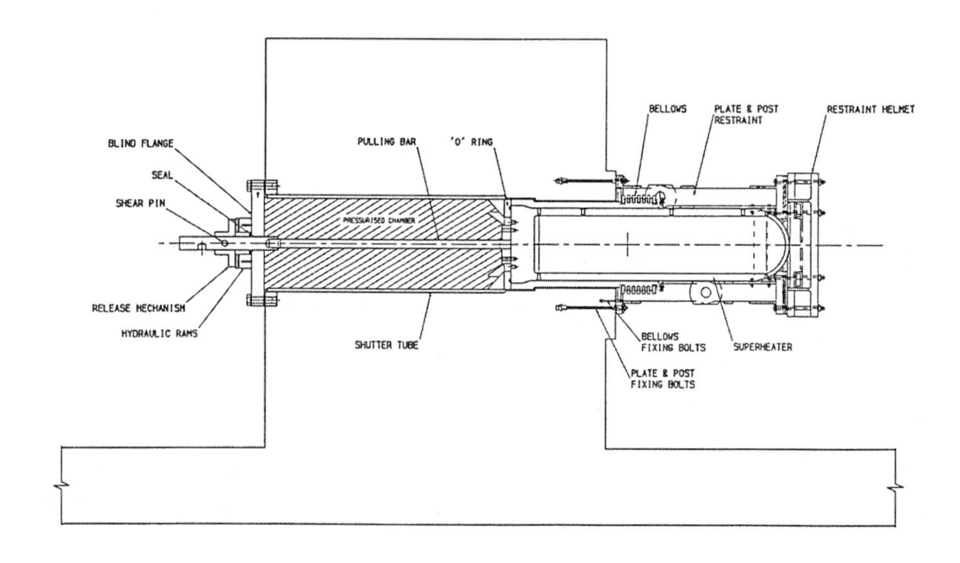

Fig. 3 Section through test structure

A loading system was required that would produce a load regime equivalent to an instantaneous exposure to a large volume of gas at pressure. The difficult feature of this loading to reproduce was the constant force on the superheater as it moved away from the PCPV wall, leading to an impact load directly followed by a steady-state load sustained for 15 hours.

At the concept design stage, several different loading systems were considered including gas pressure, a 'Kevlar rope catapult', and a spring system. For programme considerations, the test structure was initially designed to accommodate two of these systems (gas pressure and 'catapult') pending further work to determine their characteristics. Subsequent development established that the gas (air) pressure method provided the optimum loading system and was therefore selected.

Loading for the test was provided by pressurising the chamber at the rear of the superheater penetration whilst restraining it, and then suddenly removing the restraint. A pressurisation system was designed to provide an initial pressure and then, following the release and movement of the Superheater (which would lead to a rapid reduction in initial pressure), to rapidly restore and maintain the pressure at the correct design gas pressure (3.627 MPa). The system had two principal features as follows:

1. A buffer supply at higher pressure, to restore the sustained load pressure in less than 1 second following the anticipated initial fall.
2. A pressure control system to maintain the sustained pressure in the event of leakage through the bellows leak limiter (of up to 100 m3/hr at standard temperature and pressure). Sufficient air supply was available for a calculated period of 15 hours.

The initial pressure was selected such that the total energy available was the same as predicted by the analysis.

2.4 Release mechanism

A means was required to enable the full loading to be applied instantaneously to the superheater penetration.

At the design concept stage, different methods were considered including a shear pin, a hydraulically removed pin and a sacrificial (explosive) release section. A hydraulically removed pin was found during trials to be slow, and the sacrificial release section was not preferred because it involved the use of explosives. Following a series of commissioning trials, the shear pin method was adopted.

The Superheater penetration was held in position by reacting against the rear face of the wall section, and using hydraulic rams to load a tie bar (through a shear pin) connected to its rear plate. Using the rams, a nominal initial force (21 t) was placed into the tie bar to seal the pressure chamber. As the pressure in the chamber was increased to the test pressure, the rams were used to maintain a force in the tie bar above that of the chamber pressure to ensure that the chamber seal was maintained. To initiate the test, the force in the rams was increased such that the shear pin failed. At this stage, the tie bar was no longer restrained, and the superheater penetration was free to accelerate under the action of the air pressure (see 2.3 above).

3 Instrumentation

3.1 Sensors

A total of 175 sensors were used to monitor the test, including:

- Pressure transducers on pneumatic and hydraulic lines, and flow transducers;
- Displacement transducers on the superheater penetration;
- Accelerometers on the superheater and back plate (monitoring in 3 axes);
- Strain gauges on the bar, the outer face of the helmet, the back plate and the posts;
- Strain gauges on selected bolt shafts (3 in each post group, 4 of those connecting the back plate to the posts and 4 of those attaching the bellows to the wall);
- Strain gauges on the 12 post bolts to pick up vertical shear;
- Embedded strain gauges in the concrete immediately behind the anchorage and in the region of the projected failure cone;
- Strain gauges on the bellows convolutions and equalising rings.

In addition, 3 high-speed cameras (up to 10 kHz) viewed the test from different angles and, in conjunction with movement targets and motion analysis software, allowed the movement of the superheater penetration to be monitored. Additionally, two conventional video cameras monitored the full duration of the sustained test.

All sensor calibrations were traceable to UKAS (NAMAS) standards.

3.2 Data acquisition system

3.2.1 General

The data acquisition system (DAS) incorporated 214 strain gauge amplifier modules, 8 voltage input modules and 57 analogue-to-digital (A to D) modules (each capable of handling four channels of data). During the test, the DAS was required to monitor the sensors initially at a frequency of 10 kHz and then, once the dynamic phase was complete, to continue monitoring at a frequency of 1 Hz for a period of 15 hours. A hardware anti-aliasing filter was used to eliminate frequencies above 5 kHz.

Because of the critical nature of the test, all key test systems were operated from an uninterruptable power supply (UPS), which in addition to filtering the supply also provided approximately 10 minutes backup supply. Standby generators were connected, ready to provide supply after the 10-minute period. Finally, the high-speed data were protected by batteries in the DAS (for up to 15 years).

Also, selected key sensors were recorded using more than one system. For example, the chamber pressure was recorded on the DAS, a high-speed camera and a further backup monitoring system.

Selected data and video images were relayed to a visitor's suite during the test.

3.2.2 Dynamic test

The dynamic phase of the test lasted approximately 200 ms, and the DAS stored data on a 'continuous loop' with a total duration of 1.6 seconds. To ensure that the test data were successfully captured, the DAS was configured to retain data recorded between 800 ms before and 800 ms after receiving a trigger signal (see 3.3 below).

During the test, the recorded data were stored in memory on each amplifier.

3.2.3 Sustained test

After the high-speed recording was completed, the DAS continued recording at a frequency of 1 Hz. During this phase, the DAS recorded data direct to the hard disk on the controlling PC, whilst simultaneously downloading the high-speed data.

A software problem on the PC caused recording to stop approximately eight hours into the test. The recording was restarted approximately 20 minutes later and then continued until the test was completed seven hours later.

3.3 Trigger

A trigger was used to initiate the recording of the high-speed data, and the operation of the high-speed cameras and the rapid re-pressurisation buffer supply.

Two triggers were provided, comprising proximity switches attached to the displacement transducers on the superheater penetration. An additional manual trigger ('mushroom button') was provided as a backup, but not used.

A programmable logic controller (PLC) was used to provide an adequate drive capability for the signal, and also to control the logic such that the system was operated by the first of the triggers to be activated.

Since the test could not easily be repeated, an assessment was undertaken of the risks to a successful test. It was identified that an unplanned trigger signal occurring closely before the release of the Superheater would result in no data being obtained for the test. This risk was considered unacceptable, and an interlock was provided to inhibit operation of the hydraulic pump after activation of a trigger. In this event, failure of the shear pin and hence release of the Superheater would be prevented.

3.4 Data handling

The data from the test were stored as a number of ASCII data files each containing 16,000 observations for the four channels in an amplifier block. Data from each file covering the duration of the dynamic phase were imported into an Excel spreadsheet, and converted to engineering units. The average of the first 100 readings of each channel (excluding pressure measurements) was used as a zero offset.

In a number of cases, recorded data were transformed to give, for example, stresses from strains, direct forces and bending moments etc. All transformations, involving simple arithmetical operations on simultaneous readings of two or more channels of data, were carried out in excel spreadsheets.

4 Test procedure

The test procedure can be summarised as follows:

- All pre-test checks were completed and confirmed to be satisfactory before proceeding with the test.
- The air pressure and tie-bar loads were increased incrementally, until a pressure of 3.69 MPa was reached. At this stage the pressure control valve (PCV) was set, to control the upper limit during the sustained load phase of the test. A simulated check was undertaken to confirm the settings of the PCV.
- Further system checks were undertaken, especially for the triggers.
- The PCV was isolated and the buffer supply pressurised to 4.27 MPa.

Fig. 4 Test assembly.

- The test pressure was then increased incrementally to 4.7 MPa.
- The force in the tie-bar was then increased until failure of the shear pin occurred, thus simulating the sudden weld failure and initiating the test.

5 Results

The key results can be summarised as follows:

- The restraint system withstood the applied loads; the Superheater penetration was brought to rest without any observable distress to the restraint structure or its tieback system. All measured forces and stresses were well within design limits.
- There was no local failure of the concrete and, with the possible exception of one gauge close to a Liebig bolt undercut, all measured strains within the concrete were less than the tensile strain capacity.
- During the sustained phase, strains in the steel items of the restraint system remained steady, responding only to the slight variations in chamber pressure.
- The airflow into the chamber at the end of the dynamic phase of the test was 52 m^3/hr of air at STP, corresponding to a hole of approximately 1.2 mm diameter.

6 Discussion of results

The measured post forces were similar to, but greater than, the FE analysis predictions (5.45 compared to 4.85 MN), and less than the value of 7 MN total (or 1.75 MN per post) used in the design, so the outcome of the test was considered satisfactory. However, this masked significant differences between the measured and predicted values of gas pressure (higher) and distance travelled, velocity and acceleration (lower). A key part of the post-test analysis was to understand and explain these differences, as summarised in Table 1.

There were differences between the test and the reactor installation modelled in the FE analysis, which together would have increased the test load by 25% as follows:

- The stiffness of the crush pads was greater than allowed for by the design analysis due to the low temperature during the test (8 °C compared to 80 °C). This would have had the effect of increasing the pad force in the test by 18%.
- The pin clearance for the post joints on the test restraint was set at 0.25 mm, compared to a value of zero considered for the operating case and hence assumed in the FE analysis. Analysis shows that the force increases with clearance, and that a clearance at the time of loading of 0.25 mm could increase the force by 12.7%.
- The mass of the Superheater, including the mass of the tie-rod, was 5.57 t compared to the design value of 6 t. The lower mass would increase the impact velocity slightly, thereby increasing the dynamic amplification, but this effect is small.
- The free travel distance (between the superheater and the crush pads inside the helmet/plate structure) in the model was 20.6 mm, compared to a maximum design value on the reactor of 23.9 mm. A reduction in free travel reduces the impact velocity, which in turn reduces the crushing reaction of the pads and the dynamic amplification factor. It is estimated from the FE analysis that reducing the free travel distance from 23.9 to 20.6 mm would reduce the load by 5.6%.

A further significant difference concerned the behaviour of the loading system during the test. Prior to release, the pressure was 4.7 MPa with the expectation that, following an initial reduction due to the increase in pressurised area expected after release, the pressurisation system would rapidly restore and maintain the design value of 3.627 MPa. Instead the pressure remained essentially constant at 4.7 MPa, and analysis indicates that the pressurised area was smaller than expected (the available gap was insufficient to allow the gas to pressurise the bellows area). It is believed that this would also be the case for the Reactor installation.

Table 1. Comparison of analysis and test results

Parameter	FE Analysis	Scoping calc$^{\underline{n}}$	Test	Revised calc$^{\underline{n}}$
Gas pressure (MPa)	3.63	3.63	4.70	4.70
Free travel distance (mm)	22.4	21.6	20.6	20.6
Superheater mass (t)	6.50	6.00	5.57	5.57
Plate assembly mass (t)	incl.	3.85	3.85	3.85
Pressurised area (m^2)	0.666	0.666		0.273
Initial acceleration (m/s^2)	371	370	210	211
Impact velocity (m/s)	4.29	4.00	2.94	2.95
Total travel (mm)	68.4	62.8	38.0	33.6
Crush deformation (mm)	46.0	41.0	14.2	13.0
Max total post load (MN)	4.85	5.57	5.45	6.15
Final post load (MN)		2.42	2.45	
Max load per post (MN)	1.57	1.39	1.62	

The effect of the higher pressure acting on a smaller area was to produce a reduced initial acceleration (210 compared to 371 m/s^2) and impact velocity (2.94 compared to 4.29 m/s). However, the dynamic amplification factor is dependent upon the time taken to reach peak load, which was similar for both the test and the analysis. Therefore, the difference in dynamic amplification between test and analysis was small and the maximum post load observed in the test would not have changed significantly had the velocity been the predicted 4.29 m/s.

Table 1 shows predictions for the original FE analysis used in the design, a scoping calculation used to estimate the post forces and dynamic response of the system and a revised version of the calculation updated to correspond to the conditions prevailing in the test. The agreement between the measured response and the values calculated by the revised calculation is reasonably close and indicates that the effective pressurised area was significantly less than anticipated.

7 Conclusions

1. The conditions generated by the test were satisfactory.
2. The behaviour of the prototype system (including leak-limiter) was satisfactory, giving confidence in the overall robustness of the design solution.
3. A comparison between the measured test conditions and the FE analysis predictions identified differences from the expected gas flow dynamics for the test that were likely also to apply to the Reactor installation. Taking account of these differences leads to a reduction of 46% in the design dynamic force applied to the restraint structure compared to the loads predicted by the original analysis. Hence it is likely that there is a very significant conservatism in the design assessment which assumes pressurisation of the bellows area in determining the accelerating force.
4. The successful test confirmed the optimum installation methods.
5. The system tested went on to be successfully installed on a total of 16 superheaters on both PCPVs (see Fig. 5).

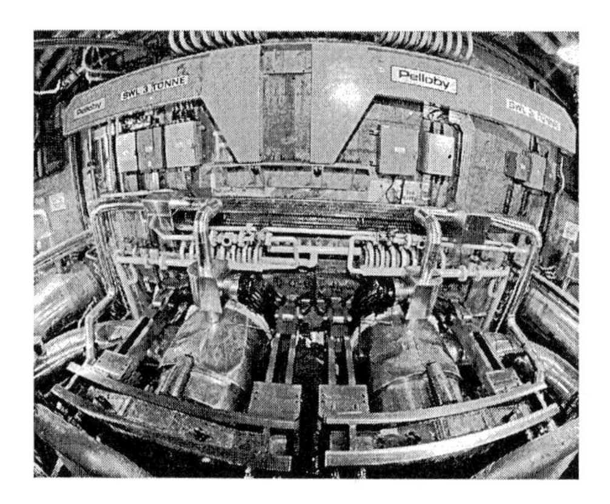

Fig. 5 Site installation.

CARBON FIBRE COMPOSITES AS STRUCTURAL REINFORCEMENT IN A NUCLEAR INSTALLATION
Bonded reinforcement for seismic strengthening

H.N. GARDEN
Taywood Engineering, Middlesex, UK

Abstract
This Paper describes the experimental evaluation of a novel system for restoring the original steel reinforcement contribution in load bearing concrete walls affected by yielding of the steel due to structural and thermal effects. Strips of a carbon fibre reinforced polymer (CFRP) laminate were bonded to the concrete in line with the reinforcement, restoring the original stiffness and strength contribution of the embedded steel. The CFRP material provided a rapidly installed reinforcement system, allowing the work to be undertaken during a strict outage period of just three weeks in this first UK application of CFRP in an operating nuclear station. The restoration of the walls to their original condition ensures the structures meet hazard loading criteria, as required by the Independent Nuclear Safety Assessors (INSA). The initial load tests of the bonded system are reviewed. A selection of the bonded laminates and the underlying concrete cracks were instrumented and are now being monitored remotely, demonstrating the effectiveness of this reinforcement system.
Keywords: Carbon fibre reinforced polymer (CFRP), epoxy adhesive, hazard loading, power station, reinforced concrete, shear load tests, tensile load monitoring

1 Introduction

1.1 The use of composites for strengthening

The rehabilitation of reinforced concrete and other structures has been achieved for many years using externally bonded steel plates. Disadvantages of this method include transporting, handling and installing heavy plates and corrosion of the plates. The use of composite materials overcomes these problems and provides equally satisfactory

Abnormal Loading on Structures edited by K. S. Virdi, R. S. Matthews, J. L. Clarke and F. K. Garas.
Published in 2000 by E & FN Spon, 11 New Fetter Lane, London EC4P 4EE, UK. ISBN 0 419 25960 0

solutions. It is both environmentally and economically necessary to apply rapid, effective and simple strengthening techniques for the upgrading of existing civil infrastructure. An excellent example of this is in the nuclear power industry which demands minimal disruption of facilities to ensure continuous production of electricity.

Composite materials, also known as fibre reinforced polymer (FRP) materials, comprise strong and stiff reinforcing fibres embedded in a polymer matrix resin. FRPs possess high strength-to-weight ratios and excellent electrochemical corrosion resistance, resulting in low maintenance costs with this method of rehabilitation. The main material types finding uses as reinforcing fibres in FRPs are glass, carbon and aramid. FRPs are generally more expensive than mild steel but the majority of the cost of a strengthening scheme is labour; the easy handling of FRPs reduces labour costs considerably. When long lengths of externally bonded reinforcement are required, the problem of having to join limited lengths of steel plate is overcome by the fact that FRPs may be delivered to site in rolls of 300 metres or more.

The use of bonded composite reinforcement, to carry only live or live and dead loads, has been reviewed in several previous publications by the author and others [1]-[11]. The technique provides a reinforcement effect by the development of tensile strain in the additional layer due to the transfer of strain through the bond by a shear action in the adhesive. The adhesive used to bond the CFRP to the concrete in the present work was a modification of the Sikadur 30 two-part (ie. base and hardener) epoxy resin manufactured by Sika Ltd., and the CFRP laminate itself was one of the Sika CarboDur range, comprising unidirectional carbon fibres embedded in an epoxy resin matrix.

1.2 The nuclear structural safety review process

The continuing operation of a nuclear power station is dependent on the submission of a satisfactory safety case by the operator to the Nuclear Installations Inspectorate (NII). The safety case is reviewed by the Independent Nuclear Safety Assessors (INSA) and includes for the integrity of the civil and structural elements of the buildings under both static and hazard loading conditions. Common forms of hazard loading to be considered are impact events (eg. steam pipe explosions) and seismic actions. This Paper concerns the case of a station at which a number of reinforced concrete structural walls had been confirmed as suitable for static loads but not hazard loads so a method of upgrading, by bonding lengths of a carbon fibre reinforced polymer (CFRP) laminate, was implemented. The strengthening works were undertaken in early 1997 but further CFRP applications have been designed and managed by Taywood Engineering (TEL) since then.

INSA conducts periodic safety reviews at which the safety case must be renewed. TEL was faced with the task of developing a method of strengthening the walls to satisfy the safety case and, therefore, allow the client to obtain INSA approval. TEL's proposed method was to be implemented during a strict outage period (ie. reactor shutdown) during which a variety of maintenance activities would be undertaken by several contractors, so logistical and programming considerations were essential. The financial consequences of a late return to service of the reactor are severe and the use of bonded composites offered the most rapid approach to upgrading the walls.

1.3 The problem to be overcome

The horizontal embedded reinforcement bars in the walls had yielded in a number of locations due to the actions of mechanical and thermal loading. This occurred due to excessive strain being transferred to the reinforcement via crack widening in the concrete. The structural assessments of the walls (also undertaken by TEL) confirmed that their as-built condition was suitable for the hazard load cases, so the objective of the additional bonded reinforcement was to restore the structures to their original condition. It was required to do this without compromising the existing static load integrity of the structures. The CFRP laminates were bonded to the concrete and oriented horizontally as 1 m lengths directly over the affected embedded reinforcement.

TEL's involvement in this work, as part of a wider scope to survey and repair defects to the civil structures, was to provide the following:

- the feasibility study to review the possible methods of upgrading the walls and the particular merits of CFRP bonding for the power station;
- the detailed design of the bonded CFRP reinforcement system;
- the installation documentation (ie. specification, method statement and quality plan);
- preliminary trials of the CFRP installation at TEL's structural testing laboratories;
- detailed surveys of the existing defects in the walls for comparison against previous and future TEL surveys;
- laboratory verification tests to confirm the suitability of the proposed CFRP system;
- site supervision of the CFRP installation works;
- quality control testing of the adhesive material from each batch applied;
- reporting on the final extent of CFRP bonding as an input to the client's return-to-service process.

The experimental verification of the proposed scheme is outlined in this Paper.

2 The strengthening works to be verified experimentally

2.1 Layout of the problem

The walls between the boiler cells and duct cells support heavy steam drum housings. The centre section of the upper part of these walls is supported by a precast beam, as indicated in Fig. 1. The figure also shows the typical form of distress in the walls; only the main cracks (for CFRP bonding) are indicated and one type of wall is shown as an example.

2.2 The widths of the cracks to be reinforced

The integrity of the walls was checked by analysis for static and seismic loads, confirming their ability to sustain seismic loads in the as-built condition only. Progressive widening of the cracks led to suspected yielding of the horizontal rebars so the contribution of these bars had to be restored. The crack widths corresponding to rebar yield were determined by analysis according to Eurocode 2 (ENV 1992-1-1).

This allowed the criterion crack widths (at rebar levels) to be identified on site to determine where the bonded reinforcement should be located.

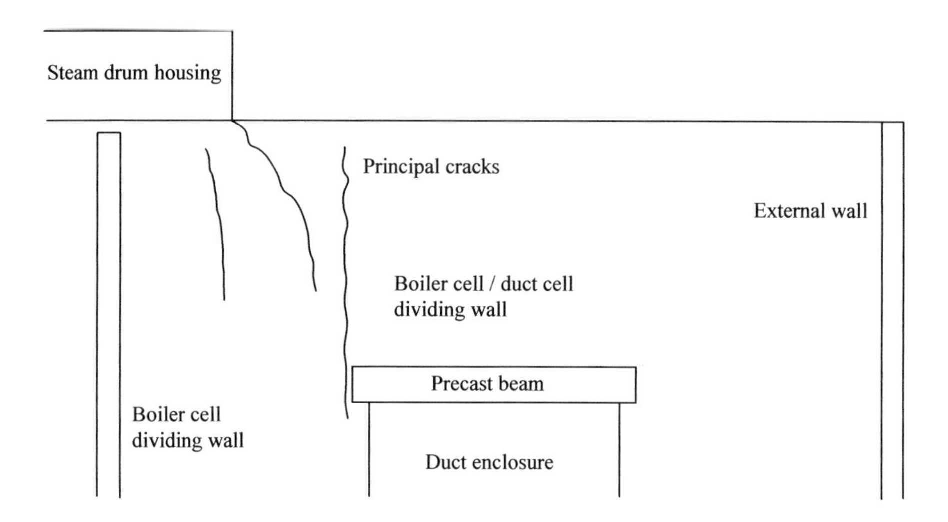

Fig. 1 Typical features of a boiler cell / duct cell dividing wall (elevation view)

2.3 The application of previous crack monitoring data

The crack width analysis considered the strain in the embedded reinforcement during operation of the reactor plant in the boiler cells and duct cells. However, the cracks are wider during operation than during outage periods due to the reduction in temperature that accompanies the shutdown of the reactor. Consequently, it was necessary to know the associated reduction in crack width caused by the drop in temperature. This was established from previous crack monitoring results collected by TEL as part of the commission to survey the condition of the civil structures.

2.4 Issues in the design of the CFRP reinforcement

The main design issues were the following:

- the load that the bonded laminate would need to carry in order to restore the contribution of the originally elastic (ie. pre-yield) rebar;
- the cross sectional area of composite laminate required to sustain this load;
- the length of bond between the CFRP and the concrete to enable the required load to be transferred into the bonded laminate without causing adhesive or cohesive failures in the bond;
- the length of bond to enable thermally induced loads to be sustained in the concrete / CFRP bond;
- the anchorage length of bond of the CFRP to the concrete to ensure a conservative design.

It is known, from the considerable research and development results collected by the first author and others, that the concrete itself generally fails (rather than the adhesive or the bond) at the ultimate limit state. The failure mode is of a shear / tensile nature so it is necessary to select an appropriate shear and tensile strength for the concrete. This value was reduced to a more conservative level by application of a material safety factor. This mode of failure and the anticipated ultimate capacity were checked by laboratory verification tests, as described below.

3 Experimental verification of the bonded system

3.1 Short term static load tests

Part of the process to obtain INSA approval was a practical demonstration of the ultimate capacity of the proposed bonded system. The state of loading of the CFRP / concrete bond was represented in a laboratory test rig designed to simulate the same shear / tensile load case as experienced in practice. Figure 2 shows this test configuration, in which a length of the CFRP laminate was loaded in tension parallel to the bonded surface of a concrete slab. The crack in the concrete was represented by a saw cut in order to encourage the wedge-type of concrete failure that was expected in the vicinity of the crack. This mode was anticipated from the results of previous research findings and was confirmed by the testing. In Figure 2, the CFRP laminate is the black strip. The CFRP strip was loaded in tension to create a shear / tensile action in the concrete and the bond.

The general form of the concrete wedge failure is shown in Figure 3. The failure started in the vicinity of the saw cut, this being the location of greatest stress along the bond. Finite element analysis [2] has shown that peak stress concentrations occur at geometric discontinuities where strain incompatibilities arise.

The load required to be carried by each bonded CFRP plate was 32 kN but the average shear / tension capacity of the system was 110 kN. However, the design of the system ensured that the stiffness imparted to the walls by the bonded CFRP was equal to that of the originally unyielded embedded reinforcement, thereby not changing the structural response of the walls. The large safety margin in ultimate capacity would be utilised in the event of a seismic load.

The magnitude of the ultimate shear / tension capacity is governed principally by the width of the bond to the concrete for a given bond length [2]. This is because the capacity depends on the strength of the wedge of concrete removed at failure. Research has shown that the capacity of the bonded system increases at a progressively slower rate, with respect to bond length, as the length increases, until a threshold length is reached beyond which no further gain in capacity is observed. Therefore, the bond length in the design (either side of the crack) was able to be limited to just 500 mm.

Fig. 2 Test configuration for laboratory load simulation

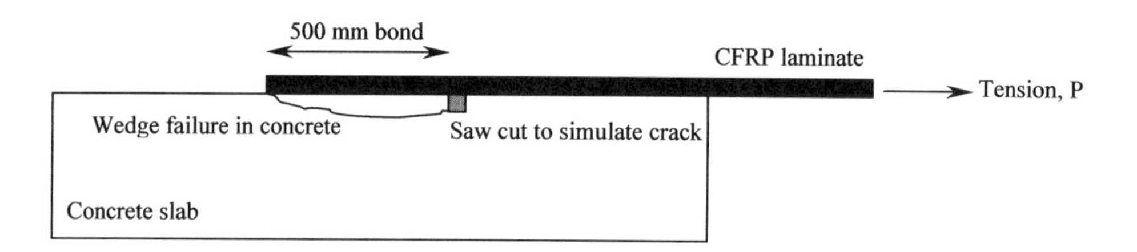

Fig. 3 Test configuration for laboratory load simulation

3.2 Long term monitoring of the bonded system
A number of CFRP / CFRP single lap shear test specimens were manufactured and placed in the boiler cell and duct cell environments so that they can be tested when entry into these areas is permitted at the next outage. The results will be compared with those from similar specimens that were not exposed to an elevated temperature environment or to atomic radiation.

In addition, the cracks in the concrete and the bonded CFRP material are now being monitored to confirm the following:

- that the bonded CFRP is displaying composite action with the concrete;
- that the tensile strain in the CFRP is of the magnitude anticipated;
- that the crack width is reduced by the amount expected and is subsequently being controlled as predicted.

Linear variable displacement transducers (LVDTs) were attached to the wall across the crack to monitor crack movement and the CFRP was instrumented with electrical resistance (ER) strain gauges over the crack and at the ends of the CFRP strip (or plate). Figure 4 shows the variations in concrete crack width and CFRP tensile strain with time for one particular wall area that was strengthened. The length of time over which these data were collected started shortly before the reactor was returned to power and continued for a period afterwards.

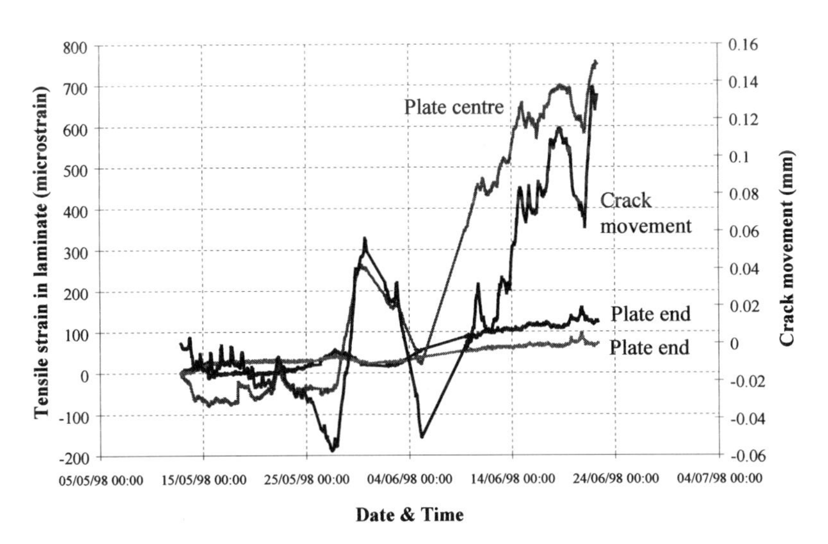

Fig. 4 Concrete crack width and CFRP tensile strain behaviours for one strengthened area of wall

The variability in the crack width and CFRP strain behaviours before 04/06/98 is due to the reactor having been returned to power and shut down during the post-outage re-commissioning trials. After 04/06/98, the crack width profile was reflected in the CFRP tensile strain profile at the centre of the plate, indicating the composite action between the CFRP and concrete due to the action of the adhesive bond. The increase in tensile strain at the ends of the CFRP layer was much smaller due to the ends being remote from the concrete crack; the ends are affected predominantly by concrete thermal movements.

4 Conclusions

The experimental testing confirmed the surface wedge type of concrete failure that was assumed in the design of the seismic strengthening scheme. For conservatism, the design was based on a uniform distribution of shear stress throughout the bond line, but the testing showed that failure is initiated locally at the position of greatest stress. The method of rehabilitation by CFRP bonding is appropriate for cracked concrete structures, as shown by the composite action generated by the adhesive at the positions of the major cracks in the walls. The design of such external strengthening systems makes certain assumptions regarding the stresses in the bond and the substrates, so experimental testing was the best way to find the true ultimate capacity and, hence, the factor of safety against failure. The use of bonded CFRP reinforcement is an effective approach for the seismic strengthening of installations containing numerous operating plant items and other congestion.

5 References

1. Barnes, R.A. and Garden, H.N. (1999), 'Time-dependent behaviour and fatigue'. In: Hollaway, L.C. and Leeming, M.B. (eds.), *Strengthening of Reinforced Concrete Structures – Using Externally Bonded FRP Composites in Structural and Civil Engineering.* Cambridge: Woodhead Publishing Ltd, 1999, pp. 183 - 221.
2. Garden, H.N. (1997), 'The strengthening of reinforced concrete members using bonded polymeric composite materials', *PhD Thesis*, University of Surrey.
3. Garden, H.N. (1998), 'Bond of strength', *Bridge Design and Engineering*, Issue 11, pp. 28-29
4. Garden, H.N. and Hollaway, L.C. (1997a), 'An experimental study of the strengthening of reinforced concrete beams using prestressed carbon composite plates', *Proc. 7th International Conference on Structural Faults and Repair*, University of Edinburgh, July 8th – 10th 1997, Vol. 2, pp. 191-199.
5. Garden, H.N. and Hollaway, L.C. (1997b), 'An experimental study of the external plate end anchorage of carbon fibre composite plates used to strengthen reinforced concrete beams', *Composite Structures*, No. 42, pp. 175 - 188.
6. Garden, H.N. and Hollaway, L.C. (1997c), 'An experimental study of the failure modes of reinforced concrete beams strengthened with prestressed carbon composite plates', *Composites, Part B*, Vol. 29, pp. 411-424.
7. Garden, H.N. and Hollaway, L.C. (1998), 'The strengthening and deformation behaviour of reinforced concrete beams upgraded using prestressed composite plates', *Materials and Structures*, Vol. 31, May 1998, pp. 247-258.
8. Garden, H.N., Hollaway, L.C. and Thorne, A.M. (1997), 'A preliminary evaluation of carbon fibre reinforced polymer plates for strengthening

reinforced concrete members', *Structures and Buildings*, Vol. 122, No. 2, pp. 127-142.

9. Garden, H.N., Hollaway, L.C., Thorne, A.M. and Parke, G.A.R. (1996), 'A parameter study of the strengthening of reinforced concrete beams with bonded composites', In: Harding, J.E., Parke, G.A.R. and Ryall, M.J. (eds.), Bridge Management 3, *Proc. 3rd International Conference on Bridge Management*, University of Surrey, Guildford, April 14th – 17th 1996, pp. 400-408.
10. Garden, H.N. and Mays, G.C. (1999), 'Structural strengthening of concrete beams using prestressed plates'. In: Hollaway, L.C. and Leeming, M.B. (eds.), *Strengthening of Reinforced Concrete Structures – Using Externally Bonded FRP Composites in Structural and Civil Engineering*. Cambridge: Woodhead Publishing Ltd, 1999, pp. 135 – 155.
11. Garden, H.N., Quantrill, R.J., Hollaway, L.C., Thorne, A.M. and Parke, G.A.R. (1998), 'An experimental study of the anchorage length of carbon fibre composite plates used to strengthen reinforced concrete beams', *Construction and Building Materials*, Vol. 12, pp. 203-219.

EXPERIMENTAL AND ANALYTICAL STUDIES OF STEEL T-STUBS AT ELEVATED TEMPERATURES

Steel stubs at elevated temperatures

S. SPYROU, J.B. DAVISON and I.W. BURGESS
Department of Civil & Structural Engineering, University of Sheffield, Sheffield, UK

Abstract

It has become apparent from careful studies of the fire tests on the Cardington composite frame building that, although connections undoubtedly have a major effect on the structural performance of a frame in fire, it is inadequate to consider simply a degradation of the ambient temperature moment-rotation characteristics without taking account of the high axial forces which also occur. Since the overall response of the connection is produced by the behaviour of different zones within it under the principal effects of horizontal tension or compression, a breakdown of the connection into components would be very advantageous. A project has recently begun in which the changes of stiffness and strength of component zones as a result of fire conditions will be examined. Combination of the components of typical beam-to-column connections used in current construction will be examined experimentally and analytically, in order to provide a satisfactory connection model for future numerical studies of frames at elevated temperatures.

An experimental programme is being undertaken in which connection components are being tested in tension and/or compression at elevated temperatures. The initial phase of the project concentrates on connections whose components are naturally idealised as T-stubs. These play a fundamental role in the behaviour of the connection, and their characteristics can be assembled and used to compute the overall behaviour under combinations of bending and axial thrust. The paper describes the use of imaging techniques, which have been adopted for the measurement of deformations in the steel T-stubs at elevated temperatures, and compares the results with an elasto-plastic model derived from beam theory.

Keywords: Component method, connections, fire engineering, image processing, T-stubs, thermal imaging, steel structures.

Abnormal Loading on Structures edited by K. S. Virdi, R. S. Matthews, J. L. Clarke and F. K. Garas.
Published in 2000 by E & FN Spon, 11 New Fetter Lane, London EC4P 4EE, UK. ISBN 0 419 25960 0

1 Introduction

The response of steel-framed structures to applied loading depends to a large degree on the behaviour of the joints between the columns and beams. Traditionally designers have assumed that these joints act either as 'pinned', with no ability to transmit moments from beam to column, or as 'rigid', providing perfect continuity between the connected members. Advances in analysis and developments in modern codes of practice permit designers to account for the real behaviour of steel joints where this is known. Even though experimental studies of joints conducted at many research centres around the world have provided a large bank of test data [1], the vast number of variables in joints (beam and column sizes, plate thicknesses, bolt sizes and spacing etc.) often means that data for a specific joint arrangement does not exist. As a result, researchers have turned their attention to ways of predicting the behaviour of such joints. One approach that has gained acceptance is based on the "component method" in which overall joint behaviour is assumed to be the sum of the responses of various simpler components. Much work has been completed in developing this approach for beam to column connections [1].

When steel-framed structures are subjected to fire, their ability to sustain loads is severely impaired and the action of the joints is of particular concern. To date, data on the response of joints at high temperatures has been gathered from full-scale furnace tests [2]. The application of the "component method" at elevated temperatures is the subject of a recently awarded research contract which aims to develop a method of predicting the behaviour of steel beam to column joints in fire conditions.

In the usual characterisation procedures, a joint is generally considered as a whole and is studied accordingly; the originality of the "component method" is to consider any joint as a set of individual basic components [3]. In the particular case of Fig. 1, which illustrates a joint with an extended end-plate connection subjected to bending, the joint is divided into the three major zones (tension, shear and compression) and then each zone is divided into the relevant components as follows:

Table 1. Zones within the joint and their components

Tension zone	Compression zone	Shear zone
End plate in bending	Column web	Column web panel
Column flange in bending	Beam flange and web	
Column web		
Beam web		
Bolts		

Each of these basic components possesses its own level of strength and stiffness in tension, compression or shear. The main objective of this study is to use the principles of the "component method" in order to calculate the effects on strength and stiffness of an individual or a group of components at elevated temperatures.

One of the advantages of this method is that the stiffness and strength characteristics of an individual or a group of components may be found by experiments (although there are limited data available at elevated temperatures), by numerical simulations using finite element programs, or by using analytical models derived from simple theory or curve-fitting.

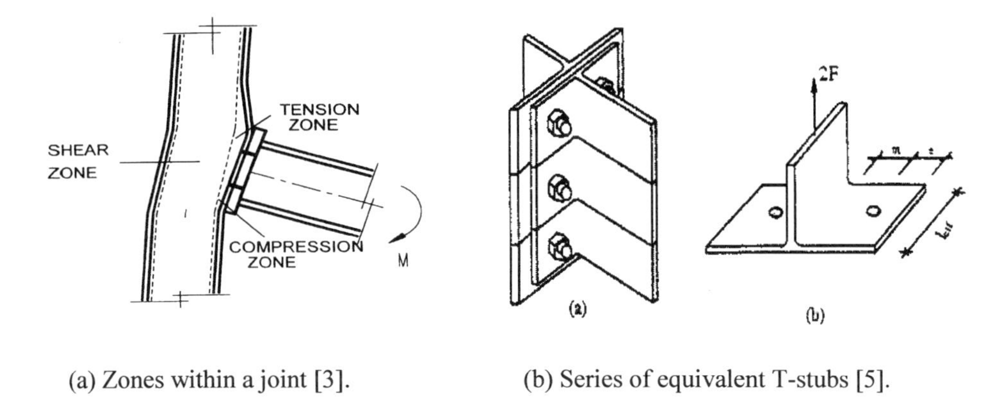

(a) Zones within a joint [3]. (b) Series of equivalent T-stubs [5].

Fig.1. The three zones in an end-plate joint, and equivalent T-stubs in the tension zone.

2 Tension zone

In the tension zone of a steel joint, the column flange and end plate at each bolt row are considered as a series of T-stub assemblies, as shown in Fig. 1(b). T-stubs play a fundamental role in the behaviour of the joint and are used to compute the stiffness contribution of the most important connection "components" which are the end plate in bending, the column flange in bending and the bolts in tension. Equivalent lengths of simple T-stubs, the effective lengths, are determined based on the recommendations of Eurocode 3, Annex J [4] to represent the more complex real arrangement found in the complete connection.

An experimental study of bolted T-stub assemblies has been undertaken in order to measure the force-displacement characteristics at elevated temperatures representative of those achieved in building fires thus extending the component approach for use in fire engineering studies.

In the past, deformations have been recorded by linear voltage displacement transducers, inclinometers and, in the more distant past, by dial gauges. A problem arises when displacements at elevated temperatures are required. An experimental method to measure deflections of a T-stub assembly by instruments, which are not attached to the specimen, has been devised and is described in this paper. Although the primary reason for developing the method was the need to record deformations at elevated temperatures, the method is also attractive for tests conducted at ambient temperatures and offers a number of advantages over more conventional methods.

2.1 Ambient temperature tests

Two pilot tests, at ambient temperature, were performed in order to compare the displacement readings of a T-stub assembly using two measurement techniques. One technique used conventional linear voltage displacement transducers (LVDTs) and the second used a digital camera. The testing arrangement is shown in Fig. 2. The specimens were made by bolting together two T-stub elements, obtained from a rolled I-beam profile (305x165x40UB), steel grade S275, by cutting it along the web plane. These T-stubs were connected through the flanges by means of four M12, Grade 4.6 bolts.

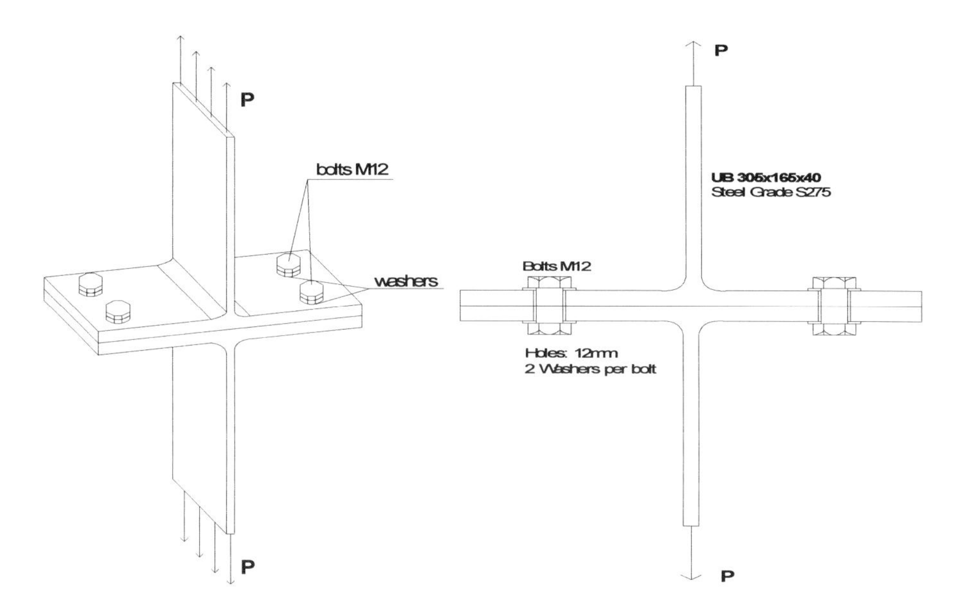

Fig. 2. The T-stub model used in ambient temperature tests.

The specimens were subjected to a tensile axial force applied to the webs by the jaws of a 1000 kN capacity Avery testing machine. The experiments were designed to check the viability of imaging techniques for measuring axial displacements under increasing tensile axial force. As these were pilot tests, readily available equipment-rather than bespoke apparatus was used to prove the techniques before purchasing more sophisticated cameras and image processing software. The axial displacements were measured in two ways:

1. S8FLP10A LVDTs with 38mm length and 10mm mechanical stroke, which provide an electrical signal directly proportional to a linear mechanical movement of the shaft, were attached directly to the specimen. This kind of arrangement gives the opportunity to use electronic logging devices, which can accurately record large amounts of experimental data automatically.
2. A Casio QV-7000SX LCD digital camera was used to capture images of the deflected shape for subsequent analysis using image-processing software. The digital camera was placed on one side of the T-stub assembly at a distance of 300mm.

When the test started the experimental data from the LVDT were automatically recorded. The LVDT was connected to an Orion Delta logging system, which converted the readings into engineering units, and stored it directly on a computer. Using the LCD Digital Camera, pictures had to be taken manually at given load intervals (every 20kN in the elastic region and every 2kN in the plastic region). At the end of the test the images were analysed using Imaging processing demonstration software in order to give the displacement readings.

The procedure to analyse the images using the software was very simple. Before the tests were carried out, reference horizontal and vertical distances were measured. Using a Vernier gauge, the flange thickness was measured in the vertical direction and the web thickness in the horizontal direction. The next step was to scale these readings according to the initial image. The two edges of the flange on the initial image were identified by eye and using the mouse the measured distance (10.55mm) was entered. As the picture was already divided into pixels this automatically generated a scaling factor between pixels and millimetres. With this conversion, any vertical distance could be measured simply by picking up with the mouse pointer any two points in the vertical direction. The software counts the number of pixels between the locations and transforms the distance into millimetres. The same procedure was used to scale the horizontal readings.

During loading of the specimens, a number of deflected images were recorded for each test. These were then analysed using the software and the results compared with the measurements obtained from the LVDTs. An image of the T-stub's deflected shape is shown in Fig. 3.

Fig. 3. Typical deformation of a T-stub specimen at ambient temperature.

2.2 Experimental results

The resolution of the recorded image at the time of Test No. 1 was 640x480 pixels. The field of view of the camera was 117mm in the vertical direction and 126mm in the horizontal direction and a total of fourteen images were captured. The relationship between the LVDT readings and the image processing results is shown in Fig. 4.

It is obvious from Fig. 4 that there is good correlation between the readings made by the camera with image processing software and the LVDTs. The sensitivity of readings taken from LVDTs was 20μm. The standard accuracy from the software, for these specific images, was up to 243.75μm, obtained by dividing the value in the vertical direction 117mm, which is the direction of displacement, by the 480 pixels in that direction. The software on the other hand uses linear sub-pixel division in order to interpolate maximum accuracy from the images. Also from Fig. 4 it can be seen that the LVDT ran out of travel distance at about 4mm. The imaging technique however is capable of reading data over a wide range of deformation.

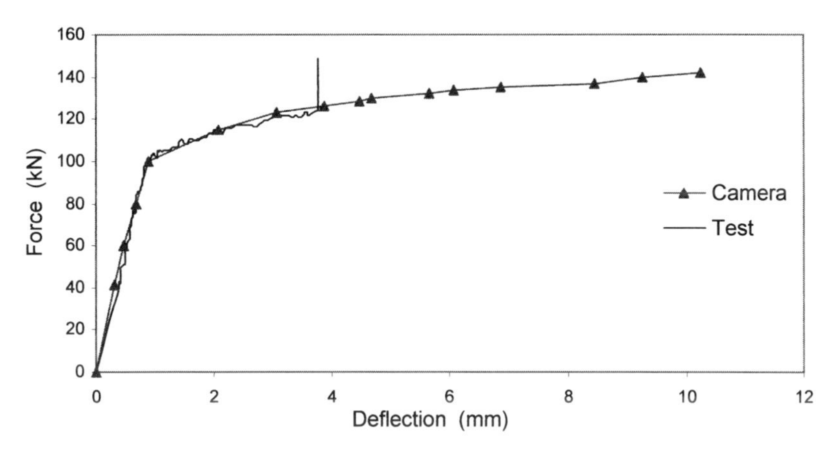

Fig. 4. Force-deflection curve for Test 1.

A second test was conducted with the image resolution set at 1280x960 pixels. This time the field of view of the camera was 74mm in the vertical direction and 121mm in the horizontal direction and a total of twelve images were captured. The relationship between the LVDT reading and the image processing results is shown in Fig. 5.

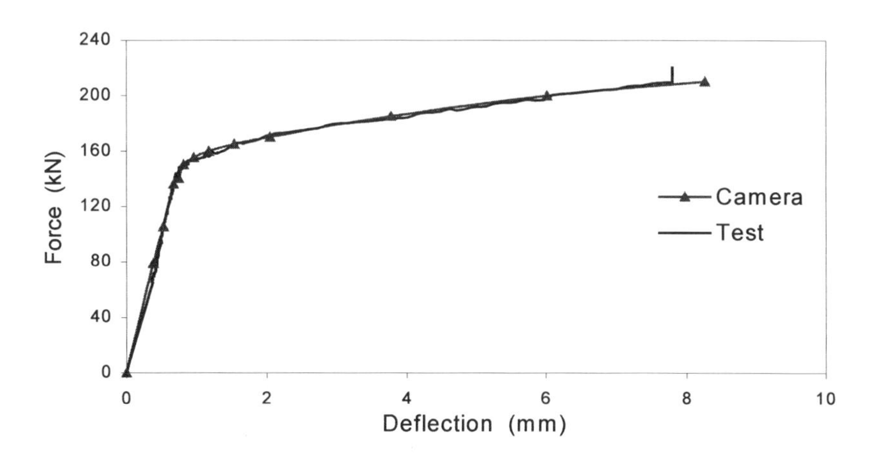

Fig. 5. Force-deflection curve for Test 2.

In this test the implied accuracy obtainable from the software was 77.08μm, but again sub-pixel division accuracy was used to get the maximum accuracy out of the images.

Three factors affect the accuracy of the tests, namely:

1) Resolution of the image taken, in pixels;
2) Field of view of the camera. The smaller the field of view, the larger the number of pixels in that field, and as a result, better accuracy;
3) A further concern, which might affect the accuracy of the tests, is the quality of the images taken at elevated temperatures, when above a certain temperature everything inside the furnace is glowing. In order to test this, another pilot test was performed by heating the already deflected T-stub specimens up to 750°C. Fig. 6 shows two images of T-stub specimen inside the furnace, one at 100°C and the other at 700°C.

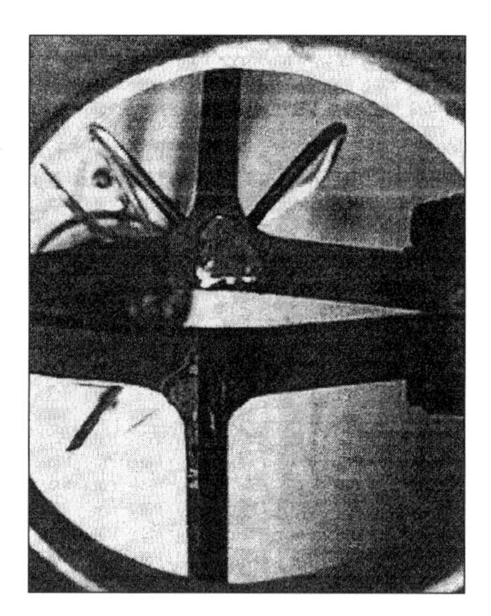

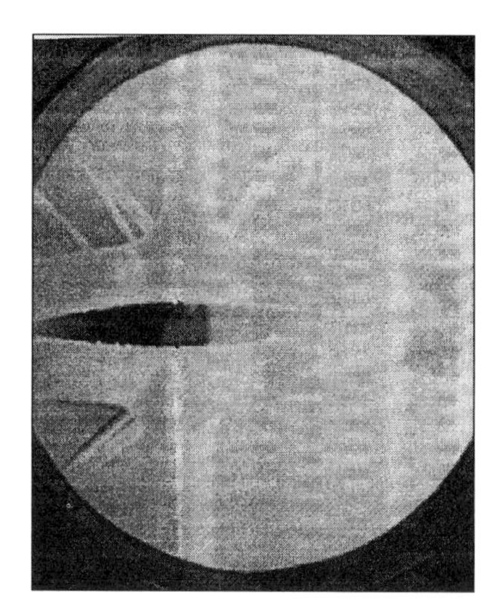

Fig. 6. T-stub specimen at 100°C and at 700°C.

From these images it is clear that well-defined edges are still visible in the image at 700°C, and these can be used to measure the displacement between the deflected flanges. Encouraged by the results from the pilot tests it was decided to use imaging techniques in the test programme to find the characteristics of components at elevated temperatures.

3 Development of the test programme

A purpose-built furnace, with an internal capacity of $1m^3$ has been designed with viewports to accommodate three video cameras. It was convenient to use an electric fan-assisted furnace in order to avoid any flames within the field of view of the video cameras. The fan also ensures a uniform distribution of atmosphere temperature up to 1100°C. Two viewports were used to accommodate the three video cameras. One is at the front of the furnace, perpendicular to the axis of loading, facing the horizontal direction where there is movement of the flanges. The other is at the top of the furnace, facing the vertical direction, in which it is believed that there will be no movement during the tests. The first viewport accommodates two video cameras, one for accurate measurements and the other for general observation of the T-stub distortion. The other viewport accommodates another video camera, again for general observation of the T-stub distortion. There are also two holes on the opposite sides of the furnace to allow a hydraulic jack to apply force to the specimens.

The image acquisition and processing system had to be carefully selected. First, the required accuracy of the deflection measurement was set at 40μm without any sub-pixel division. Then the camera field of view needed to be investigated, depending upon the position of the video cameras relative to the specimen and the type of lenses to be used; this was set at 30mm × 30mm. Another step was to investigate the resolution of the video cameras in combination with frame grabbers. The frame grabber sets, more or less, the resolution of the image taken; it can also trigger the cameras to capture images and save them to file.

Three monochrome analogue video cameras (JAI CV-M50), and a colour frame grabber (Imagenation PXC200) were selected, together giving a capture resolution of 768 x 576 pixels. The image processing software was purpose-written in order to include the change of information in an image from low to high temperatures, and the capability of controlling the sub-pixel division. Fig. 7 shows a schematic diagram of the image acquisition and processing system and Fig.8 the arrangement for the experiments.

4 Proposed T-stub deformation model

In parallel with the assembly of experimental equipment, analytical studies of T-stub behaviour have been undertaken. Using simple beam theory a deformation model for the tension zone at ambient temperatures has been developed. The deformation model adopted for the end plate, column flange system is that of the two T-stubs bolted together (Fig. 2). The tension zone deformation comes from the deformation of the column web, column flange and end plate, and the bolt elongation [5,6]. This model can been extended to predict the behaviour at elevated temperatures and will be compared with test results as they become available, and with finite element modelling.

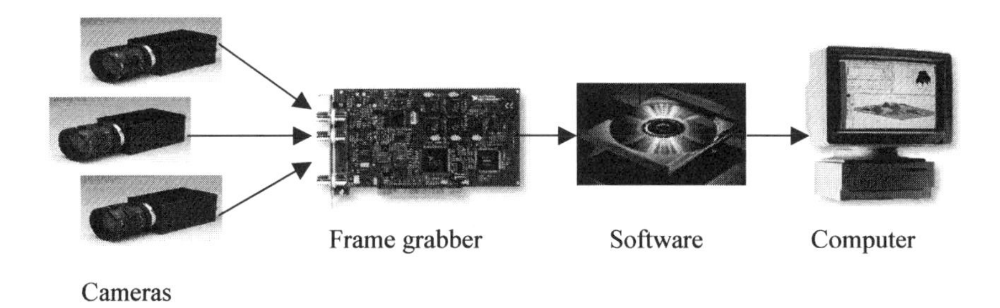

Fig. 7. Schematic diagram of the image acquisition and processing system[8].

Fig. 8. Arrangement for the experimental work.

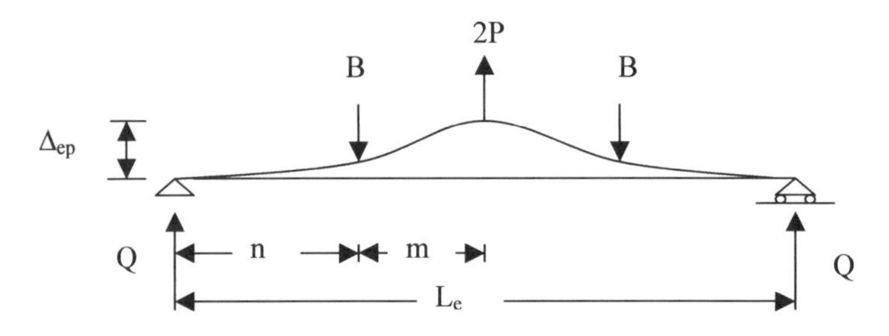

Fig. 9. Forces on T-stub assembly.

From classical beam theory, if the tension force acting on a T-stub assembly (Fig. 9) is $2P$, the elastic separation Δ_{ep} is given by:

$$\Delta_{ep} = \frac{PL_e^{\ 3}}{24EI} + \frac{\frac{P}{EI}\left(\frac{nL_e^{\ 2}}{8} - \frac{n^3}{6}\right)\left(\frac{n^3}{6} - \frac{nL_e^{\ 2}}{8}\right)}{EI\left[\frac{L_b}{E_b A_s} - \frac{1}{EI}\left(\frac{2n^3}{3} - \frac{n^2 L_e}{2}\right)\right]} \tag{1}$$

Where $I = L_{ef}\, t_f^3/12$, L_{ef} is the effective length given in Eurocode 3, Annex J; L_b is the effective length of the bolt; E_b is the Young Modulus of the bolt and A_s is the shaft area of the bolt.

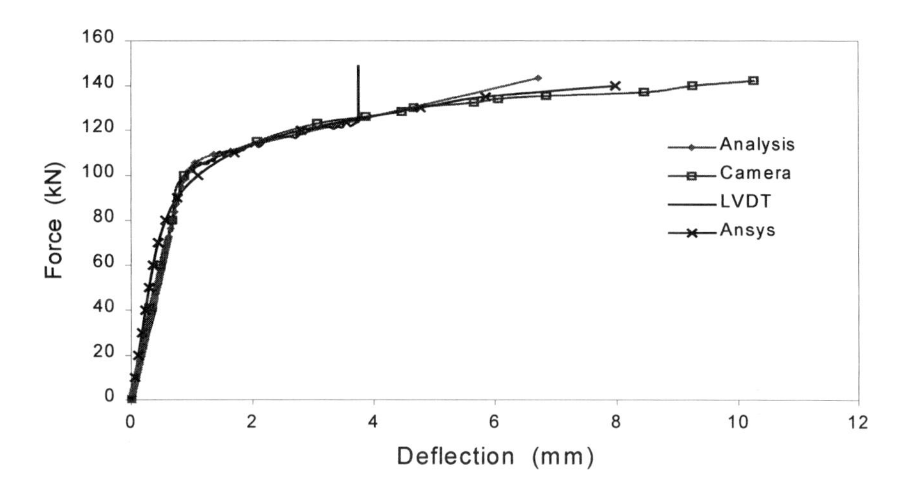

Fig. 10. Comparison of test results with finite element and simple beam model at ambient temperature.

Some preliminary two-dimensional finite element studies have been performed in order to simulate the experiments at ambient and elevated temperatures. An effective two-dimensional finite element model capable of describing plasticity and contact effects is proposed to calculate deformations at ambient[7] and elevated temperatures. Taking advantage of the two symmetry axes and with the use of 84 nodes and 136 plain stress, with thickness, quadrilateral elements were constructed to present a quarter of the T-stub specimen and solved with the finite element analysis ANSYS. The results taken from the test are compared with the finite element and the results from the simple beam model in Fig. 10.

5 Conclusions

Pilot tests have been conducted to show the feasibility of using image acquisition and processing techniques to measure deformations in joint components. This technique will be used to record data in elevated temperature tests to investigate the performance of joint components.

6 Acknowledgement

The work described herein is funded by EPSRC whose support is gratefully acknowledged.

7 References

1. Jaspart, J.P., Weynand, K. and Steenhuis, M. (1996) "The Stiffness Model of revised Annex J of Eurocode 3." *Connections in Steel Structures III,* CH.53, pp.441-452.
2. Leston-Jones, L.C., Burgess, I.W., Lennon, T. and Plank, R.J. (1997) "Elevated temperature moment-rotation tests on steelwork connections." *Proc. Instn Civ. Engrs Structs & Bldgs,* Vol. 122, pp. 410-419.
3. Jaspart, J.P., Steenhuis, M. and Anderson, D., (1998) "Characterisation of the joint properties by means of the component method." *Proceedings of the Liege COST C1 conference of semi-riged behaviuor of civil engineering structural connections,* Brussels.
4. "Eurocode 3: design of steel structures: Part 1.1, Revised Annex J.", (1992). ENV 1993-1-1, *European Committee for Standardisation.*
5. Shi, Y.J., Chan, S.L. and Wong, Y.L. (1996) "Modelling for moment-rotation characteristics for end-plate connections." *Journal of Structural Engineering,* Vol. 122, No. 11, pp. 1300-1306.
6. Agerskov, H. (1976) "High-strength bolted connections subject to prying." *J. Struct. Div.,* ASCE, Vol. 102, No. 1, pp. 161-175.
7. Mistakidis, E.S., Baniotopoulos, C.C., Bisbos, C.D. and Panagiotopoulos P.D. (1997) "Steel T-stub connections under static loading: an effective 2-D numerical model." *J. Construct. Steel Res.* Vol. 44, No. 1-2, pp. 51-67.
8. Software picture ULR, http://www.search.corbis.com, Frame Grabber and Computer picture ULR, http://www.natinst.com, Cameras picture ULR, http://www.jai.dk/uk_index.html.

NON-HOLONOMIC ELASTOPLASTIC BEHAVIOUR OF A TWO SPAN BEAM

Elasto-plastic behaviour of two span beam

R. P. WEST and A. J. O'CONNOR
Department of Civil, Structural and Environmental Engineering, Trinity College Dublin, Ireland

Abstract

The techniques of mathematical optimisation are well suited to the problem of plastic analysis and design of structural frames. Indeed, it is well known that the basic structural relations of statics, kinematics and the constitutive relations naturally formulate themselves into the Kuhn-Tucker conditions necessary for optimality. Thus in plastic limit analysis or synthesis, one proceeds directly to the point of collapse, where the collapse load is assumed to be unique, and the conventional simplex algorithm can readily produce solutions. However, in the elastoplastic region of loading, the assumption of load path independence may no longer be valid and plastic hinge unstressing may occur, even under a monotonic loading regime. While techniques exist to allow for this, the extent of this effect is not well known and even a simple two span beam can have surprising behaviour prior to collapse.

This paper will describe a parametric study of the peculiar load effect due to unstressing of hinges during elastoplastic behaviour. The results will be verified using conventional optimisation software, and a simple laboratory experiment, applied to a specific case study.

Keywords: Elastoplastic, Kuhn-Tucker criterion, non-holonomic, optimisation, plastic hinge, plastic limit analysis, simplex algorithm, unstressing,

Abnormal Loading on Structures edited by K. S. Virdi, R. S. Matthews, J. L. Clarke and F. K. Garas.
Published in 2000 by E & FN Spon, 11 New Fetter Lane, London EC4P 4EE, UK. ISBN 0 419 25960 0

1 Introduction

Plastic behaviour is observed in a material when it deforms without any increase in the applied load. Since the introduction in 1986 of limit state design in BS5950 [1], which incorporates steel design for plastic collapse, the importance of plastic theory and the study of plastic behaviour have increased greatly. Unstressing, or non-holonomic behaviour, is a phenomenon where a fully plastic section, or plastic hinge, reverts back to the elastic state, despite an increase in the force which induced plasticity. The objective of this paper is to look at the incidence of unstressing in a simple beam and loading arrangement, analysing its sensitivity to load changes and movements, whilst also looking at the frequency with which it occurs for the particular loading configuration.

2 Plasticity Theory

Plastic bending occurs in elastoplastic materials, which follow Hooke's law up to their yield stress and afterwards yield plastically without any increase in stress, the idealised and *'actual'* stress-strain diagram for such a material in tension being illustrated in Figure 1(a). Theoretically, the beam is presumed to behave elastically up to point *D'* with subsequent plastic deformation. The actual behaviour follows the path ABCD illustrated in Figure 1(a), with the corresponding stress distributions illustrated in Figure 1(b). At point A, due to low values of moment, *M*, the extreme fibre stress is below the yield stress, and the section behaves elastically with a linear stress distribution. As *M* increases, a point B is reached where the maximum stress in the section approaches the yield stress, with corresponding yield moment, M_y. Since the stress in the section may not exceed the yield stress, any further increase in load creates a plastic zone in the outer fibres. This plastic zone, extends towards the neutral axis as *M* increases, approaching point C. Finally, upon reaching point D the entire section becomes plastic.

For the range of bending moment, $M_y \leq M \leq M_P$, the section comprises an elastic core surrounded by plastic regions. As the moment increases, the elastic core takes extra stress while the outer plastic regions yield without any increase in stress. Thus the elastic core controls the deformation of the beam. This is called contained plastic flow. However, when the entire section becomes plastic, the beam can continue to deform without any increase in the bending moment being applied. This is termed unrestricted plastic flow [2], and the section has now become a, so-called, plastic hinge.

In determinate structures the formation of a single plastic hinge triggers a collapse mechanism. In an indeterminate structure, the formation of a single plastic hinge does not necessarily provide a collapse mechanism. In general, one hinge is required for the removal of each redundancy, and then another to form a mechanism and fail the structure.

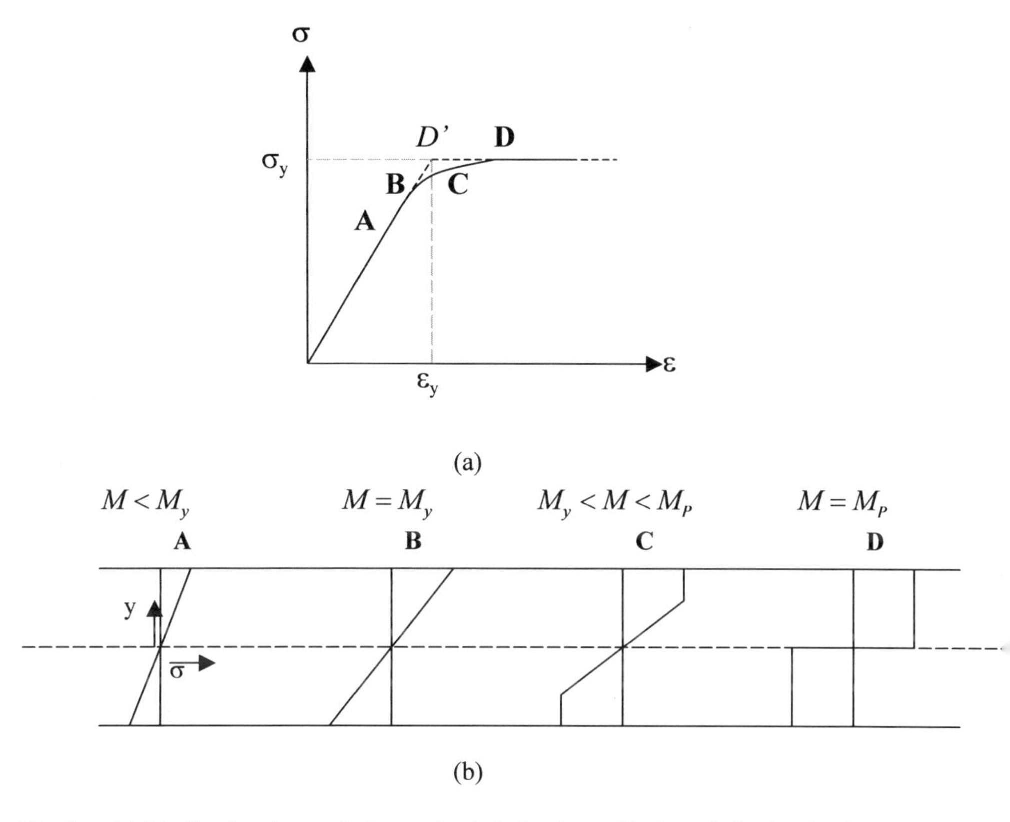

Fig. 1. (a) Idealised and actual elasto-plastic behaviour, (b) Actual elastic-plastic bending stress

2.1 Non-Holonomic Behaviour

Non-holonomic behaviour is the unstressing of a section, which has previously suffered plastic deformation, despite the application of monotonically increasing proportional loading. In non-holonomic behaviour this phenomenon occurs before the recovery of plastic deformation and, therefore, the process is irreversible [3].

Non-holonomic behaviour may only occur in indeterminate structures where more than one hinge is required for the formation of a collapse mechanism. In addition, for an indeterminate structure to collapse upon formation of a sufficient number of plastic hinges, the bending moment produced at the separate hinges throughout the structure must be in sympathy with the opening of those hinges in the collapse mechanism. When this condition, known as parity, is contravened, this implies that although a sufficient number of hinges have opened to form a collapse mechanism, a mechanism does not actually exist. Rather a *pseudo-mechanism* is formed [3,4], where this pseudo-mechanism is stabilised by the unstressing of an offending hinge to create a new equilibrium configuration. Fig. 2 illustrates the stress-strain relationship of plastic unstressing. The case shown by the dotted line indicates the formation of a plastic hinge opening in the opposite direction to that of the original, thereby recovering the original plastic deformation.

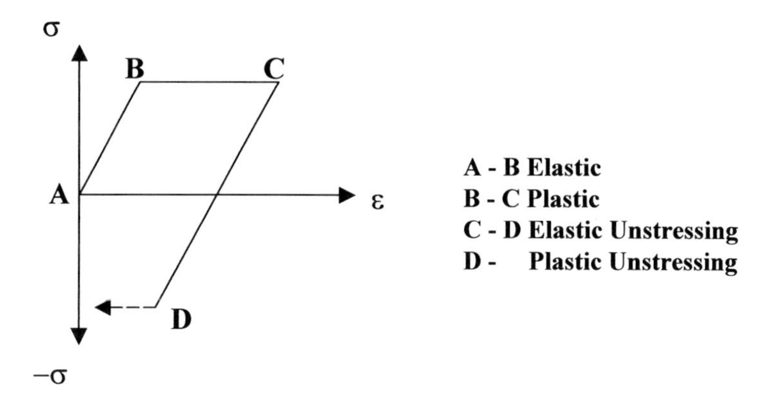

Fig. 2. Plastic unstressing stress-strain diagram [5]

2.2 Example of Non-Holonomic Behaviour

To illustrate the surprisingly little known phenomenon of unstressing, the example is provided of a two-span continuous beam, illustrated in Fig. 3, with two concentrated loads applied in one span only. For the particular case to be analysed, it is assumed that L=2.0m, Z=1.0m, $\lambda = 0.538$, say, S=$2(L-Z)/3$, R=$(L-Z)/3$, and that the plastic moment capacity is M_p throughout. Equations describing the bending moments have been developed [6] for any chosen case of L, Z, S, R, F and λ but are not included here for brevity.

Stage I: The first stage in the load path leads to the formation of a hinge at critical section T. The critical load factor for formation of the first hinge is calculated as

$$\omega_1 = \frac{M_P}{M^1_{\max}} = 5.154 \tag{1}$$

where $M^1_{\max}$ represents the maximum absolute bending moment calculated in Stage 1, that is, $M_T = 0.194F$, in Fig. 4(a). The bending moment diagram representing the end of this first stage is illustrated in Fig. 4(b).

Stage II: As the loading continues to increase, section T now behaves as a full plastic hinge. The incremental bending moment associated with this stage in loading is illustrated in Figure 5(a), leading to the formation of a second plastic hinge at Q. The critical load factor for opening of the second hinge at Q is

$$\omega_2 = \min\left(ABS\left(\frac{M_p - M_i^1}{M_i^2} \right) \right) = 0.427 \tag{2}$$

where M_i^j represents the value of bending moment calculated at location i in stage j.

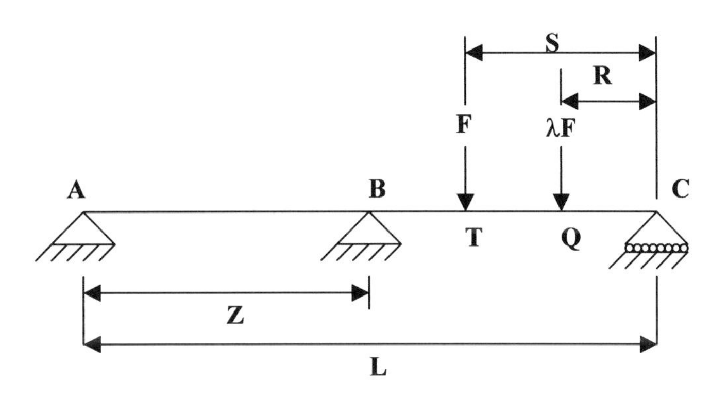

Fig. 3. Two-span structure with loading pattern as shown

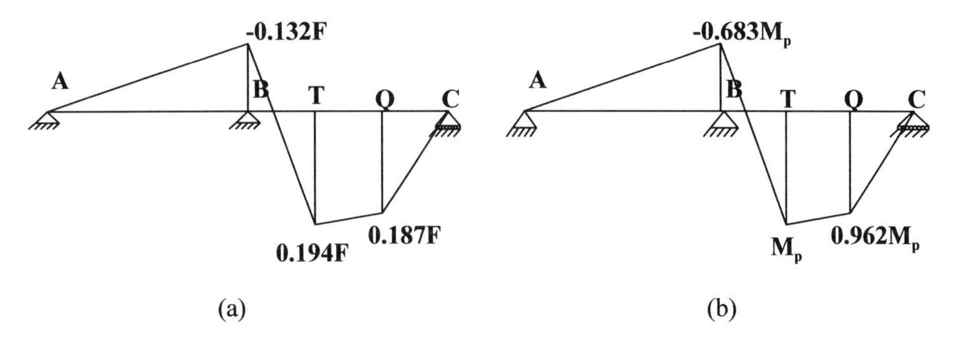

Fig. 4. Stage I: (a) Elastic bending moment diagram, (b) Bending moment at end of Stage I with full plastic hinge at T

Stage III: It may appear as though a mechanism has now formed, with the right span collapsing in a local beam mechanism (plastic hinges at T and Q). However, the positive plastic moment at T, Fig. 4(b), is incompatible with the negative sense of the hinge at T if this is the collapse mechanism, that is, the condition of parity is not satisfied. Here, joint T now unstresses under a further increasing load F and a mechanism is not formed. The incremental bending moment for the next stage is, therefore, as shown in Fig. 6(a) giving rise to a bending moment at the end of stage III as shown in Fig. 6(b), with the total bending moment as shown in Fig. 7. The critical load factor, for formation of a hinge at B is

$$\omega_3 = \min\left(ABS\left(\frac{M_p - \sum_{j=1}^{2} M_i^j}{M_i^3} \right) \right) = 0.195 \qquad (3)$$

A collapse mechanism finally occurs due to the formation of a hinge at B, in addition to the hinge at Q. The ultimate collapse load factor is calculated as, $\Omega = \sum_{i=1}^{3} \omega_i = 5.776$ (that is, collapse load $F_c = \Omega M_p$).

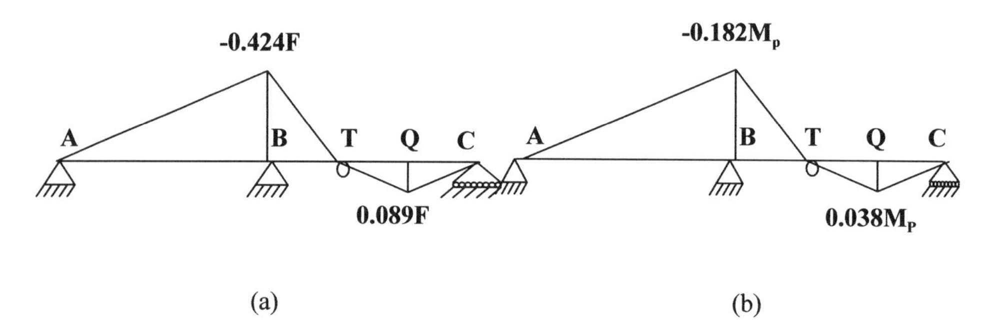

Fig. 5. Stage II: (a) Incremental bending moment diagram, (b) End of stage II incremental bending moment when plastic hinge occurs at Q

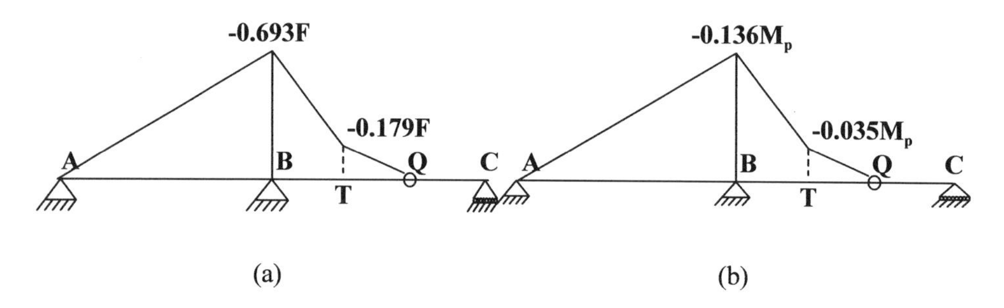

Fig. 6. Stage III: (a) Incremental bending moment diagram, unstressing at T, (b) End of stage III incremental bending moment with plastic hinge at B

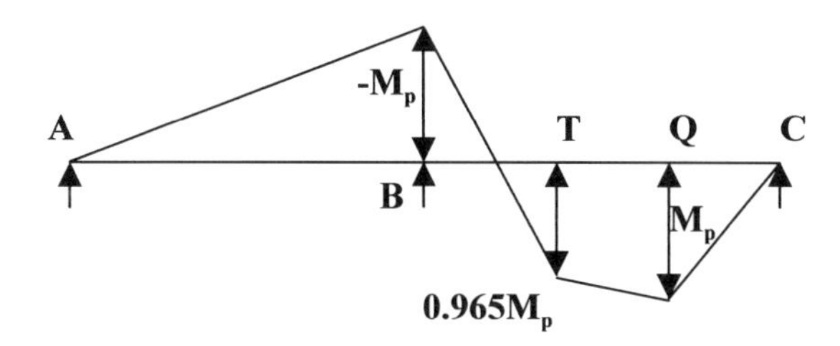

Fig. 7. Ultimate collapse bending moment diagram

3 Solution using Plastic Limit Analysis and the Simplex Algorithm

It is well-known [3] that the mesh form of the primal (kinematic) program of plastic limit analysis is given by:

$$\left.\begin{array}{l} Min\ z = \underline{m}_p^T \dot{\underline{\theta}} \\ \text{Subject to} \\ \begin{bmatrix} \underline{\mathbf{B}_0}^T & \underline{\mathrm{N}} \\ \underline{\mathrm{B}}^T & \underline{\mathrm{N}} \end{bmatrix} \dot{\underline{\theta}} = \begin{bmatrix} 1 \\ \underline{0} \end{bmatrix} \\ \text{Where} \\ \dot{\underline{\theta}} \ \geq \ \underline{0} \end{array}\right\} \quad (4)$$

in which $\dot{\underline{\theta}}$ are the mechanism deformation rates and $\underline{\mathbf{B}_0}$ and $\underline{\mathrm{B}}$ are the values of the bending moments at the critical sections in the loading and bi-action diagrams, using Static Kinematic Duality (SKD). Specifically, for the example in Fig. 3, using a moment release at the central support, one obtains the statically determinate diagrams from which $\underline{\mathbf{B}_0}$ and $\underline{\mathrm{B}}$ are derived, as illustrated in Fig. 8. From these, the input into the simplex algorithm can be derived to solve Eqn 4 above, as follows, recognising locations B, T and Q as the only critical sections:

$$\left.\begin{array}{l} Min\ z = M_p\theta_1^+ + M_p\theta_2^+ + M_p\theta_3^+ + M_p\theta_1^- + M_p\theta_2^- + M_p\theta_3^- \\ \text{Subject to} \\ \begin{bmatrix} 0 & 0.2820 & 0.2307 & 0 & -0.2820 & -0.2307 \\ 1 & 0.6667 & 0.3333 & -1 & -0.6667 & -0.3333 \end{bmatrix} \begin{bmatrix} \theta_1^+ \\ \theta_2^+ \\ \theta_3^+ \\ \theta_1^- \\ \theta_2^- \\ \theta_3^- \end{bmatrix} = \begin{bmatrix} 1 \\ 0 \end{bmatrix} \\ \\ \theta_i^+, \ \ \theta_i^- \geq 0, \quad i = 1, 2\ or\ 3 \end{array}\right\} \quad (5)$$

where θ_i^+ and θ_i^- are representative of the positive, negative (or zero) values of θ at the three critical sections.

The use of the readily available package, LINDO, for simplex algorithm analysis, yields an objective function of 5.776, that is a collapse load of $5.776M_p$ as seen previously.

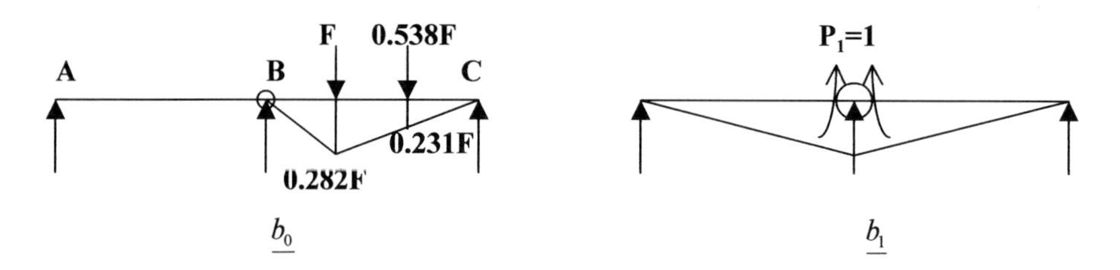

Fig. 8. $\underline{b}_0$ and $\underline{b}_1$ diagrams for Fig. 3 with a moment release at location B

The primal solution yielded is $[0 \quad 0 \quad 4.335 \quad 1.445 \quad 0 \quad 0]$ which suggests that θ_3^+ and θ_1^- are active and are in the ratio 1:3, as expected from the collapse mechanism inferred by Fig. 7. This is confirmed by the dual solution to the mathematical problem, which is also given by LINDO, namely $[5.776 \quad -1]$. This suggests that $5.776M_p \times \underline{b}_0 - M_p \times \underline{b}_1$ will yield the bending moment at collapse (that is, Fig. 7), which it does.

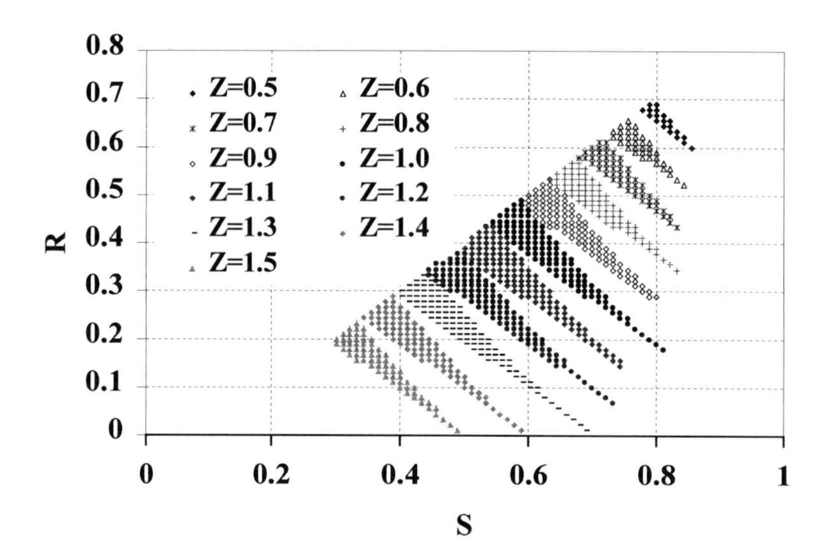

Fig. 9. Influence of variation in Z on non-holonomic behaviour

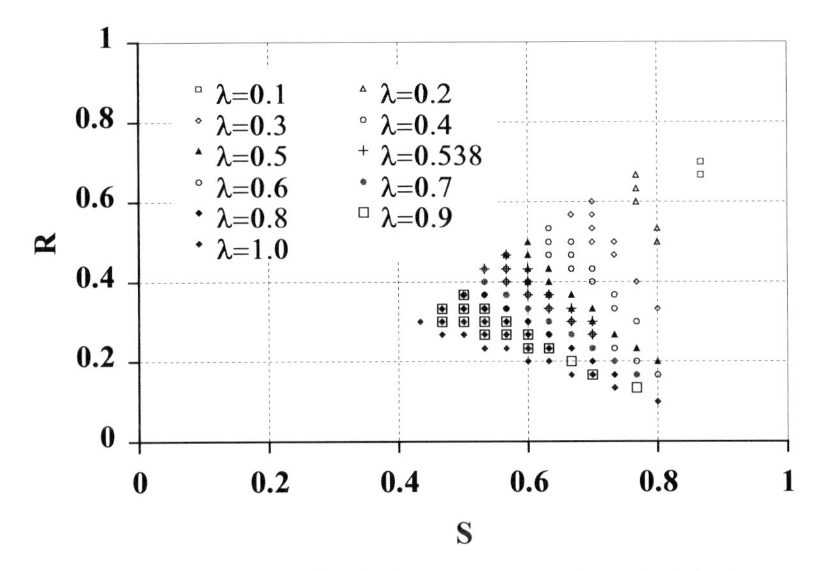

Fig. 10. Influence of variation in λ on non-holonomic behaviour

This confirms that despite the unstressing of joint T during loading as discussed earlier, plastic limit analysis yields the unique collapse load and the correct mechanism, in this case.

4 Non-Holonomic Sensitivity

The example illustrates the phenomenon of non-holonomic behaviour of a simple yet specific structural configuration (that is, span length and ratio) and loading regime (that is, load ratio, location in span, etc.). However, this surprising phenomenon in the sample structural arrangement is not unusual as evidenced by a parametric study on the parameters Z, S, R and λ. Indeed, Figs. 9 and 10 illustrate the wide range of values for Z and λ for which non-holonomic behaviour is observed [6,7].

5 Experimental Observation of Non-Holonomic Behaviour

In an attempt to observe non-holonomic behaviour experimentally, a continuous aluminium beam with configuration as illustrated in Fig. 3, with L=0.6m, Z=0.3m, S=0.14m, R=0.05m and λ=1.0 was tested to failure. The rectangular cross-section had depth d=2.94mm and breadth b=24.93mm. Deflections were measured under the load points at sections T and Q. Strain gauges were located on the top of the beam at sections B and T and on the underside of the beam at section Q. The load-strain relationship for this test is illustrated in Fig. 11 [7].

At a load of 140N, the load-strain relationship at section T became non-linear, that is the section began to behave plastically. At a load of 150N, sections B and Q also began to behave non-linearly. Once the load had reached 190N, it became obvious that the first full plastic hinge would form at section Q, as the total strain at Q was in excess of that at T. For the load increment 190-200N the rate of change of strain at B was small in comparison with that at T and Q. At this point the strains at T and Q were so large with respect to the strain at B, that it appeared as though plastic hinges may form at these sections before section B. However, during the load sequence 200-220N a change in the load distribution along the beam was observed. The rate of change of strain at section T decreased, as at section Q, but the strain at B was observed to increase dramatically. A plastic hinge had now formed at location Q. At a load of 230N section Q had rotated by such an amount that the strain gauge broke. As further load was applied, the rate of change of strain at section B increased dramatically as a hinge formed at this section, while the strain at section T decreased and the section behaved non-holonomically, that is, section T had unstressed. While it is realised that the order of formation of hinges in the experiment is different to the specific example in section 2.2, the principle is, of course, unchanged.

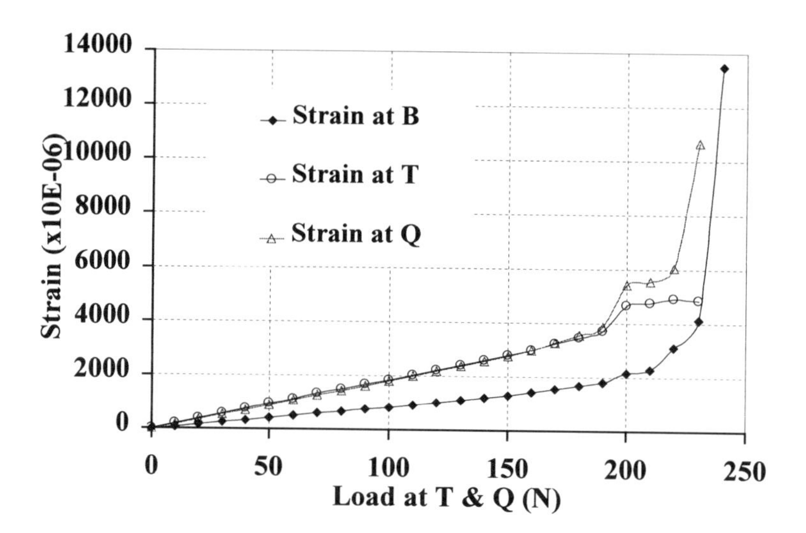

Fig. 11. Experimental results of load vs. strain at critical sections

6 Conclusions

It is worthy of note that the phenomenon of unstressing can occur unexpectedly prior to collapse even for a simple two span beam. In this paper this has been clearly verified in a step-wise approach through the elastoplastic region of behaviour prior to collapse, and, by use of a parametric study using equations developed to describe the beam's behaviour, has been shown to occur for a wide range of geometric and loading parameters. Further, the existence of unstressing in such a case has been verified using a simple laboratory experiment.

While the existence of such pseudo-mechanisms in skeletal structures has been known for some time, it is hoped that this paper will contribute, by its simplicity, to a broader understanding of the complex phenomenon.

7 Acknowledgements

The authors wish to acknowledge the theoretical and experimental work of Bermingham [5] and Duggan [6] as the background to this paper. In addition, the particular case study was originally drawn to the attention of the authors by Prof. David Smith of Imperial College in London, for which the authors are grateful.

8 References

1. British Standards Institution. (1990) *Structural Use of Steelwork in Building*, BSI, London. BS 5950: Part 1.
2. Gere, J.M. and Timoshenko, S.P., (1991), *Mechanics of Materials 3rd Ed.*, Chapman & Hall.
3. Smith, D.L. (1978) 'The Wolfe-Markowitz algorithm for non-holonomic elastoplastic analysis', *Eng. Struct.,* Vol. 1, pp 8-16.
4. Smith, D.L. and Munro, J. (1978) 'On uniqueness in the elasto-plastic analysis of frames', *J. Struct. Mech.*, 6(1), pp. 85-106.
5. Bermingham, J.N. (1992) *Optimum plastic design of planar skeletal structures by linear programming*, M.Sc theses, Trinity College Dublin.
6. Duggan, F. (1993) *An analysis of unstressing in a two-span beam*, B.A.I thesis, Trinity College Dublin.
7. Steele, D.J. (1996) *Experimental observation of non-holonomic behaviour'* B.A.I thesis, Trinity College Dublin.

SERVICEABILITY BEHAVIOUR OF A 3-DIMENSIONAL ROOF TRUSS SYSTEM

3-D Roof Truss System

T.D.G. CANISIUS, S.L. MATTHEWS and D. BROOKE
Construction Division, Building Research Establishment Ltd., Watford, England

Abstract

As part of a project on the long term study of full scale structures, a monitoring system has been installed in the trussed roof of the Meeting Hall of the Swaminarayan Hindu Mission in London. The hall is 45m × 55m in plan with the trusses spanning the 45m side at a height of 7.5m above floor level. The design of the roof trusses was governed by the deflection rising from the 'abnormal' load from the movable, hung acoustic doors. This paper details the load tests and in-service monitoring performed on this structure. Both load test results and in-service monitoring data are analysed and compared with computer-based results to understand the behaviour of the structural system. As a result some important observations are made on the design and monitoring of purpose-designed three dimensional load-sharing structural systems.

Keywords: Load tests, monitoring, serviceability, 3-D truss system

1 Introduction

The Meeting Hall of the Swaminarayan Hindu Mission in London has dimensions of 45m × 55m. The hall has roof trusses spanning the 45m dimension. The main trusses (Fig. 1), identified as T and V according to their grid lines in the building plan (Fig. 2), are 4.5m in height at centre span, diminishing to 2.5m at their ends where they are supported on stanchions 7.5m tall. As part of a project on the long term behaviour of full scale structures, a monitoring system has been installed in the roof trusses T and V. These nominally identical, unevenly spaced trusses had been designed for the case of the most heavily loaded truss. In addition to environmental loads and dead loads, the trusses have been designed to carry the substantial 'abnormal' live load of acoustic doors used to subdivide the main hall area into smaller rooms.

Abnormal Loading on Structures edited by K. S. Virdi, R. S. Matthews, J. L. Clarke and F. K. Garas. Published in 2000 by E & FN Spon, 11 New Fetter Lane, London EC4P 4EE, UK. ISBN 0 419 25960 0

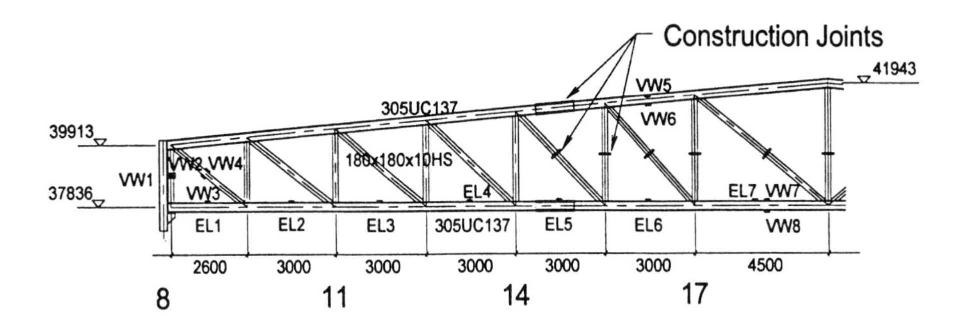

Fig. 1. Locations of construction joints and permanent transducers on Trusses T and V

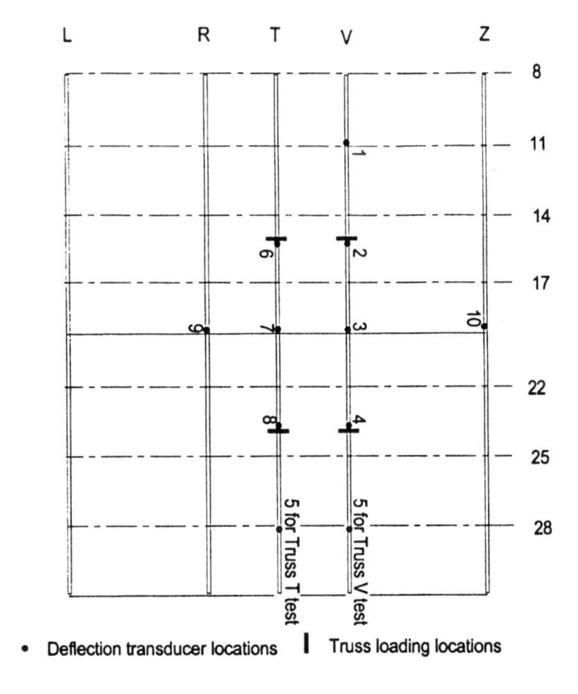

Fig. 2. Locations of displacement transducers and applied loads

The main trusses consisted of four prefabricated parts (Fig 1). The on-site joints between these parts were of bolted flanges in the truss diagonals and of splice plates in the top and bottom chords.

'Diagonal' braces, which extended from the bottom chord central nodes of Trusses V and R to the apex of acoustic door carrying Truss T, had been provided to eliminate differential deflections between these deflection-governed trusses and, hence, prevent possible damage to the ornamental ceiling plaster work.

During this research programme, load tests were carried out to verify the operation, and to calibrate the monitoring instrumentation installed on Trusses T and V. These tests, carried out prior to the installation of heavy acoustic doors, used a load approximately equal to the weight of doors to be carried.

In-service monitoring data presented here are with respect to the acoustic door loads and temperature variations in the roof. Although gauge readings were automatically taken at hourly intervals by the in-service monitoring system, in order to obtain the doors opened/closed data at a constant temperature and in a known order, this event was staged while manually operating the data acquisition system.

A structural analysis of the 3-dimensional truss system was carried out using the general purpose finite element program ANSYS [2]. The measurements and analysis combination was used to study the actual structural behaviour in relation to that intended in design.

2 Truss Instrumentation

One half of each of Trusses T and V were permanently instrumented with vibrating wire strain gauges (VW), electrolevels (EL) and LVDTs. The VWs also incorporated thermistors. The permanent instrumentation was installed *before* the on-site assembly of each truss from its four components, lying in a horizontal position. The positions of the electrolevels and VWs are shown in Fig. 1. The electrolevels were located centrally between the nodes on bottom chord members so that, assuming constant member slope, the measurements may indirectly provide the relative deflection between each pair of nodes. The LVDTs were used to measure slippage across one of the construction joints in the bottom chord of each of the two trusses. During load tests, load cells to measure the applied load and spring-loaded potentiometers to directly measure truss deflections were used as temporary instrumentation placed at ground level (Fig. 2).

3 Strains on assembly and erection of trusses

The measured strains were found to be significantly high in comparison to calculated values, suggesting that truss members had deformed appreciably from the initial positions. This is possible if the four truss components of the unassembled truss had real lack of fit and/or apparent lacks of fit due to horizontal storage of components in a deformed state prior to instrumentation. Where the deformations are due to the storage condition, part of the observed high strains could actually be strains that are *released* on assembly and erection. Due to this uncertainty the states of erected individual trusses were used as the references for subsequent measurements.

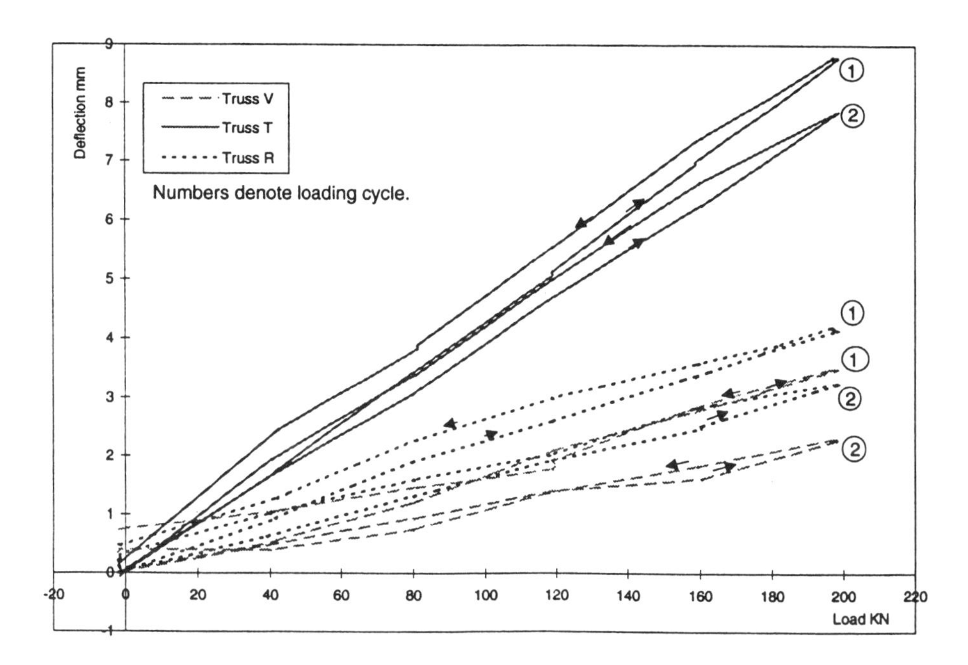

Fig. 3. Truss T load test. Central deflections - direct measurements

4 Load tests

The trusses V and T were separately loaded in two load tests, with each test being repeated immediately afterwards (giving Runs 1 and 2). A two point loading arrangement, with loads applied at approximately one-third span locations, was employed (Fig 2). The loads were applied with hand operated hydraulic jacks set-up at ground level. The maximum load for the tests was chosen as 200 kN, similar to the weight of the door system.

The directly measured central deflections of Trusses R, T and V during the Truss T load test are shown in Fig 3. The numbers 1 and 2 denote the first and second Runs, respectively, and arrows show the loading and unloading directions. These deflections have been corrected for hysteresis errors observed in the potentiometer system. Indirect deflection measurements using electrolevel readings (not presented here) showed that they are acceptable for obtaining an overall idea of the state of response of trusses, but not the detailed behaviour such as their deflected shapes.

Strains in Truss V, measured under Run 1 of Truss V load test, are given in Fig. 4. The load-strain relations for the *loaded* truss (V), were practically linear. The strains in the Truss T 'not directly loaded' showed non-linearity, but due to the small range of measurements it could be considered as practically linear. Strain gauge readings from both Runs for Truss T indicated permanent sets in all the considered trusses, probably due to movements in the bolted connections within the roof truss system.

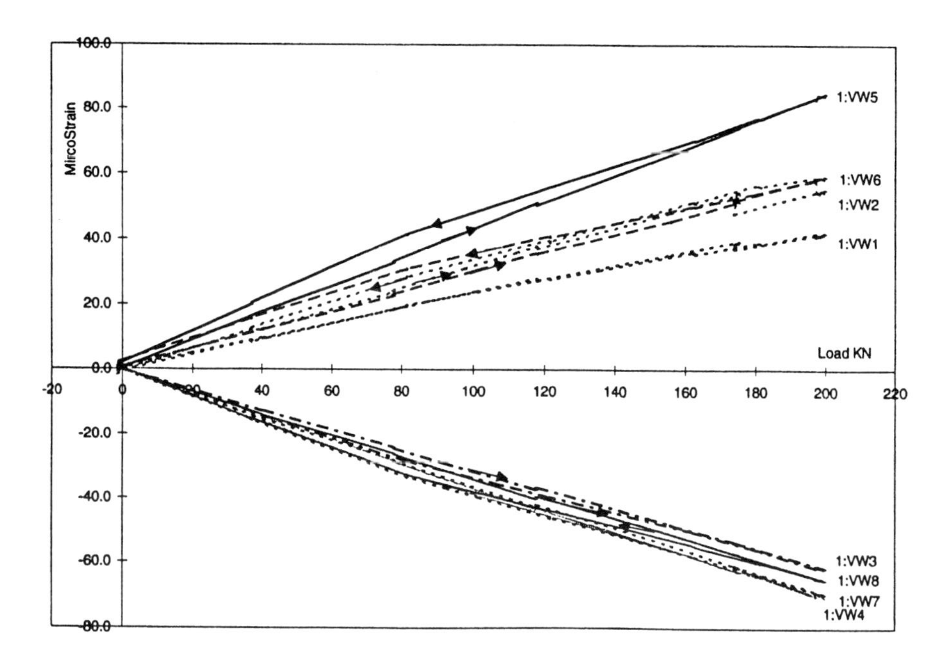

Fig. 4. Truss V load test, Run 1. Truss V strains

5 Computer simulation of load tests

The analytical model used to simulate the load tests consisted of the three main trusses R, T and V, together with associated connected steelwork or approximations of them. Assuming symmetry about the truss centre lines, only one half of the system was modelled (Fig 5, which also shows the deflection pattern under Truss T load test). The steelwork on either side of the considered three truss system, was represented by vertically acting linear springs. The equivalent stiffness of these springs were obtained by considering the force required to move the respective member, or truss, vertically by a unit value while its far end was assumed fixed. Effects of Truss Z deflections, which were not considered in obtaining the above equivalent spring stiffness values, were obtained by defining upper bounds for the deflections.

Typical deflections of the bottom chords of the trusses are shown also in Fig. 6. The central deflection of Truss T was about twice that of Trusses R and V, indicating the braces themselves compressed by an amount similar to the deflections of Trusses R and V, *i.e.* approximately 2mm. The maximum vertical deflection of Truss T was at the loading node, while those of Trusses V and R were at their centre lines. This was because, by acting as a flexible support for the truss, the diagonal braces prevented the downward deflection of Truss T, whereas they provided the (main) loads on Trusses V and R.

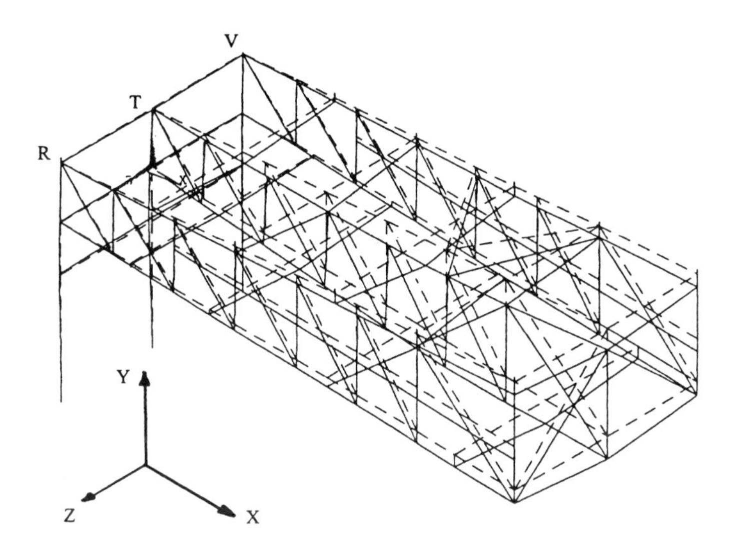

Fig. 5. Analysis model of the truss system, showing deflections under Truss T load test.

During the Truss V load test, as expected, Truss V deflected more than Truss T which in turn deflected more than Truss R. As the dominant loading of Trusses R and T occurred through the diagonal braces connected to their apexes, their deflections were a maximum at their centres. However, as Truss V was loaded at the one-third-points, away from the midspan diagonal braces and Trusses B and D (all of which restrained vertical deflections), its maximum deflection occurred at the load points.

Taking the vertical reactions at the stanchion bases as the loads resisted by each of the three trusses, the following can be observed.

- During the Truss T load test, Trusses T, V and R resisted. 40%, 12% and 12%, respectively, of the applied load. That is, 36% of the load was transferred to the steelwork adjoining this three-truss system.
- During the Truss V load test, the loads resisted by Trusses V, T and R were 33%, 14% and 5%, respectively. As determined from the equivalent spring forces, 9% and 39% of the total load were resisted by the 'external' steelwork adjoining Trusses R and V, respectively.

Thus, in both these cases, the load transferred to the supports was only a small proportion of the applied load. If the ends of Trusses B and D had not been assumed as vertically restrained at their far ends, the loads transferred to the adjoining systems would have been slightly less than calculated above. To consider upper bounds to the additional deflections resulting from the flexibility of Trusses L and Z, the analytical forces in the equivalent springs connected to Truss V, were applied on to a two-dimensional Truss Z.

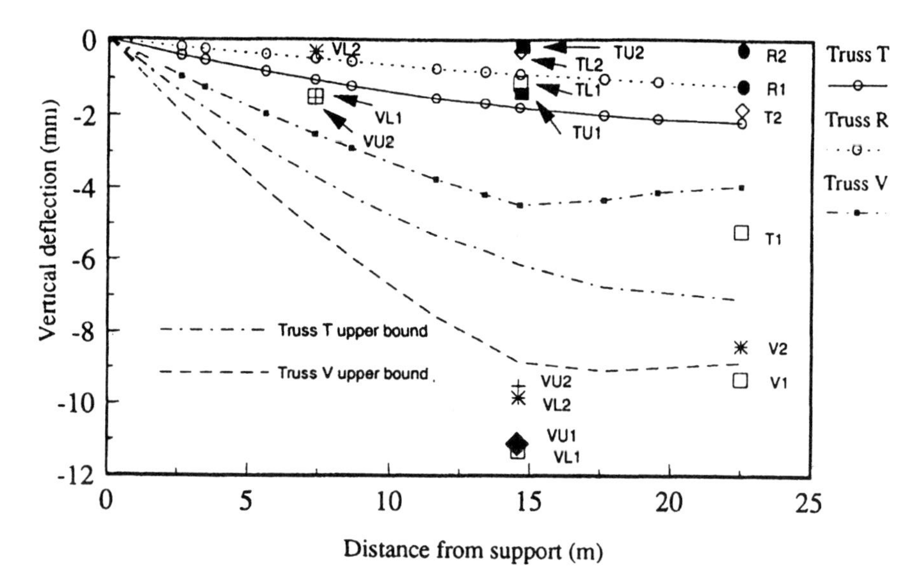

Fig. 6. Truss V load test: Measured deflections (markers only), calculated deflections (markers+lines) and upper bounds

6 Comparison of measured and calculated load test responses

Fig. 6 shows the calculated and measured truss deflections at the maximum load under the Truss V load test. The code used to identify measured values is as follows. Where there are two letters and a digit, the first letter (V or T) indicates the measured truss, the second letter (U or L) indicates the portion of the span (L for potentiometers 1, 2 and 6, and U for potentiometers 4, 5 and 8) and the digit indicates the Run number. For mid-span responses, the second letter is not required and is not used.

The following observations were made from Fig. 6 and results for Truss T load tests (not shown here).

- In the case of Truss T load tests the measured deflections are, generally, larger than the values calculated without considering the Truss Z deflections. The only exception to this is central value in Truss V during Run 2.
- Under Truss V load tests (Fig. 6), the measured deflections could be either smaller or greater than the calculated values, depending on the measurement point considered.
- The measured values at the points of loading under both load tests are appreciably higher than the calculated values. Measured Truss V central deflections under Truss V load tests also are similarly much larger than the calculated values.
- The calculated deflection upper bounds at the load points are much smaller than the measured values in all load tests. But, these measured system deflections are obviously less than those to be expected for a single truss under similar loading.

Table 1: Comparison of calculated and measured strains at 200kN load

Load Test on	Truss	Gauge	Calculated strain (TM)	Measured strain (TM)	
				Run 1	Run 2
Truss V	V	VW7	33.6	71	60
		VW8	22.3	64	72
	T	VW7	12.8	14	12
		VW8	17.7	17	16
	R	VW7	6.1	-	-
		VW8	10.7	-	-
Truss T	T	VW7	37.7	60	58
		VW8	19.1	53	49
	V(& R)	VW7	12.4	14	16
		VW8	20.7	18	20

The deflections observed at the centre of Truss Z, especially during the Truss V load tests, were much smaller than the upper bound values depicted in Fig. 6. Thus, the observed higher flexibility of the load-tested truss system could have been more due to the flexibility of the inter-truss load transfer mechanisms, than due to the flexibility of the far trusses such as Truss Z.

The measured and calculated strains corresponding to strain gauges VW7 and VW8 at the top and bottom fibres, respectively, in bottom chords (see Fig. 1) are shown in Table 1. They indicate that the strains obtained from the computer model are very similar to the strains measured in trusses not loaded directly. However, the measured strains in the loaded trusses (T or V, depending on the load test) are much greater than the calculated values and differ by a factor ranging from approximately 1.5 to 3. These observations were supported also by the deflection measurements (Fig. 6). Had there been 'slips' in the mechanisms connecting the loaded truss and the adjoining systems, higher measured strains in the loaded trusses should have been accompanied by strains lower than the calculated values in the adjoining trusses. However, the similarity of measured and calculated values of strains in adjoining unloaded trusses indicates the possibility of 'slips' in the steelwork beyond them too. It is only by having strain gauges in the load transfer paths, such as the diagonal braces and Trusses B and D, that these could have been studied in depth.

7 In-service test using actual door loads

In order to simulate an in-service condition of door closure under constant temperature, a set of data was obtained by closing the acoustic door (consisting of 35 individual panels) hung from Truss T. During door closure, the central panel is first placed at its specified central position. This gives load case LC1 here. Then two panels, one from each side, are moved from the storage position and placed next to that already in place (giving LC2). In this way, the loading starts from the truss centre and spreads towards the truss supports, giving load cases LC1 to LC18. These door panels are not left hanging freely from the truss, but are individually restrained at their bases with a mechanism located in the floor. This resulted in the transfer of an unknown amount of the weight of each panel directly to the ground, thus reducing the load on the truss.

Fig. 7 shows the measured 'm' and calculated 'c' strains at the truss upper chord (gauges VW5 and VW6) and lower chord (gauges VW7 and VW8) in Truss V. The following observations are made from the comparison between observed and calculated strains for the two load tests.

- The differences in the readings from each pair of strain gauges indicate that in Truss V local and secondary bending effects are low while in Truss T, from which the door is hung, they are large. The secondary bending effects are larger in the analytical results (for Truss T) than in the measurements. The reason for this is that the analysis assumed the panels to be hung directly from the truss bottom chord, whereas they are actually hung from an additional longitudinal member, which in turn is hung from the bottom chord, thus distributing the load more widely along the truss. As to be expected, in Truss T, secondary bending effects are more significant in the directly-loaded bottom chord (VW7 and VW8) than in the upper chord (VW5 and VW6).

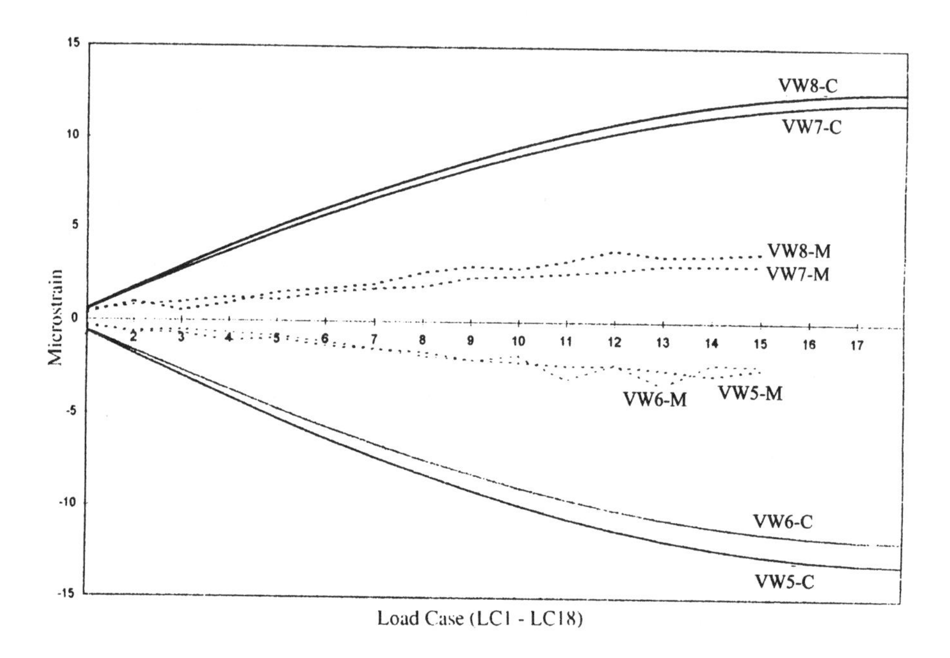

Fig. 7. Door load test. Measured 'm' and calculated 'c' strains in Truss V

- The irregular nature of the response curves, both in measurements and calculations, for bottom chord strains (VW7) should be due to the hanging of panels from the bottom chord. This strain measured at the top fibre of the member shows how during the initial loading stages, compression due to local bending dominates over the direct global tension in the chord. This points to the need to finely model the structure at positions where local effects are present.
- The measured direct strains, obtainable as the average of the pairs of strain gauge readings, indicate responses lower than that calculated.
- The measured strains are much smaller than the response from the previous load tests (see Table 1 for VW7 and VW8 of Truss T) due to the difference in the nature of the loading, and the partial support of door panels by the floor. Also, secondary local bending effects in Truss T are larger here than during the Truss T load tests.

8 Effects of temperature changes

In addition to the effect of loading from the acoustic doors, tests have also been carried out to assess the effect of temperature changes. Table 3 presents the measured temperatures at several instances when the acoustic doors were closed between 16 January to 16 June 1996. The first set of data, *viz.* at 1400hrs. of 16 January 1996, was used as the datum for obtaining response changes due to subsequent temperature changes. Table 4 presents the measured strains at the top and bottom chord positions, during some of these instances.

Electrolevel readings (not presented here) indicated large inter-nodal deflections under moderate temperature changes. Thus, as a result of their sensitivity to temperature change, the installed electrolevels were seen to be unsuitable for measuring truss response under thermal loading. This was verified by computer analyses, which considered similar, but idealised, temperature changes. These analyses showed that the relative deflections within a truss and between it and other trusses depend not only on its temperature change, but also on the states of those next to it. Thus, where significant temperature differences are expected between trusses and there are strong lateral load transfer mechanisms between them, it is important (in structural monitoring) to monitor temperatures in adjoining trusses.

Measured (temperature compensated) *elastic* strains in members, were small and similar in value to those observed in ‘non-loaded’ trusses during load tests and the door load test. However, the calculated elastic strains for the ideal temperature change cases were generally much lower than those measured. The reason for this, in addition to differences in actual temperature distributions, could be an actual restraining of the truss system against horizontal expansion. The plausibility of this was confirmed by analyses which introduced horizontal elastic restraints at the top and bottom boom nodes in the supporting columns of Trusses R, T and V.

9 Discussion

9.1 Observations on measurements and modelling

Installation of measurement gauges prior to the fabrication of trusses, constrained by Contractor's schedules, provided a measure of strains introduced on assembly of truss components. The use of additional instrumentation would have aided in the separation of the strains due to lack of fit and apparent strains due to release of strains that occurred as a result of the methods of support employed during assembly.

The considered structural system has a large amount of intentionally provided load sharing between its major elements. This gave rise to uncertainties in the interpretation of the observed behaviour of the trusses because of:

- The complexity and seemingly changing nature of the lateral load-sharing mechanism within the roof structure subject to small deflections.
- The limited numbers of measuring instruments which could be deployed and the malfunctioning during tests of some of those used.
- The difficulty of monitoring remotely (the lower boom of the truss being 7.5m above floor level) the small deflections of the deflection-governed stiff truss structure by means of electrolevels and potentiometers. A remote laser-based displacement measurement system could have been of help during load tests, but not during long term monitoring due to possible drift in readings and the currently prohibitive costs.
- The lack of instrumentation upon the diagonal bracing between trusses, and the connecting trusses that significantly influence the lateral load-sharing amongst the primary trusses.

The last of these is the most important as it would have provided valuable information upon the lateral load transfer mechanisms and, thus, helped to further understand the behaviour of the truss system. Ideally, if resources and the work programme of the Contractor had permitted, it would have been desirable to have conducted additional load tests on trusses before the inter-truss load distributing elements, such as the diagonal braces and Trusses B and D, had been introduced. This would have given a better understanding of the behaviour of the individual trusses and, subsequently, the lateral load-sharing mechanisms.

The measurements have shown that the data from the vibrating wire strain gauges (VWs) to be most useful and repeatable. The potentiometer readings, which could have been a source of reliable data, were less useful than had been hoped due to practical difficulties of monitoring the movement of a point 7.5m from the instrument location. The electrolevel readings were of limited use for precise comparison with the analytical estimates of deflection profile and central deflection. They did, however, generally give a reasonable estimate of the order of magnitude of the central deflections under load tests. The use of electrolevels in these circumstances was experimental, following on from limited laboratory trials in their use as an indirect means of establishing deflection profiles and mid-span deflection. Such procedures could provide significant advantages in situations where it is inconvenient or impracticable to use direct measurement techniques, such as LVDTs, which require instruments to be mounted upon an independent reference frame. Long span structures are an obvious application. It is now apparent that the performance of electrolevels varies significantly, especially in the long term and under temperature variation, and not necessarily within the manufacturer's stated specification. Accordingly, for sensitive applications, such as the project being discussed, it is necessary to undertake additional performance testing in order to select the most suitable instruments.

The door load test showed the importance of detailed and fine analytical modelling near positions with local effects if comparisons are to be made with strain measurements at such positions.

9.2 Observations relating to design

This work programme confirmed that the behaviour of a three-dimensional load sharing system could be very different to that assumed in simplified design or by even three-dimensional structural analysis. A simplified two-dimensional design is based on assumptions, which are expected to provide a conservative set of actions and restraints to be considered in the analysis/design process. A structural analysis would usually assume idealised conditions which may not reflect the actual behaviour, for example those of connections. Thus, depending on the type of design, and the locations of critical members and applied loads, the actual three-dimensional behaviour of a system could give rise to either a conservative or non-conservative structure.

The three-dimensional structural analysis undertaken showed that the load sharing between the different members of the system could be very significant, reducing the load transmitted to the foundations of the loaded truss to a small proportion of that applied.

The diagonal braces, provided to reduce differential deflections between trusses, also transfer large loads to the adjoining steelwork. To ensure that these lateral load distribution mechanisms remain active during overload conditions, it would be necessary for the braces to be designed more conservatively than the main trusses.

As shown by the load tests, due to the theoretically 'imperfect' nature of the connections between members, the actual behaviour of a three-dimensional structure could lie between that indicated by a two-dimensional analysis and an idealised three-dimensional analysis of the perfect structure. In the case of the current study, it is seen from the load tests that the modelling of the connections between trusses and diagonal braces should have incorporated some flexibility in their structural behaviour.

The measurements and calculated results showed that the actual behaviour could be determined only by considering the whole three-dimensional nature of the load-sharing structure. Therefore, although no load transfer may have been assumed in the design, the interpretation of results of loading tests requires consideration of the effects of possible load transfer mechanisms.

10 Conclusions

The load testing and monitoring of this unique three dimensional truss system, and the attempt to understand its in-service behaviour through structural analysis, created a large amount of information on the behaviour of the truss system. From the study of this information the following conclusions can be reached.

- The truss system has considerable load sharing/distributing properties as intended in the design. The diagonal braces between trusses (Trusses R, T and V) provided considerable load sharing and reduced potential differences in their central displacements. However, these deflections were not identical because of possible slips in the bolted joints in the trusses and the inter-truss load transfer mechanisms including the diagonal braces.
- Computer analysis gave insight into the truss behaviour intended by design. In principle, the computer model employed could have been used for further studies such as the influence of non-rigid joints.
- Although the design of a three-dimensional system can be carried out using simplifying assumptions, the actual behaviour can be determined only by considering the whole three-dimensional nature of the load-sharing structure. The use of a simplified design analysis for interpreting and understanding the actual behaviour of a complex three-dimensional structure could lead to erroneous conclusions about its performance and safety.
- It is insufficient to use only theoretical results to develop criteria for evaluating the behaviour of an actual structure, as the calculated behaviour would depend on the assumptions made for the analytical model. In load testing a three-dimensional load sharing system, it may be of considerable benefit to calibrate (at least with a small load) each major element - for example, Trusses T and V of the current structure - before the installation of lateral load transfer mechanisms.
- In some structures, temperature variations can give rise to strains comparable to service load strains.
- It is useful to have a degree of redundancy in measurement data so that erroneous measurements can be detected. Adequate time and resources are required for the proper planning and implementation of effective structural monitoring programmes.

11 Acknowledgements

BRE acknowledge the cooperation received from all those involved in the design and construction of the Mission Centre, in particular Austin Trueman, consulting structural engineers for the building, and RAKO (Design and Build) Limited who permitted the instrumentation of roof trusses.

12 References

1. Matthews S., Goodier A., Brooke D., Tsui F. and Reeves B. (1996) *In-service structural monitoring of the Meeting Hall roof, Swaminarayan Hindu Mission, Neasden;* Proc. Intl. Seminar on Structural Assessment - The role of large and full scale testing, City University, London.
2. ANSYS Version 5.3. Swanson Analysis Systems, USA, 1996.

DYNAMIC ANALYSIS OF A NARROW TURBOGENERATOR FOUNDATION ON PILES

Dynamic analysis of turbogenerator foundation

V. KARTHIGEYAN,
Offshore Safety Division, Health and Safety Executive, London, UK
G.K.V. PRAKHYA,
Design Group, Sir Robert McAlpine Ltd, Hemel Hempstead, UK

Abstract

This paper presents dynamic analysis of a thin and narrow turbo generator foundation on piles. The foundation dynamic stiffness used in the analysis varies with frequency, shear wave velocity and mode shape. Techniques for adopting suitable values of stiffness as a base case have been developed. Modal analysis was carried out using the base case. The results were used to investigate the required modifications to the base case to detect adverse or near resonance conditions. Modified stiffness values were used in the harmonic analysis to obtain the maximum amplitudes of vibration. The results are compared with those of rigid block models. It is shown that the Finite Element (FE) model predicts higher amplitudes of vibration due to flexural modes when compared to those from the rigid block model. Numerical predictions were validated using field tests and empirical methods. The approach used to account for variations in stiffness is applicable to non-piled foundations as well as to general FE analysis.

Keywords: Dynamic analysis, finite elements, foundations, harmonic analysis, piles, shear wave velocity, turbogenerator, validation.

1 Introduction

Traditionally machinery foundations are analysed as single rigid blocks with up to 4 degrees of freedom (DOF) [1][2]. The DOF are vertical and lateral translations and rotations about longitudinal and vertical axis. Stiffness and damping values required for the analysis are calculated assuming circular foundations on elastic half space or pile groups attached to rigid pile caps. In order to ensure the rigidity of the foundation and to avoid flexural vibrations, a minimum thickness of Length/10 + 600mm is generally used.

Abnormal Loading on Structures edited by K. S. Virdi, R. S. Matthews, J. L. Clarke and F. K. Garas. Published in 2000 by E & FN Spon, 11 New Fetter Lane, London EC4P 4EE, UK. ISBN 0 419 25960 0

All large generators and pumps require long and relatively narrow foundations. The thickness based on the above criteria is uneconomical for such foundations, typically longer than 25 m. Even by using the minimum recommended thickness criteria, flexural and torsional vibrations caused by high speed turbines and alternators cannot be avoided. Hence the rigid block models cannot be used.

The paper presents a method of modelling and analysis of a turbogenerator foundation system to obtain acceptable results.

2 Turbogenerator and foundation

Layout of the turbogenerator and its foundation are shown in Fig 1. The machine specifications are given in Table 1.

Table 1. Machine Parameters

	Turbine	Alternator
Operating speed (rpm)	5135	3000
Lateral critical speeds (rpm)	1860 & 3660	1570
Purge speeds (rpm)	460	270
Mass (tonnes)	128	93
Out-of-balance forces at operating speed (kN)		
Force at M_3	65	
Force at M_4		65
Force at M_9		264

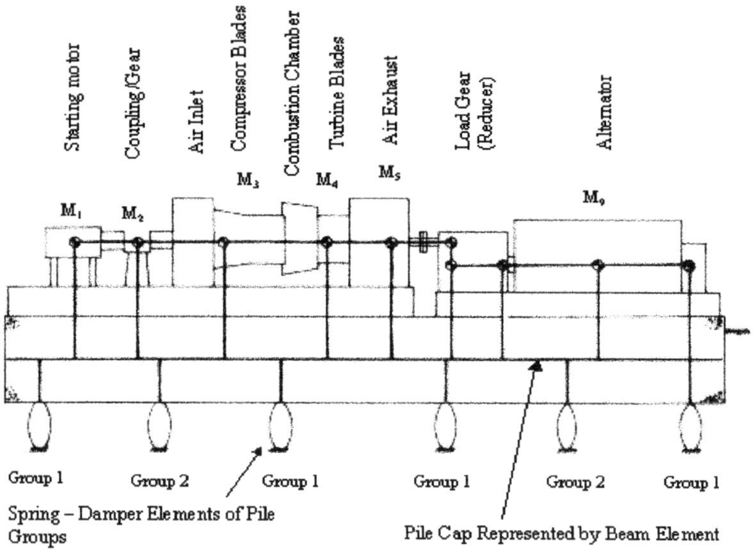

Fig 1. Layout of turbogenerator and foundation

The foundation for the machine consists of a concrete pile cap of 22m × 4m × 3m thick and 29 No, 15m long 450 mm diameter driven cast in place piles (Fig 2a).

Vibrations are caused by the centre of mass for the rotor being eccentric to the shaft centreline. The resulting amplitudes are dependent on mass/layout of machine assembly, mass/shape of the pile cap, pile geometry, and soil conditions [7][8][9].

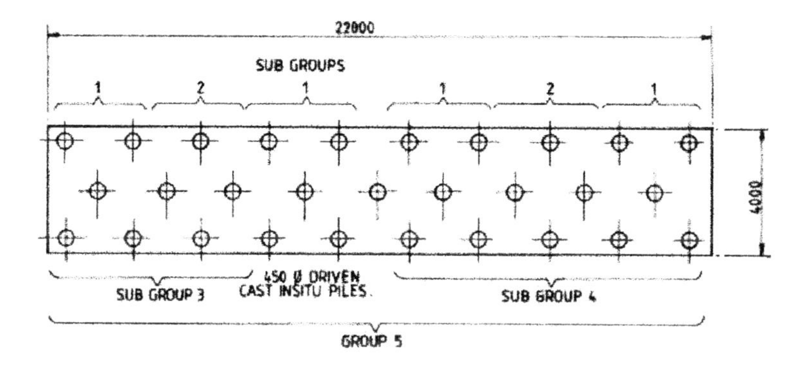

Fig. 2a. Pile cap and pile layout

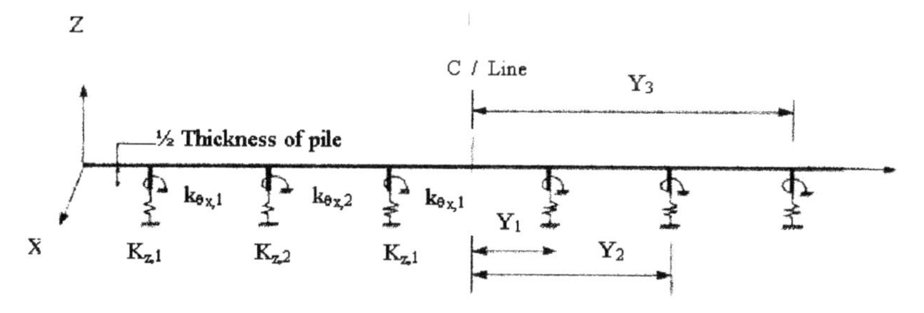

Fig. 2b. Stiffness of pile groups

3 Soil conditions

Soil conditions from the bore hole tests and measured shear wave velocity V_S using cross-hole seismic tests are given in Fig. 3. The V_s was measured for a shear strain of 10^{-4}. The measured value was validated using calculated values from soil parameters [3][4]. A correction factor of 0.9 was applied to allow for the expected higher shear strains.

Shear wave velocity V_S close to the underside of pile cap is 260 m/sec. With strain correction it is 235 m/sec. The idelaised profile of V_S is given in Fig. 3. Both the upper bound value of 340m/s (without strain correction) and a lower bound value of 235 m/s are used in the analysis. For low amplitude of vibration, strain will be low and no correction was applied to the upper bound value.

Shaker tests was carried out on two test piles with 600 mm × 600 mm × 600 mm pile caps. The results were within 5% of the calculated vertical and horizontal stiffness values, with V_S increasing linearly from 0 to 235m/sec for the top 1.5m. Calculated values were 5% higher for vertical stiffness and 25% higher for the horizontal stiffness for V_S increasing from 0 to 340m/sec for the top 1.5m. Hence V_S in the range of 235 m/s to 340 m/s is considered in the analysis.

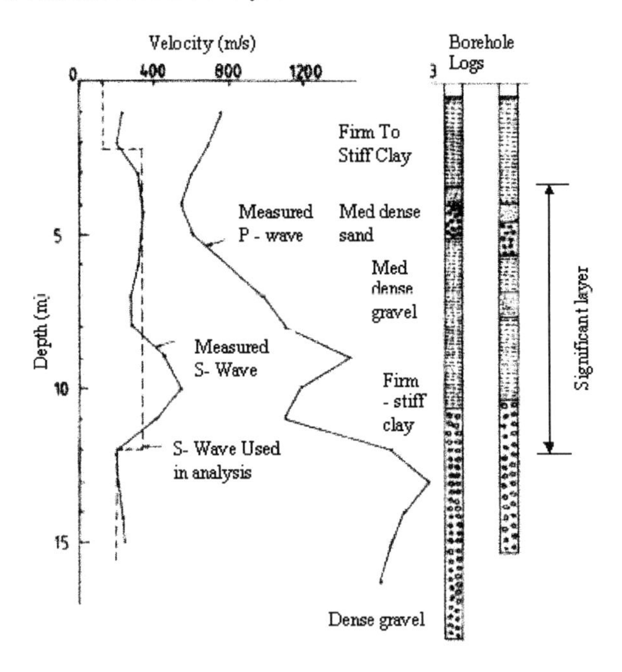

Fig. 3. Bore hole log and shear wave velocity

4 Methodolgy

The objectives are to avoid resonance or near resonance where possible, and to limit the amplitudes of vibrations. The following procedure is used to achieve the objectives.

Step 1 Identify and develop the appropriate stiffness/damping matrix to be used in the FE model. This is referred to as the base case.

Step 2 Modal analysis to compute the natural frequencies and mode shapes

Step 3 Investigate for possible variations in natural frequencies towards resonance or near resonance condition due to variations to the base case (c.f. Section 5).

Step 4 If near resonance or resonance is found, either change configuration (pile cap size, pile number and spacing) and repeat steps 1 to 3 or carry out harmonic analysis for the base case and the adverse variations.

Step 5 Carry out harmonic analysis even if resonance is not found.

5 Identification of the base case

5.1 Development of the stiffness matrix

Rigid body modes were calculated using the block model and were found to be below the operating speed. The group 5 stiffness values, which are appropriate to these modes, were used as the base case. Being low tuned, the upper bound values of stiffness give conservative results hence V_s of 340 m/sec was used.

Stiffness and damping of the pile sub-groups 1 to 5 (Fig. 2a) were computed using DYNA software, which was developed by Novak [5]. Typical results are presented in Fig. 4 for group 5. Subscripts X, Y & Z indicate direction for linear stiffness and θx, θy & θz indicate rotational stiffness about the respective axis.

Variation of stiffness K_Y and K_Z with frequency shows the effect of waves generated by piles at one end of the pile cap being at various phases with the movement of other piles. In and out-of-phase conditions create small and large apparent group factors respectively. These results, presented in Fig. 4, are valid only for rigid pile caps.

In flexural and torsional modes, groups of the piles will move out-of-phase with each other and the selective use of stiffness values for sub-groups 1 to 4 is applicable for frequencies involving these modes.

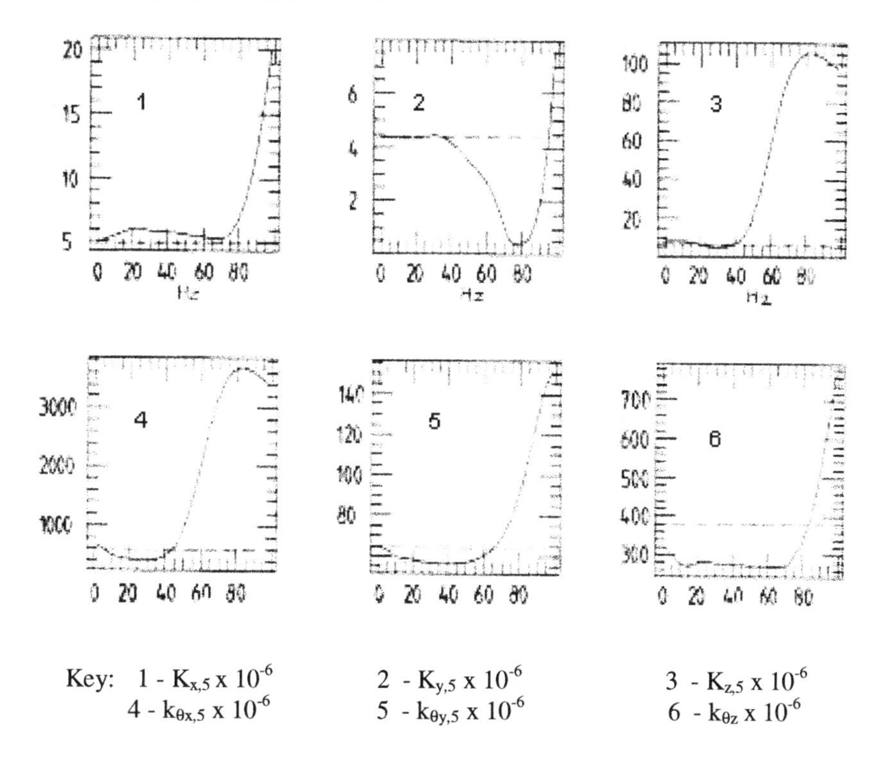

Abscissa in Hz and Ordinate in kN/m or kNm/rad

Fig. 4. Stiffness values for Group 5

In addition to the aforementioned two variabilities, the third one arises due to V_S. With this in view, one can identify two ways of enveloping the results (frequencies and amplitudes). They include:

a) analyse and examine number of combinations resulting from the above three variables;
b) analyse a base case using the most appropriate stiffness values. This should be followed by inspection and adjustment of results and/or selected re-analysis with adjusted stiffness values.

Method a) for the turbonenerator may involve up to 40 different models, even with a limited number variables. Hence, method b) is more appropriate. This approach is illustrated in this paper with the use of appropriate spring constants.

Ordinates of broken horizontal lines in Fig 4 show the stiffness values used in the dynamic analysis. These represent a reasonable upper bound for stiffness for up to 50 Hz, except for K_Z (vertical stiffness) and K_X (translational stiffness).

$K_{Z,5}$ – vertical stiffness for the whole foundation (unembedded) is listed in Table 2, together with the calculated vertical frequency.

Table 2. Vertical frequencies from block model

Stiffness kN/m	Frequencies (Hz)	Calculated Natural Frequency (Hz)
$0.87x10^7$	0 - 16	16.1
$0.60x10^7$	30	13.4
$2.0x10^7$	50	24.4
$10.0x10^7$	80	54.6

Columns 2 and 3 match up to 16 Hz. The mode corresponding to this frequency will be excited during start up and purging.

In flexural modes, the stiffness of pile cap in bending will become increasingly important in relation to that of pile groups. Hence it is adequate to use the vertical stiffness of whole group 5 at low frequency (i.e. $0.87x10^7$ kN/m) as part of the base case.

The method used to calculate vertical stiffness - K_Z (Fig. 2b) and rotational stiffness about X axis - $K_{\theta X}$ is explained below. Other spring constants are calculated in a similar manner.

Vertical stiffness for each spring (Fig. 2b) is given by:

$$K_{Z,i} = K_{Z,5} /6, i = 1,2 \tag{1}$$

where, $K_{Z,1}$ and $K_{Z,2}$ are the stiffness of vertical springs in FE model (Fig. 2b) and $K_{Z,5}$ is the vertical stiffness of group 5 shown by broken line in Fig. 4.

Rotational stiffness due to vertical spring is

$$K_{\theta V} = 2K_{Z,1}(Y_1^2 + Y_3^2) + 2K_{Z,2}Y_2^2 \quad (2)$$

where Y_i is the horizontal distances of vertical springs $K_{Z,1}$and $K_{Z,2}$ from the centroid of the pile cap (Fig. 3a).

The residue of rotational stiffness R is given by:

$$R = K_{\theta X,5} - K_{\theta V} - 4k_{\theta X,1} - 2k_{\theta X,2} \quad (3)$$

where $k_{\theta X,1}$ and $k_{\theta X,2}$ are rotational stiffness of subgroups 1 and 2 respectively.

Final rotational stiffness $K_{\theta X,1}$ & $K_{\theta X,2}$ to be used for the subgroups is given by

$$K_{\theta X,1} = k_{\theta X,1} + R/6 \quad (4)$$

$$K_{\theta X,2} = k_{\theta X,2} + R/6 \quad (5)$$

Using the above expressions, a 6 × 6 matrix was developed. Cross stiffness terms were also included in the matrix. Similar procedure is followed for developing the damping matrix. These matrices were used as the support condition for the FE model developed in the next section.

Table 3. Natural Frequencies (Hz)

Mode	Frequency (Hz)	Description
1	9.2	First coupled sliding (X) and rocking (Y)
2	11.0	Longitudinal sliding in Y
3	16.1	Vertical mode in Z
4	16.8	Rotation about vertical axis (Z)
5	21.4	Rotation about X axis
6	25.8	Second coupled sliding (X) and rocking (Y)
7	35.3	Horizontal flexure - first mode
8	54.3	Vertical flexure - first mode
9	69.7	Horizontal flexure - second mode
10	98.9	Stretching of pile cap
11	102.9	Vertical flexure - second mode

5.2 Modelling

The FE model consists of a beam element representing 3 m thick pile cap (Fig. 2b). The effects of shear and rotary inertia were included. The machinery was also modelled as beam elements with masses lumped at M_1 to M_{10}. These masses are connected to the pile cap by rigid vertical elements (Fig. 1). Springs and damper elements are located at the centroids of the pile sub-groups 1 and 2 (Fig 2a). The vertical element (Fig 2b) between the spine beam and the springs represent half the thickness of the pile cap and consists of rigid but massless elements.

6 Natural frequencies

Modal analysis carried out using ANSYS - a general FE program and the results are given in Table 3. Rigid body modes were validated using rigid block models.

7 Discussion on variations to base case

If shear wave velocity is as low as 235 m/sec, the rigid body mode frequencies can be up to 30% lower. This is illustrated in Fig. 5. Lowest possible sliding frequency (6.4 Hz) is still 42% higher than alternator speed during purge (4.5 Hz). However, the frequency of this mode coincides with purge speed of turbine (7.7 Hz) if shear wave velocity is between 235 m/sec and 340 m/sec.

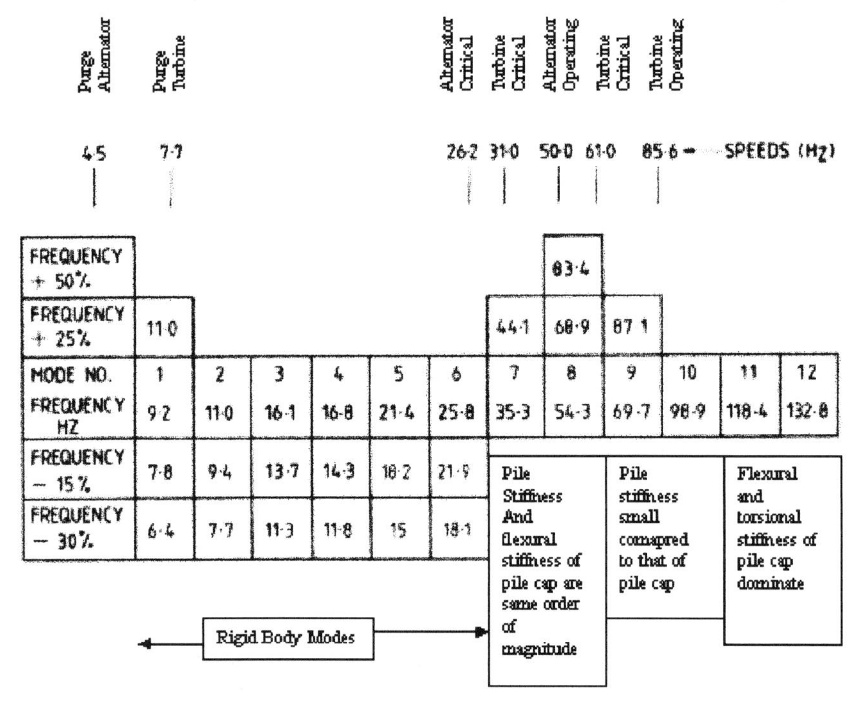

Fig. 5. Natural Frequencies and expected variations

Second natural frequency of 11 Hz, sliding in Y, can be as low as 7.7 Hz for a lower shear wave velocity. This mode couples with mode 5. Any eccentric excitation in the Z axis such as out of balance of alternator, producing moment about X, will have a secondary effect on longitudinal sliding. This is not high enough to be detuned.

Frequency for modes 3 to 5 may be lower than the calculated values. If V_s is lower than 340 m/sec these frequencies will move away from purge, critical and operating speeds.

Frequency of mode 6 is close to the alternator critical speed. The value of $K_{\theta y}$ at this frequency for un-embedded case is lower by 15%. This and the possibility of lower shear wave velocities can only improve the situation.

Modes 7 and 8 involve the first mode in flexure. Motion of the end and middle of pile cap are out of phase. Hence the spring stiffness values will be larger than those used in the analysis. Hence the use of stiffness values calculated for groups 1,2 or 3 are more appropriate. The following method is used to correct this error.

Behaviour of the pile cap in single flexure in the vertical plane (mode 8) can be modelled using half length of pile cap fixed at the centre line of full length. Deflected shape (Fig. 6) shows the two sub-groups 1 and 2 and the bending stiffness of the pile cap together resisting the deflection in this mode. The contribution of mid spring $K_{z,2}$ is small. Stiffness value for $K_{Z,1}$ and $K_{Z,2}$ calculated from Eqn. 1 for each spring and used in the finite element model is 1.45×10^6 kN/m. Whereas the stiffness of group 1 at 54 Hz (mode 8) is 7.0×10^6 kN/m. The factor for increasing this frequency is expressed as:

$$\sqrt{\frac{2 \text{ x Group 1 stiffness at 54 Hz} + \text{Pile cap stiffness}}{2 \text{ x Kz used in FE model} + \text{Pile cap stiffness}}}$$

Pile cap stiffness in the above expression is the bending stiffness of cantilever (Fig. 6) at the location of spring (at quarter point).

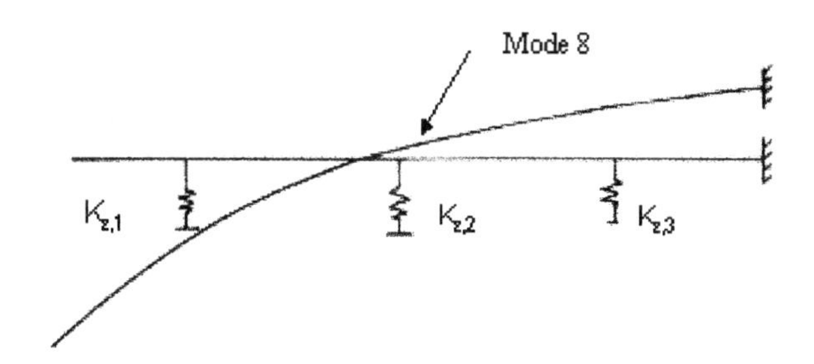

Fig. 6 Deflected shape in Mode 8

The increased frequency of mode 8 due to the above adjustment is 83.1Hz, i.e., about 50% increase on the computed frequency. The calculation illustrated can also be carried out more systematically using generalised stiffness and generalised mass [6] for the mode shape illustrated. Increase in mode 8 frequency of 50% is marked in Fig 5. A 25% increase to 68.9 Hz will be applicable if shear wave velocity in the significant layer is 235 m/sec. Similar calculations for mode 7 and second mode of flexure (mode 9) was carried out and possible increased frequencies are shown in Fig 6. It shows possible resonance with the turbine operating speed.

For modes 10 upwards the stiffness of pile groups will be small in comparison to that of pile cap and any significant variation in the frequencies is not expected.

Out of all the modes, it is clear that modes 1 and 8 may cause problems. To ensure that mode 1 frequency is substantially higher than turbine purge speed, the pile cap mass needs to be reduced (by making it thinner) and the number of piles increased. Thinner foundation will reduce the upper bound frequency (84.3 Hz) of mode 8, and the lower bound value of 54.3 Hz (Fig. 5) will move closer to the alternator speed. In order to avoid the latter it was decided to examine the results of harmonic analysis for the amplitudes before any further modifications are made to the pile cap.

8 Harmonic analysis

Three sets of harmonic analysis were carried out using;

a) spring stiffness derived from group 5 as in modal analysis,

b) using stiffness values calculated for groups 1 & 2 with V_s of 235 m/s and

c) as b) but with V_S of 340 m/s with stiffness values appropriate to the frequencies close to modes 7 and 8 .

The last two models were expected to provide conservative estimates for amplitudes caused by operating speeds of alternator and turbine.

Harmonic analysis was carried out using program ANSYS with the out of balance loads given in Table 1. Three cases were considered:

Case 1: Turbine out of balance forces at M_3 & M_4 in phase

Case 2: Turbine out of balance forces at M_3 & M_4 out of phase.

Case 3: Generator out of balance force at M_9.

Cases 1 or 2 can combine with Case 3. Each case was analysed separately for frequency range 1-100 Hz. Hence the amplitude plots are valid only at operating speeds. Values at other speeds are to be adjusted to vary with the square of speed concerned.

Table 5. Amplitudes (microns) from rigid body models and FE

	Vertical		Transverse	
	Rigid block model	FEM	Rigid block model	FEM
C.G. of pile cap	9	12.0	3.3	10.0
Alternator bearing level	7	6.5	3.6	8.0

The amplitudes were computed using rigid block models [1][2]. The comparison is given in Table 5 for load case 3 at operating speed. Typical vertical amplitude from the FE model at the centroid of pile cap is higher than that in block mass model. The increase is due to the vertical flexure mode (mode 8) at 54.3 Hz. Absence of such difference in vertical amplitude at alternator (M_9 in Fig. 1) is due to the finite element node being located near the zero crossing of the mode. If the alternator were located further away from this point, amplitudes at the centroid and at the ends of the pile cap can be much higher. Transverse amplitudes from the finite element model are higher due to the first flexural mode (mode 7) at 35.3 Hz and second flexural mode (mode 9) at 69.7 Hz.

9 Conclusions

Rigid block models cannot be used to predict the vibration amplitudes of foundations whose natural frequencies in flexural modes are within the range of operating speeds. Long flexible foundations can be modelled in finite elements using beam elements with discrete soil springs distributed along the length. Modelling such foundations require detailed understanding of soil and structural dynamics. Selective use of dynamic foundation stiffness is necessary to detect resonance or near resonance conditions. The amplitudes from the FE model are likely to be higher than predicted by rigid block models due to the contribution from flexural modes.

10 References

1 Richart et al. (1970) *Vibration of soils and foundations,* Prentice-Hall Publication, New Jersey.
2. Barkan, D.D. (1967) *Dynamics of bases and foundations*, McGraw-Hill Book Co., New York.
3. Seed, H.B and Idriss, I.M. (1978), Soil moduli and damping response analysis, *Earthquake Research Centre Report* EERC-70-10.
4. Hardin,.B.O. and Drnevich,V.P. (1979) Soil modulus and damping in soils: Design equations and curves, *Journal of the Soil Mechanics and Foundation Division, Proc ASCE*, pp 667-692, July.
5. Novak, M. (1975) Dynamic stiffness and damping of piles, *Canadian Geotechnical Journal, Vol. 11*, pp. 574-598.
6. Clough,.R.W. and Penzian,.J. *Dynamics of structures*, McGraw-Hill-Kogakusha, Japan.
7. El Sharnouby, H. and Novak, M.(1986) Flexibility coefficients and interaction factors for pile group analysis, *Canadian Geotechnical Journal, Vol. 23, No. 4*, pp. 441-450.
8. Kaynia, A. M. and Kausel, E.(1982) Dynamic behaviour of pile groups, Conference on Numerical Methods in Offshore Piling, University of Texas, Austin, Texas, pp. 509-532.
9. El Sharnouby, H. and Novak, M. (1984) Dynamic experiments with group of Piles, *Journal of Geotechnical Engineering, Proc. ASCE, Vol. 100., No. 6*, pp. 719-737.

APPENDIX

This appendix includes the abstracts of two papers which were presented at the conference, but the manuscript of which were received too late for inclusion in the proceedings.

TEMPERATURES ATTAINED BY STRUCTURAL STEEL IN BUILDING FIRES BASED UPON FULL SCALE TESTS

Building fire full scale tests

B.R. KIRBY
British Steel Swinden Technology Centre, Moorgate, Rotherham, UK

Abstract

As part of a jointly sponsored European research programme on understanding the behaviour of multi-storey steel framed buildings in fire, British Steel's Technology Centre completed the last of its major fire tests conducted on the BRE 8-storey frame at Cardington. This involved a complete burnout of a compartment fire in a fully furnished open plan office and was one of largest full-scale tests of its type conducted on an unprotected steel frame.

In the paper the design of the test arrangement is described and some of the important results given. In particular, analysis of the thermal data is presented which has been used together with results from other full scale compartment fire studies, to validate the heat transfer relationships given in Eurocodes 1 and 3 on 'Fire Actions'. The work has demonstrated that for fully developed compartment fires, the temperatures attained by steel members can be reliably predicted. These results are shown to be independent of a number of different building parameters, as well as the magnitude and type of fire loading and are an important step in the development of a performance based approach to Fire Safety Engineering.

Examples of a Nomogram are presented which can be used as an initial check in establishing the likely temperatures attained by unprotected steel members engulfed in fire. These are based upon the parametric temperature-time equations given in Eurocode 1 for post-flashover fires.

Keywords: Compartment fires, fire action, fire safety engineering, fire tests, heat transfer, multistorey frames, nomogram, steel buildings, temperatures.

Abnormal Loading on Structures edited by K. S. Virdi, R. S. Matthews, J. L. Clarke and F. K. Garas.
Published in 2000 by E & FN Spon, 11 New Fetter Lane, London EC4P 4EE, UK. ISBN 0 419 25960 0

NUMERICAL MODELLING OF AN OFFSHORE CONCRETE GRAVITY PLATFORM ON AN UNEVEN SEABED

Offshore concrete gravity platform

M. WILLFORD
Advanced Technology Group, Ove Arup & Partners, London, UK
C. HUMPHESON and R. NICHOLLS
Arup Geotechnics, Ove Arup & Partners, London, UK
P. ESPER
Arup Energy, Ove Arup & Partners, London, UK

Abstract

This paper discusses the effects of seabed-structure interaction on the design of a concrete gravity substructure supporting a conventional steel topside structure at a site having an uneven seabed. It describes the design of a levelling layer composed of gravel mounds to be placed offshore over the existing seabed to form a suitable medium on which to support the platform caisson. The paper addresses the effect of the inevitable variation in the level of individual mounds (due to surveying and placement tolerances) on the local and global design of the caisson. The methods used to predict the distribution of seabed forces on the base slab, and the resulting load effects in the caisson are described. These include:

- Non-linear finite element analysis to estimate the squashing resistance characteristics of the mounds.
- Monte-Carlo simulation of the process of survey and placement of the mounds to estimate the statistical variation of individual mound heights, mound forces and overall load effects in the caisson.
- Reliability methods to establish design values for load effects.
- The treatment of these effects in the 3-D finite element analysis of caisson for structural design verification.

Keywords: Caisson, concrete structures, finite element analysis, Monte Carlo simulation, numerical modelling, offshore structures, reliability, steel structures, topside structures.

Abnormal Loading on Structures edited by K. S. Virdi, R. S. Matthews, J. L. Clarke and F. K. Garas.
Published in 2000 by E & FN Spon, 11 New Fetter Lane, London EC4P 4EE, UK. ISBN 0 419 25960 0

AUTHOR INDEX

SUBJECT INDEX

This index is compiled from the keywords assigned to the papers, edited and extended as appropriate. The page references are to the first page of the relevant paper.